NEW HORIZONS IN TESTING

Latent Trait Test Theory and Computerized Adaptive Testing

CONTRIBUTORS

R. Darrell Bock
Linda L. Cook
Fritz Drasgow
Gerhard H. Fischer
Janice A. Gifford
Ronald K. Hambleton
Harald E. Jensen
G. Gage Kingsbury
Michael V. Levine
Frederic M. Lord
James R. McBride
Gary L. Marco
John T. Martin
Nancy S. Petersen
Mark D. Reckase
Malcolm James Ree
Fumiko Samejima
Elizabeth E. Stewart
Hariharan Swaminathan
David Thissen
Tom E. Trabin
Howard Wainer
David J. Weiss

NEW HORIZONS IN TESTING

Latent Trait Test Theory and Computerized Adaptive Testing

Edited by

DAVID J. WEISS

Department of Psychology
University of Minnesota
Minneapolis, Minnesota

1983

ACADEMIC PRESS
A Subsidiary of Harcourt Brace Jovanovich, Publishers
New York London
Paris San Diego San Francisco São Paulo Sydney Tokyo Toronto

ACADEMIC PRESS, INC.
111 Fifth Avenue, New York, New York 10003

United Kingdom Edition published by
ACADEMIC PRESS, INC. (LONDON) LTD.
24/28 Oval Road, London NW1 7DX

Library of Congress Cataloging in Publication Data
Main entry under title:

New horizons in testing.

Derived from a conference sponsored by the
U. S. Office of Naval Research.
Includes bibliographies and index.
1. Psychological tests--Congresses. I. Weiss,
David J. II. United States. Office of Naval
Research
BF176.N48 1983 153.9 82-24374
ISBN 0-12-742780-5

PRINTED IN THE UNITED STATES OF AMERICA

83 84 85 86 9 8 7 6 5 4 3 2 1

Contents

2

Robustness of Item Response Models and Effects of Test Length and Sample Size on the Precision of Ability Estimates 31

RONALD K. HAMBLETON LINDA L. COOK

3

Small *N* Justifies Rasch Model 51

FREDERIC M. LORD

4

Are We Correcting for Guessing in the Wrong Direction? 63

HOWARD WAINER

II
BEYOND THE ABILITY ESTIMATE

5
The Person Response Curve: Fit of Individuals to Item Response Theory Models 83

TOM E. TRABIN DAVID J. WEISS

6
Appropriateness Measurement: Validating Studies and Variable Ability Models 109

MICHAEL V. LEVINE FRITZ DRASGOW

III
APPLICATIONS TO TESTING PROBLEMS

7
Effects of Sample Size on Linear Equating of Item Characteristic Curve Parameters 135

MALCOLM JAMES REE HARALD E. JENSEN

8

A Test of the Adequacy of Curvilinear Score Equating Models 147

GARY L. MARCO NANCY S. PETERSEN ELIZABETH E. STEWART

9

Timed Testing: An Approach Using Item Response Theory 178

DAVID THISSEN

10

The Mental Growth Curve Reexamined 205

R. DARRELL BOCK

IV
APPLICATION TO ADAPTIVE TESTING

11
Reliability and Validity of Adaptive Ability Tests in a Military Setting 223

JAMES R. McBRIDE JOHN T. MARTIN

12
A Procedure for Decision Making Using Tailored Testing 237

MARK D. RECKASE

13
A Comparison of IRT-Based Adaptive Mastery Testing and a Sequential Mastery Testing Procedure 257

G. GAGE KINGSBURY DAVID J. WEISS

V
MODELS FOR SPECIAL APPLICATIONS

14
The Constant Information Model on the Dichotomous Response Level 287

FUMIKO SAMEJIMA

15
Some Latent Trait Models for Measuring Change in Qualitative Observations 309

GERHARD H. FISCHER

Contributors

Numbers in parentheses indicate the pages on which the authors' contributions begin.

R. Darrell Bock (205), Departments of Behavioral Sciences and of Education, The University of Chicago, Chicago, Illinois 60637

Linda L. Cook (31), Educational Testing Service, Princeton, New Jersey 08541

Fritz Drasgow (109), Department of Psychology, University of Illinois, Champaign, Illinois 61820

Gerhard H. Fischer (309), Institut für Psychologie, University of Vienna, A-1010 Wien, Austria

Janice A. Gifford (13), School of Education, University of Massachusetts, Amherst, Massachusetts 01003

Ronald K. Hambleton (31), Laboratory of Psychometric and Evaluative Research, School of Education, University of Massachusetts, Amherst, Massachusetts 01002

Harald E. Jensen[1] (135), Air Force Human Resources Laboratory, Brooks Air Force Base, Texas 78235

[1] Present address: Office of Institutional Research, U.S. Air Force Academy, Colorado Springs, Colorado 80840.

G. Gage Kingsbury (257), Department of Psychology, University of Minnesota, Minneapolis, Minnesota 55455

Michael V. Levine (109), Department of Educational Psychology, University of Illinois, Urbana, Illinois 61801

Frederic M. Lord (51), Educational Testing Service, Princeton, New Jersey 08541

James R. McBride (223), Navy Personnel Research and Development Center, San Diego, California 92152

Gary L. Marco (147), Educational Testing Service, Princeton, New Jersey 08541

John T. Martin (223), Department of Psychology, University of Minnesota, Minneapolis, Minnesota 55455

Nancy S. Petersen (147), Educational Testing Service, Princeton, New Jersey 08541

Mark D. Reckase (237), ACT, P. O. Box 168, Iowa City, Iowa 52243

Malcolm James Ree (135), Air Force Human Resources Laboratory, Brooks Air Force Base, Texas 78235

Fumiko Samejima (287), Department of Psychology, The University of Tennessee, Knoxville, Tennessee 37916

Elizabeth E. Stewart (147), Educational Testing Service, Princeton, New Jersey 08541

Hariharan Swaminathan (13), School of Education, University of Massachusetts, Amherst, Massachusetts 01003

David Thissen (179), Department of Psychology, The University of Kansas, Lawrence, Kansas 66045

Tom E. Trabin (83), Department of Psychology, University of Minnesota, Minneapolis, Minnesota 55455

Howard Wainer (63), Educational Testing Service, Princeton, New Jersey 08541

David J. Weiss (1, 83, 257), Department of Psychology, University of Minnesota, Minneapolis, Minnesota 55455

Preface

In 1968 Bert F. Green, Jr. predicted "the inevitable computer conquest of testing [Green, 1970, p. 194]." This book provides a first snapshot of the progress of the slow but deliberate revolution that psychological and educational measurement are undergoing under the influence of the computer. It provides an in-depth look at the two new frontiers in psychological measurement that have been opened up by the computer—latent trait test theory (item response theory) and computerized adaptive testing.

The availability of computers beginning in the late 1960s permitted latent trait test theory and computerized adaptive testing to develop and flourish during the 1970s and into the early 1980s. As computers became available for the processing of large data sets and the analysis of psychometric data in ways not before possible, latent trait theory—which had been developed earlier but was not implementable prior to the existence of computers—became practical. By the mid-1970s, minicomputers (and later microcomputers) made available the power of interactive computing in research laboratories and later in practical testing environments. This development led to the implementation of computerized adaptive testing.

The implementation of latent trait theory and its practical application in computerized adaptive testing have already begun to result in major changes in the way psychological testing is done. These changes, which began to occur in the late 1970s and the early 1980s, overshadowed in a span of a few years more than 60 years of prior work in test development and related areas of psychometrics.

In the United States, the development of latent trait test theory and computerized adaptive testing were spurred by the interest of the U.S. military in supporting basic and applied research in psychometrics. The financial support of the Office of Naval Research (ONR) for research in psychometrics, specifically in latent trait test theory, resulted in the publication of the classic *Statistical Theories of Mental Test Scores* (Lord & Novick, 1968). Later, ONR and other agencies stimulated the development of applications of latent trait theory by their support of a variety of research projects and investigators concerned with problems of latent trait test theory and computerized adaptive testing. The work of many of these researchers is represented in this volume.

To stimulate interaction of researchers concerned with these new psychometric problems, ONR and other agencies sponsored a series of conferences, called the computerized adaptive testing conferences. The first of these (Clark, 1976) was held in 1975 and sponsored by ONR and the U.S. Civil Service Commission (later renamed the Office of Personnel Management). The second conference, held in 1977 at the University of Minnesota (Weiss, 1978), was jointly sponsored by ONR and the Air Force Office of Scientific Research (AFOSR). The third conference, the 1979 Computerized Adaptive Testing Conference held at Spring Hill Conference Center in Wayzata, Minnesota, formed the basis for this book. This conference was jointly sponsored by ONR, AFOSR, Defense Advanced Research Projects Agency (DARPA), Military Enlistment Processing Command (MEPCOM), and Navy Personnel Research and Development Center (NPRDC). The majority of the chapters in this book are derived from those presented at that conference, with a few chapters based on papers presented at the conference or discussions that occurred at the conference. Additional papers presented at the conference but not included in this book can be found in Weiss (1980).

Because of the rapid development in the last decade of latent trait test theory, its applications, and computerized adaptive testing, very little of this research has appeared in the journal or other archival literatures. Although this is somewhat less true of the research in latent trait theory and its applications, it is certainly true of the research in computerized adaptive testing. Most of the current research in both fields, however, has been distributed via the proceedings of conferences and through technical reports that are interchanged among active researchers in these fields. Consequently, it was appropriate to bring together into one volume the papers that define the state of the art in these two areas.

Much of the work presented in this book presupposes a basic understanding of latent trait test theory. For readers not familiar with latent trait theory, papers by Hambleton *et al.* (Hambleton & Cook, 1977;

Hambleton, Swaminathan, Cook, & Eignor, 1978) provide excellent introductions. Allen and Yen (1979) also provide a basic introduction to latent trait test theory. For a somewhat more advanced treatment see Lord (1980), and for a mathematical treatment see Chapters 16–20 of Lord and Novick (1968).

Thus, this book represents the first collection of state-of-the-art research on latent trait test theory and computerized adaptive testing. It is designed to bring to the reader an appreciation of the current state of research in these fields as well as specific information on important aspects of latent trait test theory and computerized adaptive testing, their applications, and the underlying models. It should help readers appreciate the range of applicability of latent trait theory and computerized adaptive testing to specific psychometric problems and give them an appreciation for the utility of both of these fields for the solution of important psychometric problems.

This book should be useful to both psychometricians who are new to latent trait test theory and computerized adaptive testing, as well as those with more advanced knowledge. Both applied and theoretical readers will find chapters of interest. Although some of the chapters are relatively technical and presuppose a basic knowledge of the ideas of latent trait test theory and computerized adaptive testing, the introduction to the book and the introductions to each of the sections are designed to help give the newcomer an awareness of what follows and the relationships of the chapters to each other.

Although this book is directed toward the psychometrician who is concerned with the development of psychometric models and procedures and their implementation, it should also be of interest to those who use or construct psychological tests in any environment. While it is difficult to change the way applied psychometricians have constructed and implemented tests for the last 60 years, the last three chapters on psychometrics in the *Annual Review of Psychology* (Bock & Wood, 1971; Lumsden, 1976; Weiss & Davison, 1982) have all made a strong case for the abandonment of classical test models in favor of the latent trait test theory models discussed in this book. It is hoped that this book will spur the development and application of latent trait test theory models and the application of computerized adaptive testing to practical testing problems.

The book is organized into five parts. The first, on parameter estimation, addresses basic problems in estimating the parameters of the item response theory models that constitute a class of latent trait test theory models. These parameters include the constants that describe the test items and the parameters that describe the trait levels of individuals. In the second part, the implications of item response theory for measuring individuals using more than just simply a trait level (e.g., ability) score are discussed. Part III describes the application of item response theory models to specific applied problems, including the problem of equating tests or linking items into a pool, a latent trait model for timed tests, and the problem of measuring growth using scores derived from the application of item response theory models. Part IV is concerned with the application of item response theory to computerized adaptive testing. It describes the application of computerized adaptive testing to obtain measurements with equal precision across individuals, and to classifications such as those used in mastery decisions. Finally, Part V includes discussion of two special models beyond the standard models used in the rest of the book. One of these models, the constant information model, is a simplification of the general latent trait models, whereas the other is an extension of latent trait models to the problem of measuring change.

REFERENCES

Allen, M. J., & Yen, W. M. *Introduction to measurement theory*. Monterey, California: Brooks/Cole, 1979.

Bock, R. D., & Wood, R. Test theory. *Annual Review of Psychology*, 1971, *22*, 193–224.

Clark, C. L. (Ed.). *Proceedings of the first conference on computerized adaptive testing* (U.S. Civil Service Commission, Personnel and Development Center PS-75-6). Washington, D.C.: U.S. Government Printing Office, 1976. (Superintendent of Documents Stock No. 006-00940-9.)

Green, B. F., Jr. Comments on tailored testing. In W. H. Holtzman (Ed.), *Computer-assisted instruction, testing, and guidance*. New York: Harper & Row, 1970.

Hambleton, R. K., & Cook, L. L. Latent trait models and their use in the analysis of educational test data. *Journal of Educational Measurement*, 1977, *14*, 75–96.

Hambleton, R. K., Swaminathan, H., Cook, L. L., Eignor, D. R., & Gifford, J. L. Developments in latent trait theory: Models, technical issues, and applications. *Review of Educational Research*, 1978, *48*, 467–510.

Lord, F. M. *Applications of item response theory to practical testing problems*. Hillsdale, N.J.: Lawrence Erlbaum Associates, 1980.

Lord, F. M., & Novick, M. R. *Statistical theories of mental test scores*. Reading, Mass.: Addison-Wesley, 1968.

Lumsden, J. Test theory. *Annual Review of Psychology*, 1976, *27*, 251–280.

Weiss, D. J. (Ed.) *Proceedings of the 1977 Computerized Adaptive Testing Conference*. Minneapolis: University of Minnesota, Department of Psychology, Psychometric Methods Program, 1978.

Weiss, D. J. (Ed.), *Proceedings of the 1979 Computerized Adaptive Testing Conference*. Minneapolis: University of Minnesota, Department of Psychology, Psychometric Methods Program, Computerized Adaptive Testing Laboratory, 1980.

Weiss, D. J., & Davison, M. L. Test theory and methods. *Annual Review of Psychology*, 1981, 629–658.

Acknowledgments

As indicated in the preface, the conference on which this book was based and the preparation of this book would not have been possible without the financial support of ONR, AFOSR, DARPA, MEPCOM, and NPRDC (under ONR Contract N00014-79-C-0196, NR150-432). Particular thanks are due to Marshall J. Farr of ONR, who arranged the funding for the conference. Preparation of this book was also facilitated by contract N00014-79-C-0172, NR150-433, also from ONR with additional funding from AFOSR and the Army Research Institute. Thanks are also due to the *British Journal of Mathematical and Statistical Psychology* for permission to reproduce portions of Chapter 6 from a paper by Michael Levine and Fritz Drasgow entitled "Appropriateness measurement: Review, critique, and validating studies," which appeared in the *British Journal of Mathematical and Statistical Psychology*, 1982, *35*, 42–56.

NEW HORIZONS IN TESTING

Latent Trait Test Theory and Computerized Adaptive Testing

Introduction

DAVID J. WEISS

LATENT TRAIT THEORY

The term *latent trait theory* embraces a family of mathematical models. These models are statements of functional relations between observable variables and the underlying hypothetical trait constructs that give rise to the observable variables. Latent trait models consist of several elements:

1. There must be a stimulus variable, or set of variables, that is presented to individuals. These variables can be, for example, test items on an ability test or an achievement test, personality questionnaire items, or items on an attitude scale.
2. The items are presented to an individual, and they elicit certain responses that are observed and recorded.
3. To enable the psychometrician to infer a person's status on the trait based on the observed responses to a specified stimulus variable, or set of stimulus variables, the hypothesized relationships between the observed responses and the underlying trait levels are formalized by an equation that describes the functional form of that relationship.

NEW HORIZONS IN TESTING
Latent Trait Test Theory and
Computerized Adaptive Testing

ISBN 0-12-742780-5

Thus the core of latent trait models is the equation expressing the relationship between the observable response to a stimulus (such as a test item) and the underlying trait level as a function of the characteristics of the stimulus variable.

When the characteristics of the stimulus variables are known elements in the equation, the purpose of latent trait models is to estimate the unobservable trait levels for individuals based on their observable responses to the specified stimuli, using the mathematical function that is assumed to describe that relationship. Thus, given an individual's set of observable responses to test items of known characteristics, latent trait models enable the inference of the person's trait level from the observable responses to those items.

When latent trait theories are applied to tests of ability or achievement, they have been known as item characteristic curve theory, or most recently, item reponse theory. These terms will be used interchangeably throughout this book. The *item characteristic curve* is the curve that portrays the probability of a correct response to a test item as a function of trait levels that would give rise to those probabilities. The term *item response theory* emphasizes both the role of the test item and the responses of examinees.

At first glance, item response theory may appear to be quite different from classical test theory, but the idea of item response theory is implicit in the classical test theory that has been used in psychological measurement since the 1920s (e.g., Gulliksen, 1950; Lord & Novick, 1968, Chaps. 1–11). In classical test theory, the observed score on a test is not taken to be an exact measurement on an individual, but is assumed to include error, and thus functions as an imperfect estimator of an unobservable "true" score. As a consequence, classical test theory is concerned with the reliability of measurements, and more specifically with the standard error of measurement that reflects the degree of error associated with an observed score as an estimator of true score. A "true" score is analogous to the trait levels used in latent trait theory, since neither true score in classical test theory nor trait level (usually designated by the Greek letter theta, θ) in item response theory is observable.

Also similar to item response theory, classical test theory assumes a functional mathematical relationship between the observed variable "observed score" and the latent or unobservable variable "true score." In classical test theory this relationship is assumed to be linear—that is, the index of reliability (the square root of the reliability coefficient) describes the linear functional relationship between observed and true scores. Thus classical test theory incorporates a simple linear mathematical model that enables the estimation of a latent variable (true score) from an observable variable (observed score, or total number correct). In this sense, then, even classical test theory is a latent trait model, although a very simple one.

While the ideas of latent trait theory were implicit in much of early test theory, they were also implicit in other aspects of psychological measurement. The attitude scaling work done by Thurstone also includes ideas of latent trait theory (Lumsden, 1980). In Thurstone's models, as in other latent trait models, the observable variables were not considered important in themselves except as representations and estimators of an unobservable trait status. Thurstone's scaling models thus related the observed responses of individuals on an attitude questionnaire to an inferred latent trait status. Brogden (1977) also describes relationships between latent trait models and Thurstone's law of comparative judgment, as well as relationships between latent trait models and additive conjoint measurement. Relationships between item response models and Likert scaling are described by Andersen (1977) and Andrich (1978). Thus latent trait models are not new in psychological measurement, but what is new is the capability to apply them in practical measurement situations and to apply models that are more complex, and therefore more informative, than the simple linear model implicit in classical test theory.

The formal roots of latent trait theory appear in a number of places. Relationships between latent trait concepts and psychophysics were described by Mosier (1940, 1941), with basic statistical methodology having been proposed by Lawley (1943). At the same time that Mosier was working within the field of test theory and Lawley was approaching the problem from a statistical point of view, Guttman (1941, 1944) developed the basics of latent trait theory to solve scaling problems of attitude measurement. Lazarsfeld (1950) developed some of the same ideas into latent structure analysis, a class of latent trait models concerned with estimating categorical group membership of individuals rather than levels of an assumed continuous variable.

The item response theory version of latent trait theory, that is, latent trait theory applied to tests of ability and achievement and specifying a class of latent trait models primarily for dichotomous item responses (e.g., correct or incorrect) was formalized by Lord (1952). Important work on the models that now define item response theory was done by Birnbaum and incorporated (as Chapters 17–20) into Lord and Novick's (1968) formalization of latent trait models. Lord has continued to develop and refine these models (Lord & Novick, 1968; Lord, 1980), as well as procedures for estimating model parameters (e.g., Wingersky, Barton, & Lord, 1982; Wood, Wingersky, & Lord, 1978).

A parallel development in the European countries was the development of the one-parameter logistic model by Rasch (1960, 1980), which is a special case of the more complex two- and three-parameter models developed by Lord and others. Research extending latent trait models beyond the dichotomous (correct–incorrect) item response to the use of additional information in individuals' responses to test or attitude scale items was done by Samejima (e.g., 1969) and Bock (e.g., 1972). Other current contributors to the development and application of latent trait models will be met later in this book.

While latent trait theories were originally formalized based on ideas that had been implicit in psychological measurement since its inception, their practical implementation was not feasible until the availability of computing facilities in the late 1960s. As scientific computing facilities became available to psychometricians, procedures became operational for estimating the parameters of the mathematical models on which latent trait theories are based. The first problem was that of determining the characteristics of test items that define the set of constants in the latent trait mathematical models. Related to this problem was that of estimating trait levels of individuals from their observable responses to test items with known characteristics. As techniques for solving these two problems became available, latent trait theories became practical. One exception to this trend was the one-parameter Rasch model, which enjoyed popularity during the 1960s because a subset of its procedures did not require the use of computers for estimating item parameters or trait levels of individuals (e.g., Wright & Stone, 1979). As procedures became available for estimating the second and third parameters of test items and more complex procedures for estimating individual trait levels became practical, however, the more realistic two- and three-parameter models began to replace the more simplified one-parameter model.

In addition to providing powerful solutions to a variety of applied psychometric problems, latent trait models have an additional advantage over the classical test models that preceded them. Because the models specify a functional relationship between observable responses and unobservable trait levels, this functional relationship can be used in computer simulation to investigate the behavior of the models and their applicability to a wide range of measurement problems. This capability has provided a powerful tool for the rapid evaluation of a variety of circumstances that may affect the utility of a specific latent trait model. Monte Carlo simulations enable the psychometrician to evaluate rapidly the effects of a variety of conditions upon the performance of various models and to compare procedures for estimating the parameters of both items and individuals. Therefore a substantial amount of the research in this book is based on computer simulations.

ADAPTIVE TESTING

Adaptive testing, also sometimes called tailored testing, involves the selection of test items during the process of administering a test so that the items administered to each individual are appropriate in difficulty for that individual. The result is a test that is "adapted" or "tailored" to each individual's ability or trait level during the process of test administration, so that the test is neither too easy nor too difficult for the individual. In an adaptive test, items are selected out of a larger pool using a set of rules, or "strategies," that may operate differently. While adaptive testing is not dependent on item response theory (Weiss, 1973, 1974; Weiss & Betz, 1973), item response theory is useful in the efficient implementation of adaptive testing.

Like item response theory (IRT), adaptive testing has its roots in early psychological measurement as well as in the psychophysical tradition of psychology. The first adaptive test was developed by Alfred Binet. While Binet's "strategy" was relatively simplistic in comparison to IRT-based adaptive tests, his test had all the characteristics of current adaptive tests:

1. The Binet test used variable entry, in that the first set of items to be administered to an individual depended on the examinee's ability level as estimated by the test administrator prior to the start of testing.
2. Test items were scored as they were administered, and the correctness of the test items was used to determine the branching, or selection of items to be administered next to an individual.
3. Binet's test had a variable termination criterion, which meant that different individuals could be administered tests of different lengths. In the case of the Binet tests, testing was terminated when a "ceiling level" was reached at which no items were answered correctly by an examinee.

Current adaptive tests (e.g., Weiss, 1982) are considerably more complex, but still maintain the characteristics of the Binet tests.

Adaptive procedures have been used in psychophysical laboratories for many years for measuring thresholds (e.g., Mayer, Hooper, Nagle, & DeMayo, 1982). Experimental psychologists have called these adaptive convergence procedures by various names, including the Method of Adjustment (Method of Average Error, Production Method) and the Method of Limits (Method of Minimal Change, Method of Serial Exploration, Staircase Method). All these methods use variations of adaptive procedures to converge on perceptual thresholds for individuals.

In psychometrics, after Binet's implementation of the individually administered adaptive test, adaptive testing lay dormant until the early 1950s. At that time, the idea of adaptive testing by paper-and-pencil was briefly examined, but abandoned because of its complexity. Other simple procedures of adaptive testing (e.g., two-stage tests in which the second test given to an individual is dependent on the examinee's performance on a first test) were investigated in studies that simulated the responses of individuals to adaptive tests based on conventional tests that had already been administered (see Weiss & Betz, 1973, for a review of this early work).

As the availability of interactive computers increased in the early 1970s, adaptive testing became computerized adaptive testing (CAT), in which test items are administered by interactive computers on cathode-ray terminals, and examinees respond on the terminal keyboard. The computer is used as a means of selecting the next item to be administered, and early research was based on mechanical branching rules not using item response theory (e.g., Betz & Weiss, 1973, 1974, 1975; Larkin & Weiss, 1974; Vale & Weiss, 1975). As procedures for item response theory became practical, CAT and IRT merged into the current IRT-based strategies.

There are currently two primary IRT-based item selection strategies for implementing CAT (Weiss, 1982). In the maximum information item selection strategy (used by Kingsbury and Weiss in Chapter 13) values of item information are calculated for each item at an individual's current estimated level of the trait ($\hat{\theta}$). All items in the pool are searched, and the item having the maximum value of information at the individual's current $\hat{\theta}$ level is selected for administration. The item is scored, and θ is reestimated. This procedure continues until a predetermined termination rule is invoked. Related to this item selection strategy is the Bayesian item selection strategy (used by McBride and Martin in Chapter 11), which has the capability to use prior information about the examinee more completely than does the maximum information item selection strategy. In this item selection approach, an item pool is searched to find the one item that when administered will maximally reduce the posterior variance of an individual's ability estimate. Closely associated with these two strategies are two IRT-based methods for estimating ability, maximum likelihood and Bayesian (Bejar & Weiss, 1979). While the maximum likelihood scoring method is most frequently used with maximum information item selection and the Bayesian scoring approach is used with Bayesian item selection, it is possible to combine the scoring methods and item selection strategies in a reverse combination (e.g., see Chapter 13).

Similar to Binet's application of adaptive testing, much of the work in adaptive testing was originally developed for the problem of measuring traits for individuals with a wide range of possible trait levels. With some modifications, however, adaptive testing can also be used for the problem of mastery or dichotomous classifications. Both these applications are represented in this book.

Both item response theory and computerized adaptive testing represent new horizons in psychological testing that have developed for practical application during the late 1970s and early 1980s. Both these areas, singly and together, promise to have profound implications for the improvement of psychological measurement and for the solution of a variety of applied problems that have not been adequately solved by over a half century of classical psychometrics. At the same time, they have opened a variety of new research questions and new ways of doing psychometric research. This book provides a description of the state of the art in both those fields as a road map for future developments in psychological testing.

REFERENCES

Andersen, E. B. Sufficient statistics and latent trait models. *Psychometrika*, 1977, *42*, 69–81.

Andrich, D. Application of a psychometric rating model to ordered categories which are scored with successive integers. *Applied Psychological Measurement*, 1978, *2*, 581–594.

Bejar, I. I., & Weiss, D. J. *Computer programs for scoring test data with item characteristic curve models* (Research Report 79-1). Minneapolis: University of Minnesota, Department of Psychology, Psychometric Methods Program, February 1979.

Betz, N. E., & Weiss, D. J. *An empirical study of computer-administered two-stage ability testing* (Research Report 73-4). Minneapolis: University of Minnesota, Department of Psychology, Psychometric Methods Program, October 1973.

Betz, N. E., & Weiss, D. J. *Simulation studies of two-stage ability testing* (Research Report 74-4). Minneapolis: University of Minnesota, Department of Psychology, Psychometric Methods Program, October 1974.

Betz, N. E., & Weiss, D. J. *Empirical and simulation studies of flexilevel ability testing* (Research Report 75-3). Minneapolis: University of Minnesota, Department of Psychology, Psychometric Methods Program, July 1975.

Bock, R. D. Estimating item parameters and latent ability when responses are scored in two or more nominal categories. *Psychometrika*, 1972, *37*, 29–51.

Brogden, H. The Rasch model, the law of comparative judgment and additive conjoint measurement. *Psychological Bulletin*, 1977, *42*, 631–634.

Gulliksen, H. *Theory of mental tests*. New York: Wiley, 1950.

Guttman, L. A basis for scaling qualitative data. *American Sociological Review*, 1944, *9*, 139–150.

Guttman, L. The quantification of a class of attributes: A theory and method of scale construction. In P. Horst *et al.* (Eds.) *Prediction of personal adjustment*. New York: Social Science Research Council, 1941.

Larkin, K. C., & Weiss, D. J. *An empirical investigation of computer-administered pyrami-*

dal ability testing (Research Report 74-3). Minneapolis: University of Minnesota, Department of Psychology, Psychometric Methods Program, July 1974.

Lawley, D. N. On problems connected with item selection and test construction. *Proceedings of the Royal Society of Edinburgh, Series A,* 1943, *61,* 273–287.

Lazarsfeld, P. F. The logical and mathematical foundation of latent structure analysis. In S. A. Stouffer, L. Guttman, E. A. Suchman, P. F. Lazarsfeld, S. A. Star, & J. A. Clausen. (Eds.), *Measurement and prediction.* Princeton, New Jersey: Princeton University Press, 1950.

Lord, F. M. A theory of test scores. *Psychometric Monograph,* 1952, No. 7.

Lord, F. M. *Applications of item response theory to practical testing problems.* Hillsdale, New Jersey: Lawrence Erlbaum Associates, 1980.

Lord, F. M., & Novick, M. R. *Statistical theories of mental test scores.* Reading, Massachusetts: Addison-Wesley, 1968.

Lumsden, J. Variations on a theme by Thurstone. *Applied Psychological Measurement,* 1980, *4,* 1–7.

Mayer, M. J., Hooper, K. S., Nagle, M., & DeMayo, R. *Demonstrations of sensory perception.* New York: Wiley, 1982.

Mosier, C. I. Psychophysics and mental test theory: Fundamental postulates and elementary theorems. *Psychological Review,* 1940, *47,* 355–366.

Mosier, C. I. Psychophysics and mental test theory. II. The constant process. *Psychological Review,* 1941, *48,* 235–249.

Rasch, G. *Probabilistic models for intelligence and attainment tests.* Copenhagen: Denmarks Paedagogiske Institut, 1960.

Rasch, G. *Probabilistic models for some intelligence and attainment tests (expanded ed.).* Chicago, Illinois: University of Chicago Press, 1980. [Original edition: Copenhagen: Denmarks Paedagogiske Institut, 1960.]

Samejima, F. Estimation of latent ability using a response pattern of graded scores. *Psychometrika Monograph Supplement,* 1969, *34* (4, Pt. 2, Monograph, No. 17).

Vale, C. D., & Weiss, D. J. *A study of computer-administered stradaptive ability testing* (Research Report 75-4). Minneapolis: University of Minnesota, Department of Psychology, Psychometric Methods Program, October 1975.

Weiss, D. J. *The stratified adaptive computerized ability test* (Research Report 73-3). Minneapolis: University of Minnesota, Department of Psychology, Psychometric Methods Program, September 1973.

Weiss, D. J. *Strategies of adaptive ability measurement* (Research Report 74-5). Minneapolis: University of Minnesota, Department of Psychology, Psychometric Methods Program, December 1974.

Weiss, D. J. Improving measurement quality and efficiency with adaptive testing. *Applied Psychological Measurement,* 1982, *6,* 473–492.

Weiss, D. J., & Betz, N. E. *Ability measurement: Conventional or adaptive?* (Research Report 73-1). Minneapolis: University of Minnesota, Department of Psychology, Psychometric Methods Program, February 1973.

Wingersky, M. S., Barton, M. A., & Lord, F. M. *LOGIST user's guide: Logist, version 1.0.* Princeton, New Jersey: Educational Testing Service, 1982.

Wood, R. L., Wingersky, M. S., & Lord, F. M. *LOGIST: A computer program for estimating examinee ability and item characteristic curve parameters* (Research memorandum 76-6). Princeton, New Jersey: Educational Testing Serivce, 1976. (Modified January 1978.)

Wright, B. D., & Stone, M. H. *Best test design: Rasch measurement.* Chicago, Illinois: Mesa, 1979.

I

ESTIMATING ITEM PARAMETERS AND ABILITY

As indicated, IRT models specify the probabilistic relationship between the observed responses of an individual to a test item and the individual's level on the latent trait. The model is an equation having constants and variables. Equation (1.1) in Chapter 1 and Eq. (6.1) in Chapter 6 are examples of the mathematical models used in item response theory to express this relationship.

In the simplest case, the variables in the equation are the difficulty of a test item and the ability or trait level of an individual. This is usually referred to as the one-parameter model. When the assumed functional relationship between the unobservable trait levels and observed responses for an item is a curve called a logistic curve, the model is the one-parameter logistic model, which has been described by Rasch and others. While this simple model allows items to differ in their difficulties, it does not allow items to differ in terms of other characteristics. Thus additional variables or parameters are needed in the equation to describe more fully the characteristics of test items and to allow them to vary in terms of these characteristics.

In the two-parameter case, two parameters are used to describe test items. In addition to difficulty levels, items are permitted to vary in terms of how quickly the probability of a correct response changes as a function of underlying trait level or in terms of item discrimination (see, e.g., Figure 7.1 in Chapter 7). When test items can be answered correctly by chance guessing, such as is the case with a multiple-choice test item or a true-or-false test item, a third parameter is used to describe the characteristics of the test items more completely. This parameter, known as the *c* or pseudoguessing parameter, increases the number of parameters for the items to three, resulting in the three-parameter model. Where the equation describing the probability of a correct response as a function of the underlying trait level is the logistic curve, the three-parameter logistic model is a generalization of the one-parameter logistic model. In some cases, the curve or function relating probability of correct response to an underlying trait level is assumed to be a normal ogive, a close relative of the logistic curve, resulting in the three-parameter normal ogive model.

To implement IRT models, it is necessary to estimate the values of the variables in the equation, which are the characteristics of the test item (i.e., the item parameters) and the trait levels of individuals (i.e., the person parameter). This is usually done in two steps. In the first step, the focus is on estimating the parameters of single items. This phase of estimation is usually referred to as item calibration, or item parameterization. The result is a set of values that characterize each of a number of test items, with the number of parameters to be estimated depending on the model being used (i.e., one- versus two- versus three-parameter models). Although the process of estimating item parameters also involves the estimation of the trait level parameters of the individuals to whom the items were administered, interest at this stage is primarily in estimating the parameters of the items, and the trait level parameters are only incidental. Therefore, they are generally not used to characterize individuals in that group for measurement purposes.

There are several ways of estimating the parameters of a set of test items, given the responses scored correct or incorrect of a group of individuals to those items. In Chapter 1, Swaminathan and Gifford compare two estimation procedures—ANCILLES and LOGIST—on a variety of criteria and under different conditions. Computer simulation is used here, as in many of the chapters of this book, since it permits the generation of item response data with known trait levels of individuals and known parameter values characterizing each of the items. The question at issue is which estimation procedure can better reproduce the known values of the item parameters—usually referred to as *a* for discrimination, *b* for difficulty, and *c* for the pseudoguessing parameter—with different sample sizes, numbers of items, and other characteristics of the data.

The second stage of estimation in IRT models is to estimate the trait levels of individuals (usually referred to as "ability" for convenience) when the item parameters for the items used in a test have been previously estimated with other data. In this application, the item parameters become constants in the mathematical model, and person responses vary, as do the person trait levels that give rise to those responses. In the process of trait estimation, however, different numbers of parameters can be used, and the parameters themselves can have errors in them, as shown by Swaminathan and Gifford in Chapter 1. In Chapter 2, Hambleton and Cook address these problems concerning the effects of different numbers of item parameters under different conditions on the estimation of trait levels for individuals. In addition, they examine the effects of errors in item parameters on the standard errors of estimate conditional on θ that are associated with different situations in which θ is being estimated.

In Chapter 3, Lord addresses a similar problem. Lord is concerned with whether the one- or two-parameter model is better for estimating θ, as a function of sample size, even though the one-parameter model may be technically inappropriate since items differ in their discriminations. The one-parameter model permits the use of the number-correct score as a means of estimating ability, whereas in the two-parameter model ability is estimated by a score weighted by the discriminations of the items. If the items differ in discrimination, the use of the one-parameter model is technically incorrect. When item discriminations are estimated with relatively small sample sizes, however, the errors of the item discrimination parameter estimates may be so large that using those estimates may be inefficient. Lord investigates this problem to determine under what circumstances the use of the one-parameter model is appropriate even though technically incorrect if small sample sizes have been used to estimate the item discrimination parameters.

When ability is being estimated using the three-parameter model, the pseudoguessing (c) parameter of the items is used to correct for chance successes due to guessing. The effect of this correction is most obvious for low-ability examinees, but there is some effect through most of the ability range, by artificially raising the probability of a correct response to an item as a function of guessing. Only for examinees of very high ability, where the probability of a correct response is 1.0 or close to it, is there no effect as a result of the pseudoguessing parameter. As Wainer points out in Chapter 4, the effect of correcting for guessing this way is to lower the ability estimates. Wainer argues in his chapter that instead of always lowering ability estimates, the effect of guessing on ability estimates should be based on the entire response pattern and its deviation from the model, and in some cases an increased ability estimate might be appropriate if guessing is properly taken into account. Wainer then proposes a way to improve IRT ability estimates when guessing is a problem.

1

Estimation of Parameters in the Three-Parameter Latent Trait Model

HARIHARAN SWAMINATHAN   JANICE A. GIFFORD

The successful application of latent trait theory to practical measurement problems hinges on the availability of procedures for the estimation of the parameters. Hence, investigations of the adequacy of the available procedures for estimating parameters in latent trait models are necessary and, indeed, play a crucial role when assessing the usefulness of latent trait theory.

Although the problem of estimating parameters in the one-parameter latent trait model appears to be solved, some degree of controversy seems to surround the estimation of parameters in the two- and three-parameter models (Andersen, 1973; Wright, 1977). Lord (1975) has empirically evaluated the maximum likelihood procedure for estimating the parameters in the three-parameter model and has provided answers to some of the questions that arise with respect to estimation of parameters. Jensema (1976) has compared the efficiency of a heuristic procedure suggested by Urry (1974) for estimating the parameters in the three-parameter model with the maximum likelihood procedure. Ree (1979) has compared the properties of the Urry estimators and the maximum likelihood estimators and has investigated the effect of violating the underlying assumptions on the estimates, fixing the test length (80 items) and the number of examinees, however. Despite these efforts, little is known regarding the statistical

NEW HORIZONS IN TESTING
Latent Trait Test Theory and
Computerized Adaptive Testing

ISBN 0-12-742780-5

properties of the estimators in the three-parameter model and the effect of test length and examinee population size on the estimates.

PURPOSE

The purpose of this study was to investigate the efficiency of the Urry (1976) procedure and the maximum likelihood procedure for estimating parameters in the three-parameter model, to study the properties of the estimators, and to provide some guidelines regarding the conditions under which they should be employed. In particular, the issues investigated were (*a*) the "accuracy" of the two estimation procedures; (*b*) the relations among the number of items, examinees, and the accuracy of estimation; (*c*) the effect of the distribution of ability on the estimates of item and ability parameters; and (*d*) the statistical properties, such as bias and consistency, of the estimators.

DESIGN OF THE STUDY

In order to investigate the issues mentioned above, artificial data were generated according to the three-parameter logistic model

$$P_{ij}(\theta) = c_i + (1 - c_i)\{1 + \exp[-1.7a_i(\theta_j - b_i)]\}^{-1} \qquad (1.1)$$

using the DATGEN program of Hambleton and Rovinelli (1973). Data were generated to simulate various testing situations by varying the test length, the number of examinees, and the ability distribution of the examinees. Test lengths were fixed at 10, 15, 20, and 80 items. Since the accuracy of maximum likelihood estimation with large numbers of items has been sufficiently documented by Lord (1975), tests with small numbers of items—10, 15, and 20—were chosen so that the accuracy of the estimation procedure could be ascertained for short tests. This is particularly important if latent trait theory is to be applied to criterion-referenced measurement. Similarly, the sizes of examinee population were set at 50, 200, and 1000 in order to study the effect of small sample size on the accuracy of estimation.

In the Urry (1976) estimation procedure, the relationships that exist for item discrimination and item difficulty between the latent trait theory parameters and the classical item parameters are exploited (Lord & Novick, 1968, pp. 376–378). These relationships are derived under the assumption that ability is normally distributed and that the item characteristic curve

(ICC) is the normal ogive. In order to study how the departures from the assumption of normally distributed abilities affect the Urry procedure, three ability distributions were considered: normal, uniform, and negatively skewed. The normal and uniform distributions were generated with mean 0.0 and variance 1.0. (The uniform distribution was generated in the interval −1.73 to 1.73 to ensure unit variance.) A beta distribution with parameters 5 and 1.5 was generated to simulate a negatively skewed distribution, and then rescaled so that the mean was 0.0 and the variance 1.0. The distributions were standardized to remove the effect of scaling on the estimates of the parameters.

The three factors—test length (4 levels), examinee population size (3 levels), and ability distribution (3 levels)—were completely crossed to simulate 36 testing situations. Test data arising from these situations were subjected to the Urry estimation procedure using the computer program ANCILLES and to the maximum likelihood estimation procedure using the computer program LOGIST (Wood, Wingersky, & Lord, 1978).

Lord (1975) has emphasized that simulated data should in some way resemble real data; otherwise, results obtained through simulation studies will not generalize to real situations. An attempt was therefore made to generate test data as realistically as possible. In order to accomplish this, item difficulty parameters were sampled from a uniform distribution defined in the interval $b = -2.0$ to 2.0, and item discrimination parameters were sampled from a uniform distribution in the interval $a = .6$ to 2.0. Since data were generated to simulate item responses to multiple-choice items with four choices, the pseudochance level parameters were set at $c = .25$. It should be noted, however, that this does not ensure close approximation of the generated data to real data. Combinations of item difficulty and discrimination that may not occur in constructed tests may occur with simulated tests and hence may affect the estimation procedures, limiting the generalizability of the findings in simulated studies to real situations. On the other hand, since the purpose of this study was to compare two estimation procedures and to study the statistical properties of estimators, the possible lack of correspondence between simulated and real data may not be a serious problem

RESULTS

Accuracy of Estimation

Comparisons between ANCILLES and LOGIST across various test lengths, examinee population sizes, and ability distributions are indicated in Tables 1.1–1.3. The statistics reported are (a) the mean μ of the popula-

TABLE 1.1
Comparison of Estimates of Item and Ability Parameters from LOGIST and ANCILLES Based on a Normal Distribution of Ability

		Discrimination (a)					Difficulty (b)					Chance-level parameter (c)					Ability (θ)				
			ANCILLES		LOGIST			ANCILLES		LOGIST			ANCILLES		LOGIST			ANCILLES		LOGIST	
No. of items	No. of exam- inees	μ	$\bar{X}$	ρ	$\bar{X}$	ρ	μ	$\bar{X}$	ρ	$\bar{X}$	ρ	μ	$\bar{X}$	SD	$\bar{X}$	SD	μ	$\bar{X}$	ρ	$\bar{X}$	ρ
10	50	1.46	3.47	.21	1.53	.43	−.15	−.87	.92	−.60	.95	.25	.34	.38	.12	.04	.02	−.10	.63	.00	.71
	200	1.18	2.82	.08	1.72	.46	.46	.41	.87	.22	.99	.25	.36	.18	.25	.00	.13	.07	.77	−.13	.76
	1000	1.46	2.97	.36	2.00	[a]	−.15	−.45	.95	−.15	.99	.25	.36	.28	.23	.02	−.00	.11	.71	.09	.75
15	50	1.17	2.08	−.25	1.67	−.02	.32	.61	.92	.29	.89	.25	.36	.25	.23	.00	.01	.04	.83	−.23	.78
	200	1.17	2.12	.38	1.59	.47	.32	.35	.97	.19	.96	.25	.35	.14	.23	.03	.11	−.00	.77	−.10	.77
	1000	1.40	2.61	.42	1.72	.86	−.09	−.04	.97	−.05	1.00	.25	.33	.17	.25	.00	.02	−.01	.86	−.04	.85
20	50	1.35	2.09	.40	1.60	.37	.16	.22	.95	.04	.96	.25	.29	.14	.18	.04	−.08	−.04	.87	−.00	.87
	200	1.35	2.17	.33	1.41	.46	.16	.22	.97	.08	.97	.25	.30	.12	.24	.01	−.02	.05	.88	−.12	.88
	1000	1.35	1.99	.66	1.59	.76	.16	.37	.98	.16	.99	.25	.36	.11	.25	.02	.00	.05	.89	−.06	.88
80	50	1.28	1.48	.40	1.40	.62	.15	.06	.85	.13	.88	.25	.20	.13	.22	.02	−.08	.12	.96	−.00	.97
	200	1.28	1.42	.64	1.46	.81	.15	.20	.96	.15	.98	.25	.22	.09	.25	.01	−.04	.09	.98	−.00	.97
	1000	1.28	1.36	.84	1.37	.88	.15	.21	.99	.12	1.00	.25	.23	.08	.25	.01	−.00	.08	.98	−.02	.97

[a] Indicates that correlation coefficient was not computed, since estimates of discrimination parameters attained the upper bound, 2.00.

TABLE 1.2

Comparison of Estimates of Item and Ability Parameters from LOGIST and ANCILLES Based on a Negatively Skewed Distribution of Ability

No. of items	No. of exam-inees	Discrimination (a)					Difficulty (b)					Chance-level parameter (c)					Ability (θ)				
			ANCILLES		LOGIST			ANCILLES		LOGIST			ANCILLES		LOGIST			ANCILLES		LOGIST	
		μ	$\overline{X}$	ρ	$\overline{X}$	ρ	μ	$\overline{X}$	ρ	$\overline{X}$	ρ	μ	$\overline{X}$	SD	$\overline{X}$	SD	μ	$\overline{X}$	ρ	$\overline{X}$	ρ
10	50	1.18	2.67	.13	1.81	−.38	.46	.79	.91	.68[a]	.78[a]	.25	.39	.28	.25	.02	.05	.16	.70	−.30	.78
	200	1.46	2.52	−.58	1.96	−.31	−.15	.31	.98	−.19	.99	.25	.56	.26	.20	.01	−.07	−.01	.71	−.14	.78
	1000	1.46	2.98	−.06	1.95	−.31	−.15	.62	.97	−.26	.98	.25	.41	.40	.22	.02	−.01	.03	.57	−.17	.77
15	50	1.40	2.05	.10	1.11	−.01	−.09	.25	.96	−.31	.94	.25	.42	.16	.22	.01	−.17	−.13	.80	−.06	.82
	200	1.40	2.58	.23	1.61	.45	−.09	−.09	.94	−.37	.99	.25	.44	.27	.23	.01	.03	.08	.80	−.12	.91
	1000	1.40	2.37	.18	1.79	.87	−.09	.15	.95	−.10	1.00	.25	.48	.24	.25	.01	.00	.14	.79	−.06	.87
20	50	1.35	2.16	.49	1.25	.27	.16	.28	.92	.16[a]	.72[a]	.25	.34	.22	.19	.03	.02	.03	.81	−.14	.89
	200	1.35	2.03	.03	1.54	.10	.16	.21	.96	.05	.98	.25	.41	.19	.25	.00	.08	.15	.77	−.05	.87
	1000	1.35	2.10	.49	1.59	.52	.16	.51	.96	.08	.99	.25	.39	.13	.24	.01	.01	.06	.86	−.05	.91
80	50	1.29	1.49	.22	1.19	.28	.18	.10	.85	1.82[a]	.30[a]	.25	.21	.16	.21	.01	.07	.22	.93	−.30	.97
	200	1.28	1.26	.69	1.13	.61	.15	.22	.94	1.72[a]	.27[a]	.25	.21	.16	.24	.01	.14	.06	.96	−.04	.96
	1000	1.28	1.27	.68	1.24	.82	.15	.38	.97	.16	.99	.25	.20	.11	.25	.01	−.03	.08	.95	−.06	.97

[a] Indicates that the difficulty estimate for an item has taken on an extreme value.

TABLE 1.3
Comparison of Estimates of Item and Ability Parameters from LOGIST and ANCILLES Based on a Uniform Distribution of Ability

		Discrimination (a)					Difficulty (b)					Chance-level parameter (c)					Ability (θ)				
			ANCILLES		LOGIST			ANCILLES		LOGIST			ANCILLES		LOGIST			ANCILLES		LOGIST	
No. of items	No. of exam-inees	μ	$\bar{X}$	ρ	$\bar{X}$	ρ	μ	$\bar{X}$	ρ	$\bar{X}$	ρ	μ	$\bar{X}$	SD	$\bar{X}$	SD	μ	$\bar{X}$	ρ	$\bar{X}$	ρ
10	50	1.18	2.50	.33	1.26	.02	.46	.64	.68	.40	.81	.25	.43	.20	.18	.04	.06	.06	.59	.00	.71
	200	1.46	2.86	.60	1.74	.70	−.15	−.28	.90	−.49	.94	.25	.36	.19	.21	.00	.15	.09	.66	−.02	.75
	1000	1.46	2.52	.22	2.00	[a]	−.15	.06	.98	−.13	.99	.25	.33	.14	.29	.02	−.02	−.04	.74	−.10	.77
15	50	1.40	2.85	.33	1.90	.47	−.09	−.13	.91	−.04	.96	.25	.22	.16	.25	.01	.07	−.04	.90	−.00	.90
	200	1.40	2.70	.13	1.52	.03	−.09	−.04	.92	−.03	.91	.25	.22	.12	.20	.02	−.04	−.04	.89	−.00	.88
	1000	1.40	2.43	.37	1.61	.11	−.09	.16	.95	.20	.87	.25	.25	.09	.23	.01	−.03	−.02	.88	−.03	.87
20	50	1.35	2.35	.09	1.69	.47	.16	.52	.94	.24	.91	.25	.26	.30	.25	.01	−.07	.02	.89	−.14	.88
	200	1.35	2.08	.46	1.68	.34	.16	.40	.92	.07	.98	.25	.35	.24	.25	.00	.09	.05	.91	−.10	.88
	1000	1.35	1.98	.43	1.64	.56	.16	.34	.99	.06	1.00	.25	.46	.34	.24	.02	.04	.03	.90	−.02	.89
80	50	1.29	1.38	.30	1.38	.29	.18	.51	.88	.53	.86	.25	.20	.14	.21	.03	−.30	.09	.95	.00	.96
	200	1.28	1.32	.54	1.38	.73	.15	.29	.93	.20	.95	.25	.30	.15	.23	.01	−.04	.08	.97	.00	.97
	1000	1.28	1.26	.83	1.34	.94	.15	.22	.98	.12	1.00	.25	.36	.18	.25	.00	.02	.08	.97	−.00	.97

[a] Indicates that correlation coefficient was not computed, since estimates of discrimination parameters attained the upper bound, 2.00.

tion item parameters for each population size; (*b*) the mean $\bar{X}$ of the estimated item parameters; and (*c*) the correlation ρ between the true parameters and their estimates. These statistics are reported for the estimates obtained by employing both ANCILLES and LOGIST.

A comparison of the mean of the generated item parameters, μ, and the mean of the estimates, $\bar{X}$, for each of the item parameters—discrimination (*a*), difficulty (*b*), pseudochance level (*c*)—and ability (θ) provides some indication of the accuracy of estimation. This comparison is rather weak when carried out alone, however, since the means do not contain all the essential information. Simultaneous comparisons of the means and examination of the correlations between the parameters and estimates, on the other hand, provide more complete information regarding the accuracy of estimation. If the correlation is high and the means differ, then it can be concluded that the estimation was not sufficiently accurate.

Lord (1975) has implied that if heteroscedasticity exists, it may not be meaningful to compute correlations between true and estimated values, and, in general, we agree. However, since in the strict sense heteroscedasticity will invalidate the computation of a least squares regression line—the more appropriate criterion to employ is the generalized least squares criterion—and hence will rule out the use of simple, interpretable statistics for the evaluation of the accuracy of estimation, heteroscedasticity (when it occurred) was ignored; correlations and least squares regression equations were computed.

ESTIMATION OF THE DISCRIMINATION PARAMETER

Examination of the results in Tables 1.1–1.3 indicates that the *a* parameter was poorly estimated for short tests. The highest correlation between true values and estimates for a test with 10 items and normally distributed ability was .36, with the mean of the estimates exceeding the mean of the true values. The correlations improved with increasing sample size and test length, with the mean of the estimated values approaching the mean of the true values from above. The highest correlation between the estimated and true values was .88 for an 80-item test with 1000 examinees. This trend was also evident for the uniform and negatively skewed distributions of ability. In general, the *a* parameter was poorly estimated by ANCILLES, with the estimation improving more rapidly with increasing test length than with increasing examinee population size.

The least squares regression lines (for normally distributed ability) for predicting the estimates from true values, given in Table 1.4, were plotted (not shown) and compared with the line $y = x$ in order to determine the extent of the bias in estimation. The regression lines for all the test-length and sample-size combinations fell above the line $y = x$, indicating that

TABLE 1.4
Regression Coefficients and Standard Errors for Predicting the Estimates from True Values Based on a Normal Distribution of Ability

		Discrimination (a)								Difficulty (b)								Ability (θ)							
		ANCILLES				LOGIST				ANCILLES				LOGIST				ANCILLES				LOGIST			
No. of items	No. of exam-inees	b_0	SE	b_1	SE	b_0	SE	b_1	SE	b_0	SE	b_1	SE	b_0	SE	b_1	SE	b_0	SE	b_1	SE	b_0	SE	b_1	SE
10	50	2.55	1.38	.63	.95	.19	.90	.92	.62	−.71	.17	1.06	.15	.78	.13	1.20	.12	−.11	.17	.48	.11	−.02	.09	.77	.11
	200	2.57	.91	.21	.77	1.17	.36	.45	.27	−.04	.20	.97	.18	−.20	.06	.91	.04	−.01	.06	.59	.04	−.25	.04	.97	.06
	1000	1.52	1.22	.99	.83		[a]			−.29	.14	1.06	.11	.00	.07	1.03	.04	.11	.03	.58	.02	−.09	.02	.94	.03
15	50	2.89	.80	−.69	.68	1.72	.42	−.04	.35	.30	.14	.98	.11	−.01	.17	.94	.13	.03	.09	.78	.08	−.24	.07	1.00	.12
	200	1.23	.55	.76	.47	.96	.32	.54	.26	.09	.11	.80	.06	−.13	.11	1.01	.07	−.07	.05	.68	.04	−.20	.04	.91	.06
	1000	.85	.99	1.26	.71	.24	.23	1.06	.16	.05	.08	.95	.07	.05	.02	1.08	.00	−.03	.02	.78	.02	−.06	.01	.96	.02
20	50	1.14	.49	.70	.36	.83	.44	.57	.33	.07	.09	.95	.07	−.10	.08	.88	.06	−.10	.08	.82	.07	.07	.07	.87	.07
	200	.70	.99	1.09	.73	.49	.41	.68	.30	.04	.06	1.11	.06	−.09	.06	1.08	.07	.07	.03	.82	.03	.09	.03	1.07	.04
	1000	.36	.41	1.21	.30	.23	.27	1.01	.20	.21	.06	1.02	.05	.00	.04	1.02	.04	.05	.02	.80	.01	−.06	.01	.98	.02
80	50	.46	.27	.80	.21	.26	.16	.89	.13	−.06	.07	.83	.06	−.01	.06	.94	.06	.18	.07	.74	.03	.06	.05	.80	.03
	200	.11	.17	1.02	.14	.23	.10	.96	.08	.05	.04	.97	.03	.04	.03	.74	.02	.12	.02	.89	.01	.04	.02	.93	.02
	1000	.12	.09	.97	.07	.11	.08	.98	.06	.05	.02	1.03	.02	−.02	.01	.96	.00	.08	.01	.91	.01	−.02	.01	.96	.01

[a] Regression coefficients and their standard errors were not computed, since all estimates of discrimination parameters attained the upper bound, 2.00.

ANCILLES systematically overestimated the a parameter, with the regression lines approaching the line $y = x$ with increasing test length. Again, the convergence to the line $y = x$ was more rapid with increasing test length than with increasing sample size.

Trends similar to that observed with ANCILLES were also observed with LOGIST. Although the estimation of a was poor, the LOGIST estimates were consistently better than those from ANCILLES in that the correlations between true values and estimates were higher and the means of the estimates were much closer to the means of the true values. Comparison of the plots of the regression lines, given in Table 1.4, with the line $y = x$ showed that although there was a general tendency for the parameters to be overestimated, this tendency was not as marked as with ANCILLES; the convergence of the regression lines to the line $y = x$ was more rapid. These trends—the higher correlations between true and estimated values than for ANCILLES estimates, the tendency for the means of the estimates to be closer to the means of the true values, and the rapidity of convergence of the regression line to the line $y = x$—were also observed with the uniform and negatively skewed distributions of ability.

ESTIMATION OF THE DIFFICULTY PARAMETER

ANCILLES was very successful in providing accurate estimates of the b parameter. The correlations between estimates and true values ranged from .85 to .99. Comparison of the regression lines for normally distributed ability, given in Table 1.4, with the line $y = x$ indicated that the b parameter was generally overestimated for tests with 15 and 20 items. With larger numbers of items, there was a tendency for difficult items to be overestimated and for easy items to be underestimated. The bias was slight, however, in that the convergence of the regression line to the line $y = x$ was rapid with increasing items and sample size.

In general, the LOGIST estimates of the b parameters were better than the estimates produced by ANCILLES. The correlations between true and estimated values ranged from .88 to 1.00, whereas ANCILLES yielded correlations ranging from .85 to .99. The means of the estimates were, in general, closer to the means of the true values than they were with ANCILLES. Comparisons of the regression lines, given in Table 1.4, with the line $y = x$ revealed that with increasing test length and sample size, the regression line approached the line $y = x$ rather rapidly, demonstrating that there was no bias in the estimation. No clear trends were visible with 10, 15, and 20 items, although the test with 10 items and 50 examinees produced overestimates of the b parameter. These results appeared to hold for both the uniform and negatively skewed distributions of ability, although with the skewed distribution there were two instances when the estimates of difficulty went out of bounds. These cases are indicated with a superscript in Table 1.2. However, with 80 items and 1000 examinees, the agreement between estimated values and true values was comparable to that obtained with normally distributed ability.

In general, the b parameter was estimated rather well by both LOGIST and ANCILLES. LOGIST fared surprisingly well with small numbers of items and examinees in comparison with ANCILLES, and in general produced better estimates (as determined by the correlations) than did ANCILLES.

CHANCE-LEVEL PARAMETER

The true value of the chance-level parameter was set at $c = .25$ for all the items. Given this lack of variation among the true values, correlations between estimates and true values were not computed. Hence, only the mean of the true values, the mean of the estimates, and the standard deviation of the estimates are reported in Tables 1.1–1.3.

ANCILLES clearly produced very poor estimates of the c parameter. The means of the estimates were consistently higher than the mean of the true values, with relatively large standard deviations. LOGIST estimates, on the other hand, were close to the true values, with small standard deviations. The mean LOGIST estimates ranged from .12 to .25 for normally distributed ability, from .19 to .25 for skewed distribution of ability, and from .18 to .25 for uniformly distributed ability. In comparison, ANCILLES yielded estimates that ranged from .20 to .36, .20 to .56, and .22 to .46, respectively, for the three distributions of ability.

ESTIMATION OF ABILITY

An examination of Tables 1.1–1.3 indicates a consistent pattern in the estimation of ability (θ) for both LOGIST and ANCILLES. The correlations between true values and estimates did not seem to be affected by increasing sample sizes for fixed test lengths. On the other hand, increasing the lengths of the test greatly affected the magnitude of the agreement between true values and estimates. This unsurprising trend held for the three distributions of θ.

In general, it appears that although no differences existed between the ANCILLES and LOGIST estimates of θ for tests with 15 items or more, the LOGIST estimates fared better than the ANCILLES estimates for short tests with 10 items. This effect was more pronounced with the skewed ability distribution.

Comparison of the regression lines (that obtained by regressing the estimates on the true values with the line $y = x$) indicated that, in general, ANCILLES underestimated θ for examinees with high true abilities and overestimated θ for examinees with low true abilities. This may partly be attributed to the fact that the c parameters were overestimated. No such trends were evident with the LOGIST estimates. These regression lines rapidly converged to the line $y = x$ with increasing test length.

Effect of Ability Distribution

A χ^2 test was used to determine if the uniform and the beta distributions deviated from the normal. The beta distribution yielded a χ^2 value of 63.5 when the tails of the normal distribution were excluded and a value of 193.1 when the tails were included. The uniform distribution yielded a χ^2 value of 69.6 when tails were excluded and 307.7 when the tails were included. This indicates that both distributions deviated sufficiently from the normal, with the uniform distribution deviating even more than the beta distribution.

Comparisons of the results in Tables 1.1–1.3 reveal that, in general, the beta distribution affected both estimation procedures, while the uniform distribution produced results similar to those obtained using a normal ability distribution. Although the beta distribution affected the estimation of a for both procedures and c and θ for ANCILLES, the estimation of b did not seem to be affected in either case. ANCILLES fared poorly with the skewed distribution in comparison to LOGIST in the estimation of the a, c, and θ parameters.

The estimates for the a parameter, resulting from both procedures, were negatively correlated with the true values for short tests. For longer tests, although estimates from both procedures improved, ANCILLES produced poor estimates in comparison to LOGIST. For an 80-item test with 1000 examinees, a correlation of .68 was obtained using ANCILLES, as compared to a correlation of .82 obtained from LOGIST.

The estimates of the c parameters resulting from ANCILLES were extremely high for all tests except those of 80 items. The mean values ranged from .20 to .56 with the beta distribution, as compared to a range of .20–.36 for the normal distribution of ability. LOGIST, on the other hand, underestimated the c parameter, although the estimates were comparable to those obtained using a normal distribution of ability.

The LOGIST estimates of ability resulting from a skewed distribution of ability were as good as, and in some cases better than, the estimates obtained with a normal distribution. In contrast, ANCILLES with a skewed distribution resulted in poorer estimates. This effect held true even as sample size and test length increased.

Thus ANCILLES estimates of θ, a, and c parameters seemed to be affected more dramatically than the LOGIST estimates when ability had a skewed distribution. It should be noted that although the uniform distribution had a larger χ^2 value than the beta distribution, the results obtained with the uniform distribution of ability were similar to those obtained with the normal distribution. It is then not departures from normality but departures from symmetry and the unavailability of examinees in the lower tail of the ability distribution that affected the estimation procedure.

Statistical Properties of Estimators

BIAS

If g is an estimator of γ, then g is an unbiased estimator of γ if

$$E(g) = \gamma, \tag{1.2}$$

where $E(\cdot)$ is the expectation operator. This is a desirable property of estimators.

Schmidt (1977) has pointed out that the Urry procedure, developed by Urry in 1974, systematically overestimated the a parameter and underestimated the b parameter. Urry (1976) suggested a correction for this and incorporated this into the ANCILLES program, employed to estimate parameters in this study. Since it appears that for large numbers of items and examinees the estimates are unbiased (Lord, 1975), in order to study the effect of this correction on the estimates and to examine if the LOGIST estimates were unbiased, a relatively short test of 20 items with 200 examinees was selected, response data were generated, and item parameters were estimated; this was replicated 20 times. Since the replications were obtained by generating sets of random examinees, the bias in the estimator of ability was not investigated.

The results of the replications are presented in Table 1.5, in which the true value μ of the 20 item parameters is given together with the mean estimate $\overline{X}$ of the item parameters over 20 replications. The standard error and the t value obtained as

$$t = (\overline{X} - \mu)/SE \tag{1.3}$$

are also given to indicate the degree of departure of the mean estimate from the true value.

ANCILLES clearly overestimated the a parameter, as did LOGIST. The bias in the LOGIST estimates, however, did not appear to be as severe as the bias in the ANCILLES estimates. This finding is borne out in Figure 1.1, where the regression line for predicting $\overline{X}$ from μ is plotted for both ANCILLES and LOGIST and compared with the line $y = x$. The LOGIST regression line is closer to the line $y = x$ and shows that small

TABLE 1.5

Bias in the Estimation of Item Parameters Based on Twenty Replications for Twenty Items and Two Hundred Examinees

	Discrimination (a)							Difficulty (b)							Chance-level parameter (c)						
		ANCILLES			LOGIST				ANCILLES			LOGIST				ANCILLES			LOGIST		
Item	μ	$\overline{X}$	SE	t	$\overline{X}$	SE	t	μ	$\overline{X}$	SE	t	$\overline{X}$	SE	t	μ	$\overline{X}$	SE	t	$\overline{X}$	SE	t
1	.77	1.17	.067	5.9	.85	.073	1.9	1.63	1.90	.119	2.3	1.65	.097	.2	.25	.37	.020	6.0	.24	.007	−1.4
2	.84	1.60	.069	10.9	1.09	.129	1.9	−1.49	−1.28	.079	2.6	−1.52	.107	−.3	.25	.39	.017	8.2	.23	.004	−5.0
3	1.79	2.24	.056	8.0	1.89	.067	1.5	1.82	2.05	.112	2.1	1.75	.005	−1.3	.25	.33	.039	2.1	.23	.006	−3.3
4	1.11	1.91	.054	14.8	1.19	.119	.7	−1.54	−1.57	.076	−.4	−1.91	.188	−2.0	.25	.44	.021	9.0	.23	.004	−5.0
5	1.28	2.35	.202	5.3	1.57	.077	3.8	−.47	−.38	.206	.4	−.50	.057	−.5	.25	.33	.036	2.2	.23	.004	−5.0
6	1.53	2.42	.151	5.9	1.77	.088	2.6	−1.26	−1.24	.046	.4	−1.30	.063	.6	.25	.35	.036	2.8	.23	.004	−5.0
7	1.31	1.72	.129	3.2	1.62	.105	2.9	1.17	1.33	.079	2.0	1.06	.049	−2.2	.25	.30	.018	2.8	.23	.004	−5.0
8	1.31	1.89	.184	3.2	1.68	.098	3.8	1.47	1.75	.139	2.0	1.37	.067	−1.5	.25	.33	.036	2.2	.23	.005	−4.0
9	1.45	2.48	.117	8.8	1.54	.086	1.0	−1.78	−1.89	.089	−1.2	−1.96	.067	−2.7	.25	.48	.049	4.7	.23	.004	−5.0
10	1.48	2.23	.173	4.3	1.69	.088	2.4	−1.02	−.94	.053	1.5	−1.03	.044	−.2	.25	.34	.028	3.2	.23	.004	−5.0
11	1.58	2.26	.198	3.4	1.78	.081	2.5	.71	.83	.054	2.2	.67	.026	−1.5	.25	.29	.024	1.7	.23	.007	−2.8
12	1.43	1.89	.163	2.8	1.70	.106	2.6	.84	.97	.076	1.7	.74	.034	−2.9	.25	.30	.023	2.2	.23	.006	−3.3
13	1.97	3.06	.192	5.7	1.96	.027	−.4	.19	.07	.062	−1.9	.17	.023	−.9	.25	.18	.030	−2.3	.22	.007	−4.3
14	1.52	2.54	.133	7.7	1.62	.104	.9	−1.64	−1.72	.629	−.1	−1.89	.059	−4.2	.25	.41	.042	3.8	.23	.004	−5.0
15	.73	1.24	.105	4.9	.93	.112	1.8	.01	.22	.059	3.6	.01	.043	0.0	.25	.34	.019	4.7	.23	.006	−3.3
16	1.49	2.07	.221	2.7	1.75	.089	2.9	1.07	1.21	.073	1.9	1.03	.045	−.9	.25	.28	.026	1.2	.23	.005	−4.0
17	1.15	2.05	.129	6.9	1.47	.108	2.9	−.25	−.04	.045	4.7	−.23	.029	.7	.25	.35	.024	4.2	.22	.006	−5.0
18	1.89	2.29	.237	1.7	1.77	.089	−1.4	1.53	1.86	.135	2.4	1.57	.068	.6	.25	.31	.035	1.7	.23	.007	−1.4
19	1.23	1.58	.096	3.6	1.41	.127	1.4	1.28	1.77	.125	3.9	1.31	.086	.3	.25	.36	.016	6.9	.23	.004	−5.0
20	1.20	1.71	.076	6.7	1.56	.094	3.8	.94	1.05	.589	.2	.81	.068	−1.9	.25	.33	.010	8.0	.23	.007	−2.8

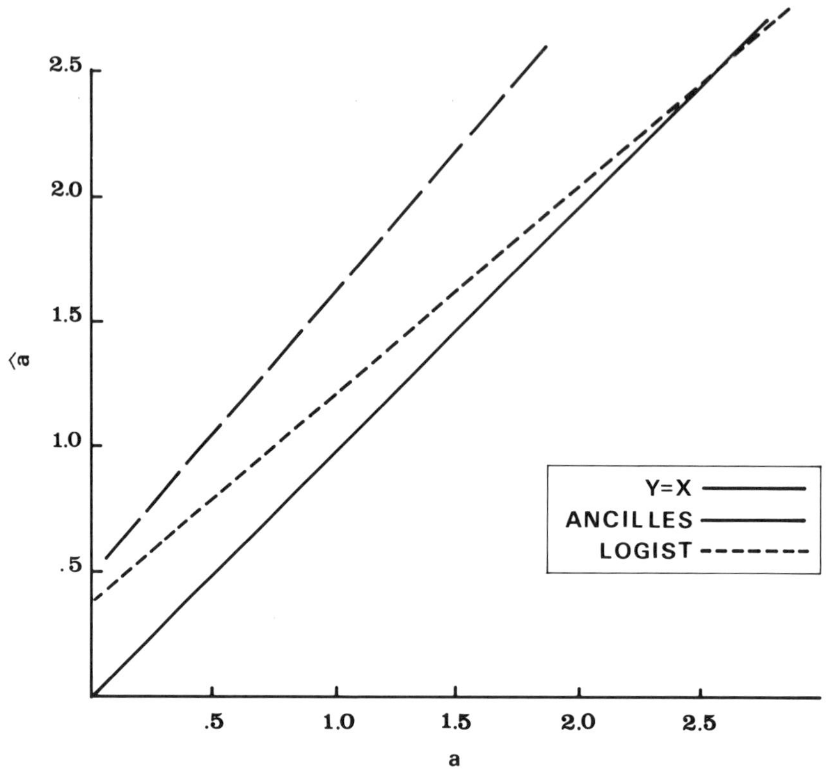

FIGURE 1.1. Bias in the estimation of the discrimination parameter of the three-parameter logistic model.

values of a were overestimated, whereas very large values tended to be estimated accurately, partly because an upper limit was imposed on the estimates. On the other hand, ANCILLES tended to overestimate large values, even more than small values, of a.

With item difficulty, LOGIST tended to underestimate easy items while producing relatively accurate estimates of very difficult items (Figure 1.2). ANCILLES, on the other hand, tended to overestimate items with high b levels and to underestimate items with negative b levels. In general, ANCILLES seemed to produce biased estimates of b throughout the entire range.

CONSISTENCY

If g_n is an estimator of γ, g_n is a consistent estimator of γ if for any positive ε and η there is some N such that

$$\text{Prob}\{|g_n - \gamma| < \varepsilon\} > 1 - \eta, \qquad n > N. \tag{1.4}$$

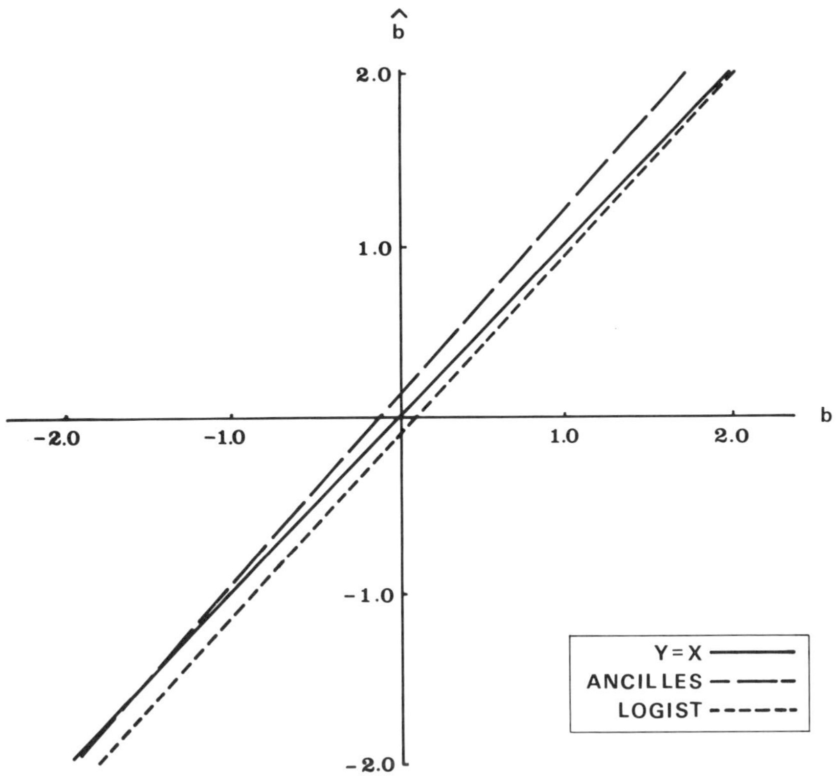

FIGURE 1.2. Bias in the estimation of the difficulty parameter of the three-parameter logistic model.

Consistency is a desirable property in that it ensures that an estimator tends to a definite quantity, which is the true value to be estimated.

The problem of consistency has raised several questions concerning the estimation of parameters in the latent trait models. Andersen (1973) has argued that a consistent estimator of the discrimination parameter does not exist and, hence, has questioned the meaningfulness of the two- and three-parameter models.

In order to investigate whether or not the LOGIST and ANCILLES estimators were consistent, the regression equations for predicting the estimates from the true values of the various parameters were examined. The definition for a consistent estimator given earlier implies that an estimator is consistent if it is asymptotically unbiased and its variance tends to 0 with increasing sample size. Consequently, in order for the estimators of the latent trait parameters to be consistent (*a*) the slope of the regression equation must approach 1.0 and the intercept must approach

0.0; and (*b*) the variance, and hence the standard errors of the estimate of the slope and intercept, must approach 0.0. If these conditions are met, then the estimator is consistent.

The regression coefficients and the standard errors are reported in Table 1.4. The results indicate that when both the number of items and the number of examinees increase, the slope and intercept coefficients approach 1.0 and 0.0, respectively, with the standard errors approaching 0.0. This tendency is evident for both ANCILLES and LOGIST estimators for the a, b, and c parameters and for θ. In all these cases, the LOGIST estimator converged in probability to the true value more rapidly than the ANCILLES estimator. It should be pointed out, however, that the results reported here do not conclusively support this. It is clearly necessary to examine the standard errors and the regression coefficients with a greater number of items and examinees.

DISCUSSION

The purpose of this study was to compare two methods for estimation of parameters in the three-parameter logistic model, the Urry method of estimation, and the maximum likelihood procedure. The computer programs that were used were the ANCILLES program and the LOGIST program (Wood *et al.*, 1978). The efficiency of the procedures were compared with respect to the accuracy of estimation, the effect of violating underlying assumptions (for ANCILLES), and the statistical properties of the estimators. The factors that were controlled were test length (4 levels), examinee population size (3 levels), and ability distribution (3 levels).

The results indicate that, in general, the maximum likelihood procedure was superior to the Urry procedure with respect to the estimation of all item and ability parameters. The differences were pronounced in the estimation of the discrimination and chance-level parameters, but with respect to the estimation of ability and difficulty parameters, the differences were less remarkable. Differing θ distributions had little effect on the estimation of b and θ. With a skewed distribution of θ, however, ANCILLES produced poorer estimates of a and c parameters than with normal or uniform θ distributions. LOGIST, although faring better than ANCILLES (with the exception of the 10-item test), produced slightly poorer results with the skewed distribution than with the normal or uniform distribution.

The number of examinees had a slight effect in improving the accuracy of estimation of the b and c parameters and θ. Increasing the number of items and the number of examinees, however, considerably improved the accuracy of the a estimates with both procedures. Surprisingly enough, a 20-item test with 1000 examinees produced excellent estimates of the b and c parameters and reasonably good estimates of a and θ. Tests with 80 items and 1000 persons fared considerably better, providing good estimates of all parameters.

Although these results are encouraging, they should be interpreted cautiously. As pointed out earlier, the LOGIST program requires that limits be set on the range of values the estimates of the a parameter can take. The default lower and upper limits are set at 0.0 and 2.0, respectively. The lower limit is clearly appropriate. Recent studies by the authors indicate that in some situations the upper limit poses a problem. The estimates of a for some items have been observed to reach this limit and even higher limits that have been set. This outward drift does raise some questions regarding the estimability of the a parameter. In general, however, this outward drift rarely occurs with large numbers of items and examinees.

Limits are also imposed for the c parameters in the LOGIST program. Moreover, in general, easy items are grouped together, and an average value for the c parameter is obtained for these items. The generation of data with a common c value may have therefore favored the LOGIST procedure.

Tests with 15 items or fewer, while yielding good estimates of b and c parameters and reasonable estimates of θ, yielded poor estimates of the a parameter. This severely limits the application of the three-parameter latent trait model to criterion-referenced measurement situations, since criterion-referenced tests typically have fewer than 10 items. It should be pointed out, however, that this limitation exists only if the item parameters and ability parameters are estimated simultaneously. If item banks with known item characteristics are employed to estimate ability, or if the one-parameter model is employed, this limitation may not exist.

Although the LOGIST estimates were superior to the ANCILLES estimates, especially in the case of short tests, the difference between them was negligible when the number of items and the number of examinees increased. This is of particular importance since ANCILLES requires considerably less computer time than LOGIST. The computer time taken by LOGIST, especially with large numbers of items and examinees, may become forbidding enough to warrant the use of ANCILLES in this situation. It should be noted that, in fairness to the maximum likelihood procedure, the Urry procedure, in general, deletes more items and examinees during estimation than does the maximum likelihood procedure. This may explain the rapidity of convergence and indicate a weakness in ANCILLES.

The bias and consistency results indicate that for small numbers of items, the estimates of the item and ability parameters are biased, with the ANCILLES more biased than the LOGIST estimates. As the number of examinees and the number of items increase, it appears that the estimators are unbiased and, hence, are consistent. This, in a sense, supports a conjecture of Lord (1968) and shows that the three-parameter model may be statistically viable.

ACKNOWLEDGMENTS

The project was performed pursuant to a contract from the United States Air Force Office of Scientific Research; however, the opinions expressed here do not necessarily reflect their position or policy, and no official endorsement by the Air Force should be inferred.

REFERENCES

Andersen, E. B. Conditional inference in multiple-choice questionnaires. *British Journal of Mathematical and Statistical Psychology,* 1973, *26,* 31–44.

Hambleton, R. K., & Rovinelli, R. A FORTRAN IV program for generating examinee response data from logistic test models. *Behavioral Science,* 1973, *18,* 74.

Jensema, C. A simple technique for estimating latent trait mental test parameters. *Educational and Psychological Measurement,* 1976, *36,* 705–715.

Lord, F. M. An analysis of the Verbal Scholastic Aptitude Test using Birnbaum's three-parameter logistic model. *Educational and Psychological Measurement,* 1968, *28,* 989–1020.

Lord, F. M. *Evaluation with artificial data of a procedure for estimating ability and item characteristic curve parameters* (Research Bulletin 75-33). Princeton, New Jersey: Educational Testing Service, 1975.

Lord, F. M., & Novick, M. R. *Statistical theories of mental test scores.* Reading, Massachusetts: Addison-Wesley, 1968.

Ree, M. J. Estimating item characteristic curves. *Applied Psychological Measurement,* 1979, *3,* 371–385.

Schmidt, F. L. The Urry method of approximating the item parameters of latent trait theory. *Educational and Psychological Measurement,* 1977, *37,* 613–620.

Urry, V. W. Approximations to item parameters of mental test models and their uses. *Educational and Psychological Measurement,* 1974, *34,* 253–269.

Urry, V. W. Ancillary estimators for the item parameters of mental tests. In W. A. Gorham (Chair), *Computerized testing: Steps toward the inevitable conquest* (PS-76-1). Washington, D.C.: U.S. Civil Service Commission, Personnel Research and Development Center, 1976.

Wood, R. L., Wingersky, M. S., & Lord, F. M. *LOGIST: A computer program for estimating examinee ability and item characteristic curve parameters* (Research Memorandum 76-6) (revised). Princeton, New Jersey: Educational Testing Service, 1978.

Wright, B. D. Solving measurement problems with the Rasch model. *Journal of Educational Measurement,* 1977, *14,* 97–116.

2

Robustness of Item Response Models and Effects of Test Length and Sample Size on the Precision of Ability Estimates

RONALD K. HAMBLETON LINDA L. COOK

Although item response models are potentially very useful, there remain many practical problems at the application stage. For example, how should an item response model be selected? It is tempting to use the more general models, since these models will provide "better" fits to the available test data. Unfortunately, the more general item response models (for example, the three-parameter logistic test model) require more computer time to obtain satisfactory parameter estimates, larger samples of examinees, and longer tests, and are more difficult for practitioners to work with. Clearly, more needs to be known about the goodness of fit and robustness of item response models. Such information would aid practitioners in the important step of selecting a test model.

There has been research on the goodness of fit of different item response models to a variety of test data sets (e.g., Hambleton & Murray, 1983; Hambleton & Swaminathan, 1983; Lord, 1975; Tinsley & Dawis, 1977; Wright, 1968), and generally the results suggest that the models fit many data sets at least reasonably well (Hambleton, Swaminathan, Cook, Eignor, & Gifford, 1978). Only a few studies, however, compared the fit of more than one item response model to the same data sets (e.g., Hambleton & Traub, 1973). In the Hambleton and Traub study, improvements

NEW HORIZONS IN TESTING
Latent Trait Test Theory and
Computerized Adaptive Testing

ISBN 0-12-742780-5

were obtained in predicting test score distributions (for three tests) from the two-parameter model as compared to the one-parameter model.

On the question of model robustness (i.e., the extent to which the assumptions underlying the test model can be violated to a greater or lesser extent by the test data and be fitted by the model), the results of several studies have been reported (Dinero & Haertel, 1977; Hambleton, 1969; Hambleton & Traub, 1976; Panchapakesan, 1969). These results have been mixed, perhaps because of the confounding of results with sample sizes.

One problem with most of the goodness-of-fit studies and the robustness studies reported to date is that they provide no indication of the practical consequences of fitting a less-than-perfect model to a data set. It is of little interest to the practitioner to know that 15 out of 20 items failed to be fitted by a test model when the range of discrimination parameters reached a value of .80. If the size of the examinee sample is large enough, probably all items can be identified by a χ^2 statistic of goodness of fit as *not* fitting the model. On the other hand, if the size of the examinee sample is small enough, perhaps none of the 20 items would be misfit by the model. It would be more useful to compare item response models using a criterion measure that can be clearly interpreted and has practical significance. This desirable goal is addressed in Study 1 in this chapter.

One of the features of using any item response model in test development work is the opportunity to specify a *target information curve* and then to select test items from an item pool to produce a test with a similar test information curve to the one specified in the target information curve. A target information curve describes the desired level of information at each point on the ability scale underlying examinee test performance. Information, in turn, is directly related to the degree of precision of ability estimation at points on the ability continuum (Lord, 1980). In fact, as long as a test is not too short, the standard error of ability estimates at a particular ability level is equal to 1 divided by the square root of information provided by the test at the ability level in question. In practice, since the contribution of each test item to the test information curve (referred to as a *score information curve* when item parameter estimates are used instead of the item parameter values) is known—that is, once the item parameter values or the item parameter estimates are specified—it is possible to select test items from a pool of calibrated test items (i.e., a pool of test items with associated parameter estimates) to produce a score information curve that approximates a desired target information curve.

One of the problems with the paradigm offered above for test development is the imprecision associated with the item parameter estimates. Score information curves—and therefore the associated standard errors of ability estimates—will depend on the precision of item parameter estimates. In turn, precision of item parameter estimates is influenced by the examinee sample size used to estimate the item parameters, and in the case of the item discrimination parameter, estimates are influenced by the length of the test also. Study 2 was designed to address this issue.

STUDY 1

The purpose of Study 1 was to study systematically the goodness of fit of the one-, two-, and three-parameter logistic models. Using computer-simulated test data, the effects of four variables were studied: (*a*) the variation in item discrimination parameters; (*b*) the average value of the pseudochance-level parameters; (*c*) the test length; and (*d*) the shape of the ability distribution. Artificial or simulated data representing departures of varying degrees from the assumptions of the three-parameter logistic test model were generated and the goodness of fit of the three test models to the data was studied. The goodness-of-fit measures used were chosen for their practical significance.

In choosing an appropriate goodness-of-fit measure, attention must be directed to the intended use of the test scores. In this study, one of the most common uses of test scores was assumed, that of rank ordering of examinees on the ability measured by a test. All aptitude tests and many standardized achievement tests are used in this way. Developers of these tests expend considerable energy to enhance the validity of the rank orderings of examinees obtained from administering these norm-referenced tests: They make the tests as long as is practical; they ensure that the items are of high quality; they select test items that increase test score variability; and they standardize the test administrations and testing environments. In this investigation, interest was centered on another factor that has impact on the validity of rank orderings of examinees: choice of logistic test model. It seemed reasonable, therefore, to compare the three logistic test models in terms of how well they recovered the true rankings of examinee ability with several diverse test data sets.

Method

SIMULATING THE TEST DATA

The simulation of item response data for examinees was accomplished using the three-parameter logistic model. First, the number of examinees

(N), shape of the ability distribution, and values of the ability parameters ($\theta_i = 1, 2, \ldots, N$) were specified. Next, the number of items in the test (n) and values of the three item parameters (a_g, b_g, c_g, $g = 1, 2, \ldots, n$) were specified. Then, the examinee and item parameters were substituted in the equation of the three-parameter logistic model to obtain p_{ij} ($0 \leq p_{ij} \leq 1$), representing the probability that examinee i correctly answered item j. The probabilities were arranged in a matrix **P** of order $N \times n$ whose (i, j)th element was p_{ij}. **P** was then converted into a matrix of the item scores for examinees (1 = correct response, 0 = incorrect response) by comparing each p_{ij} with a random number obtained from a uniform distribution in the interval 0–1. If the random number was less than or equal to p_{ij} (which would happen on the average p_{ij} of the time), p_{ij} was set equal to 1; otherwise, p_{ij} was set to 0. The matrix **P** of 0s and 1s was the simulated test data. Three statistics, which corresponded to those used in the estimation of examinee ability in the respective models, were calculated:

One-parameter logistic score, $\Sigma_{g=1}^{n} u_g$, the number-correct score
Two-parameter logistic score, $\Sigma_{g=1}^{n} a_g u_g$
Three-parameter logistic score, $\Sigma_{g=1}^{n} w_g(\theta) u_g$

For the three-parameter model statistic, since the item weights $[w_g(\theta)]$ depend on examinee ability, three-parameter model estimates of ability

TABLE 2.1
Test Lengths, Range of Discrimination Parameters, and Pseudochance-Level Parameters for Each Data Set

Data set	Test length	Variation in discrimination parameters	Pseudochance level parameters
A	20	0.00	.00
B	20	0.00	.25
C	20	.81–1.43	.00
D	20	.81–1.43	.25
E	20	.50–1.74	.00
F	20	.50–1.74	.25
G	40	0.00	.00
H	40	0.00	.25
I	40	.81–1.43	.00
J	40	.81–1.43	.25
K	40	.50–1.74	.00
L	40	.50–1.74	.25

were obtained for each examinee from LOGIST (Wood, Wingersky, & Lord, 1976).

The values of the examinee and item parameters, summarized in Table 2.1, were chosen in the following way:

Item Parameters. Two test lengths (20 and 40 items) were used in the simulations. Item difficulty parameters b were selected at random from a uniform distribution in the interval $[-2, 2]$. An analysis of the difficulty parameters reported by Lord (1968) suggested that this decision was reasonable.

The discrimination parameters a were selected at random from a uniform distribution with mean 1.12. The range of the discrimination parameters was a variable under investigation. The range was varied from 0.0 to a maximum of 1.24 [.50 to 1.74], and an intermediate value of .62 [.81 to 1.43] was also studied. The maximum value of discrimination was similar to the range and distribution of the discrimination parameters reported for the Verbal Section of the Scholastic Aptitude Test (SAT-V; Lord, 1968).

The extent of guessing in the simulated test data was another variable under study. Two values of the average guessing parameter were considered: $c = 0.0$ and $c = .25$. All pseudochance-level parameters were set equal to the mean value of the c parameter under investigation.

Examinee Parameters. The number of examinees was set to 500. This number was found to be sufficient to produce stable goodness-of-fit results. Two distributions of ability were considered: uniform $[-2.5, 2.5]$ and normal $[0, 1]$.

Factor Structure. For all of the tests simulated in this study, it was assumed that the tests were unidimensional.

GOODNESS OF FIT

Twelve data sets, denoted by the letters A–L, were formed for each of two ability distributions (uniform and normal). The data sets were produced from the combinations of two test lengths, two levels of pseudochance parameters and three levels of variation in discrimination parameters. With each data set, three scoring methods were used to estimate ability based on the one-, two-, and three-parameter logistic models. Since simulated data were used, the true examinee ability scores were known; these true ability scores served as the criterion against which the ability score estimates derived from the three test models could be judged. The rankings of examinees based on ability estimates obtained from each model (for each set of test data) were then compared to examinee "true" abilities using Spearman rank-difference correlations and the average discrepancy in ranks. Both statistics describe the extent to which a set of ability scores obtained from a particular logistic model results in a ranking of examinees that matches the true ranking of examinees on the ability measured by the test.

Because of the arbitrariness of the scale on which $\hat{\theta}$ is measured, summary statistics such as $\Sigma_{i=1}^{N}|\theta_i - \hat{\theta}_i|/N$ were not studied. To facilitate the interpretation of results further, they are reported separately for each half of the ability distribution as well as for the total ability distribution.

TABLE 2.2
Spearman Rank-Order Correlations (r) and Average Absolute Difference in Rank Orders (AAD) for the Two Halves of the Uniform Ability Distribution

Data set	True vs. 1-P Model		True vs. 2-P Model		True vs. 3-P Model	
	r	AAD	r	AAD	r	AAD
Lower half (θ = −2.5 to 0.0)						
A	.88	54.24	.88	54.24	.88	54.24
B	.77	76.61	.77	76.61	.83	64.98
C	.88	56.07	.88	56.41	.88	56.40
D	.76	77.14	.76	76.90	.83	64.28
E	.87	56.50	.87	56.56	.87	56.56
F	.75	80.08	.75	79.92	.83	65.77
G	.94	36.48	.94	36.48	.94	36.48
H	.87	58.58	.87	58.58	.91	48.70
I	.95	36.50	.95	36.47	.95	36.47
J	.87	57.66	.88	56.86	.91	48.01
K	.94	37.86	.95	36.96	.95	36.74
L	.87	57.82	.88	56.87	.91	48.22
Upper half (θ = 0.0 to +2.5)						
A	.88	54.45	.88	55.62	.88	55.62
B	.84	63.68	.83	65.35	.83	65.73
C	.89	52.23	.88	55.38	.88	55.38
D	.85	63.80	.83	65.02	.84	63.19
E	.87	56.99	.88	55.38	.88	55.47
F	.80	71.57	.80	70.72	.80	69.16
G	.94	39.03	.94	40.50	.94	40.50
H	.90	50.19	.90	51.05	.90	50.85
I	.94	40.65	.93	41.83	.93	41.85
J	.91	49.14	.90	50.55	.91	50.27
K	.93	40.79	.94	38.93	.94	38.94
L	.89	52.88	.89	52.90	.89	52.68

Results

Results are summarized in Tables 2.2–2.5.

LEVEL OF VARIATION IN DISCRIMINATION PARAMETERS

For the values investigated in the study, using discrimination parameters as item weights contributed very little to the correct ranking of examinees.

LEVEL OF PSEUDOCHANCE-LEVEL PARAMETERS

With the 20-item tests the three-parameter model was considerably more effective at ranking examinees correctly in the lower half of the ability distribution. Correlations were about .08 higher (about .75 to .83) in the uniform distribution of ability and about .08 higher in the normal distribution (about .65 to .73). When the one-parameter model results for the average absolute difference in rank order are considered as a baseline, it is seen from Table 2.2 that the three-parameter model reduced the ranking discrepancies between true and estimated abilities by about 17%.

With the 40-item tests, the three-parameter model was again more effective at ranking examinees correctly in the lower half of the ability distribution. Correlations were about .04 higher in both ability distribu-

TABLE 2.3
Spearman Rank-Order Correlations (*r*) and Average Absolute Difference in Rank Order (AAD) for the Full Uniform Ability Distribution ($\theta = -2.5$ to $+2.5$)

Data set	True vs. 1-P Model		True vs. 2-P Model		True vs. 3-P Model	
	r	AAD	*r*	AAD	*r*	AAD
A	.97	28.26	.97	28.37	.97	28.37
B	.93	41.85	.93	41.97	.95	36.97
C	.97	28.81	.97	29.14	.97	29.14
D	.93	42.40	.93	43.93	.94	38.59
E	.97	30.83	.97	30.14	.97	30.14
F	.93	42.20	.93	42.73	.94	39.02
G	.98	20.44	.98	20.61	.98	20.61
H	.96	30.13	.96	30.26	.97	27.02
I	.98	21.09	.98	21.25	.98	21.25
J	.96	30.69	.96	30.75	.97	27.74
K	.98	22.48	.98	21.81	.98	21.81
L	.96	31.49	.96	30.50	.97	27.30

tions. The three-parameter model again reduced the ranking discrepancies by about 17%. The reduction in effectiveness of the three-parameter model weights was to be expected with the longer tests. Gulliksen (1950) noted the insignificance of item scoring weights when a test is of substantial length and test items are positively correlated.

For examinees in the upper half of the ability distribution and for the data sets studied, the number-correct score was about as effective as the more complicated scoring weights used in the two- and three-parameter models.

TABLE 2.4
Spearman Rank-Order Correlations (*r*) and Average Absolute Difference in Rank Order (AAD) for the Two Halves of the Normal Ability Distribution

Data set	True vs. 1-P Model		True vs. 2-P Model		True vs. 3-P Model	
	r	AAD	*r*	AAD	*r*	AAD
Lower half (θ = 0.0, SD_θ = 1.0)						
A	.82	65.58	.82	65.58	.82	65.58
B	.65	94.93	.65	94.93	.74	82.54
C	.84	62.72	.83	63.26	.83	63.31
D	.65	95.18	.65	95.77	.73	83.49
E	.80	70.65	.80	69.43	.80	69.41
F	.66	94.63	.64	95.80	.73	63.38
G	.91	46.03	.91	46.03	.91	46.03
H	.81	68.70	.81	68.70	.85	61.63
I	.90	48.23	.91	67.28	.91	47.28
J	.81	68.08	.82	67.05	.85	60.09
K	.90	48.22	.91	46.58	.91	46.58
L	.81	69.01	.81	68.66	.85	61.58
Upper half (θ = 0.0, SD_θ = 1.0)						
A	.84	60.51	.84	60.81	.84	60.81
B	.76	75.75	.76	76.16	.77	75.08
C	.85	61.09	.85	61.60	.85	61.61
D	.76	76.41	.76	78.02	.77	75.63
E	.83	64.79	.85	63.08	.85	63.08
F	.75	78.69	.75	79.92	.77	77.01
G	.90	50.71	.90	50.75	.90	50.75
H	.82	65.18	.82	65.45	.83	64.24
I	.89	51.25	.90	50.21	.90	50.23
J	.82	65.92	.83	64.84	.84	63.16
K	.89	51.01	.90	49.95	.90	49.95
L	.81	67.60	.82	64.51	.83	63.96

TABLE 2.5
Spearman Rank-Order Correlations (r) and Average Absolute Difference in Rank Order (AAD) for the Full Normal Ability Distribution ($\bar{\theta}$ = 0.0, SD_θ = 1.0)

Data set	True vs. 1-P Model		True vs. 2-P Model		True vs. 3-P Model	
	r	AAD	r	AAD	r	AAD
A	.94	36.84	.94	36.91	.94	36.91
B	.88	53.94	.88	53.90	.91	47.55
C	.94	35.87	.94	35.99	.94	35.98
D	.88	54.31	.88	54.34	.91	48.61
E	.93	41.11	.93	40.96	.93	40.96
F	.87	55.73	.87	57.94	.88	53.13
G	.97	26.60	.97	26.62	.97	26.62
H	.95	36.44	.95	36.46	.96	33.03
I	.97	25.20	.97	25.54	.97	25.53
J	.94	38.86	.94	37.65	.95	34.15
K	.97	27.04	.97	25.88	.97	25.87
L	.94	38.79	.94	37.33	.95	34.68

SHAPE OF THE ABILITY DISTRIBUTION

As expected, correlations tended to be higher for the uniformly distributed ability scores because score variability was increased.

TEST LENGTH

Increases in correlations were observed because of the doubling of the length of the test. The increases were small as would be predicted from the well-known formulas linking test length to test validity.

Conclusions

It is clear from the data that with modest test lengths (n = 20) there are some sizable gains to be expected in the correct ordering of examinees at the lower end of the ability continuum when three-parameter model estimates are used (as opposed to the number-correct score). The gains were cut roughly in half when the test lengths were doubled (n = 40). It was surprising that item discrimination parameters as weights had so little effect on the results. Gulliksen (1950) summarized the research on item weights, however, and came to essentially the same conclusion. Consequently, to the extent that these simulated data sets are typical of real data, it would appear that the application of item response models to the problem of ranking examinees is probably not essential except in those situations where gains of the size noted for lower ability examinees are important. (One situation might arise when there is interest in selecting candidates from the lower portion of an ability distribution for special programs.) The number-correct score ranks examinees nearly as well as the most complicated scoring method.

The results of this single study should be generalized with caution for the following reasons. First, the values of the item parameters used may not be typical of real data sets. Second, the criterion measures of goodness of fit seem suitable for the situation in which a user desires to make norm-referenced interpretations of test scores. There are many other test situations (e.g., those involving adaptive tests, test score equating, and criterion-referenced tests) where a different criterion to judge the quality of a solution would be more suitable. Third, these results provide a somewhat unfair comparison of the two-parameter model with the other two models because the item discrimination parameters used in the weighting process to derive statistics for ability estimation would have been somewhat different had the ''best-fitting'' two-parameter curves to the three-parameter item characteristic curves been used. The item discrimination parameters in the best-fitting two-parameter curves would have differed somewhat from those defined in the three-parameter curves to which they were fitted. Finally, the correlation results for the one-parameter model and, to a much lesser extent, the two-parameter model, are inflated (to an unknown extent) because of tied scores. Therefore, the true differences in the reported correlations are somewhat larger than those reported in Tables 2.2–2.5.

STUDY 2

This second study was designed to investigate two practical questions that are important to test developers:

1. What are the effects of examinee sample size and test length on the standard errors of ability estimation (SEE) curves?
2. What effects do the statistical characteristics of an item pool have on the precision of standard errors of ability estimation curves?

A study of the use of SEE curves and factors that affect their stability was motivated by item response model test development procedures. When item statistics are available they are commonly used by test developers for selecting test items from a pool of test items to produce a test with a desired SEE curve. Because of random errors associated with the item statistics, however, the SEE curve obtained from the examinee group in which the new test is administered may be different from that based on error-free item parameters. When the curve is considerably different, the new test may be of limited value, for it does not have the desired characteristics. Therefore it is desirable for test developers to have an idea of the amount of change to be expected in an SEE curve because of random errors associated with the item parameter estimates. Since the size of these errors is influenced by both the sample size of examinees used in item calibration and the length of the test being constructed, both of these variables were considered in this study.

Method

VARIABLES

Tests of three lengths were considered: 10, 20, and 80 items. Since a test with 10 items is about as short a test as is ever used in practice, the 10-test item length was selected to be studied; the 80-item test was selected because the length represents about as long a test as is used in practice.

Ability scores were simulated to be normally distributed (mean = 0, SD = 1). This assumption was made to conform with an assumption made in Urry's (1974) item parameter estimation method, which was used (with slight modification) in this study.

Three examinee sample sizes were used: 50, 200, and 1000. The smallest sample size (N = 50) is considerably smaller than should be used in practice. It was chosen to identify the poorest results that could be expected. The other two sample sizes define minimum and maximum sample sizes typically used in test development work with item response models. Ranges of parameter values for items in the two pools are given in Table 2.6. An examination of Table 2.6 shows that items in Item Pool 1 had a

TABLE 2.6
Range of Item Parameter Values for the Two Simulated Item Pools

Item parameter	Range of values	
	Item Pool 1	Item Pool 2
Difficulty (b)	−2.00 to 2.00	−1.00 to 1.00
Discrimination (a)	.60 to 2.00	.60 to 1.50
Pseudochance (c)	.25 to .25	.25 to .25

wider range of difficulty and discrimination values than those in Item Pool 2.

DATA SIMULATION

The eight steps in the data simulation were as follows:

1. Item Pool 1 was selected for study.
2. A test length (10, 20, or 80 items) and a sample size (50, 200, or 1000 examinees) were selected. A sample of examinee ability scores was drawn from a normal distribution (mean = 0, SD = 1).
3. Computer program DATAGEN (Hambleton & Rovinelli, 1973), produced (*a*) item parameters, given the constraints of the item pool under investigation, and (*b*) examinee item scores. The computer program used the three-parameter logistic model, the ability scores from Step 2, and item parameters generated at this step to produce probabilities of correct responses for examinees to the test items. These probabilities, in turn, were converted to examinee item scores (0 or 1) by a random number generator.
4. The examinee item scores from Step 3 were used in Urry's computer program to estimate item and ability parameters. Only the item parameter estimates, however, were used further in this particular study.
5. The item parameter estimates were used to obtain the standard errors of estimate for estimating θ [SEE(θ)]. The values of SEE(θ) at seven ability levels ($\theta = -3.00, -2.00, -1.00, 0.00, 1.00, 2.00, 3.00$) were calculated and reported.
6. Steps 3–5 were repeated three times to obtain three estimates of SEE(θ). All item and ability parameter values for the three runs were identical. The particular examinee item scores varied from one run to the next because of the probabilistic nature of the score outcomes.
7. Steps 3–6 were repeated for each combination of test length and sample size ($3 \times 3 = 9$).
8. Steps 2–7 were repeated with Item Pool 2. In all, 54 sets of simulated test data were considered in the study.

Results

Tables 2.7–2.9 contain the SEE curves from Item Pool 1 obtained for three replications of three examinee sample sizes (N = 50, 200, and 1000) and three test lengths (n = 10, 20, and 80) and for seven ability levels. Test lengths and sample sizes given under the column headed "Actual" are the number of items and examinees remaining after a satisfactory set of item and ability parameter estimates were obtained from Urry's computer program.

EFFECT OF EXAMINEE SAMPLE SIZE

The data for a test length of 10 items, given in Table 2.7, clearly show the lack of stability of the SEE curves for all examinee sample sizes. There was little improvement, if any, because of increasing sample size. This result, however, may be due to the limited amount of data considered, since improvements were obtained in Item Pool 2 and at other test lengths.

Table 2.8 contains the results for 20-item test lengths and shows that the SEE curves were beginning to stabilize. Except at extreme values of the ability continuum, the results for the smaller examinee sample sizes were nearly as good as those obtained with the larger examinee sample size ($N = 1000$).

At a test length of 80 items, the SEE curves were highly stable, as clearly indicated by the data in Table 2.9. Similar to the effect noted with test lengths of 20 items, the expected decrease in variation of the SEE with increase in sample size was apparent only at ability levels −1, +1, and +2.

TABLE 2.7
Standard Error Estimates (SEE) Adjusted to Correspond to Ten-Item Tests for Various Sample Sizes and Ability Levels with a Heterogeneous Item Pool

Sample size and replication	Actual: Test length	Actual: Sample size	Ability level: −3.0	−2.0	−1.0	0.0	1.0	2.0	3.0
50									
1	10	34	.66	.33	.67	.22	.75	1.60	2.19
2	10	34	2.40	1.88	.56	1.04	.20	1.34	1.37
3	9	34	.73	.57	1.03	.22	.58	.43	2.19
200									
1	10	172	.64	.21	.52	2.15	1.60	1.50	1.48
2	10	137	.22	.51	.36	1.30	.37	.96	2.45
3	10	174	2.63	2.14	.27	2.75	.92	.76	1.91
1000									
1	10	841	.98	.26	.58	1.43	3.33	.57	1.18
2	10	833	1.03	1.03	.67	1.05	.45	1.01	1.06
3	10	892	2.44	.49	.67	.30	.29	.89	1.33

TABLE 2.8
Standard Error Estimates (SEE) for Various Sample Sizes and Ability Levels with a Heterogeneous Item Pool

Sample size and replication	Actual		Ability level						
	Test length	Sample size	−3.0	−2.0	1.0	0.0	1.0	2.0	3.0
50									
1	20	50	2.84	.70	.35	.30	.31	.44	1.23
2	20	50	1.93	1.53	.39	.32	.24	.45	1.19
3	20	46	2.07	.83	.58	.31	.36	.68	1.48
200									
1	20	193	—	.57	.26	.39	.33	.50	.77
2	20	196	—	1.51	.37	.34	.25	.53	.86
3	20	196	—	1.03	.22	.49	.34	.40	1.15
1000									
1	20	955	—	1.05	.48	.33	.33	.45	.82
2	20	969	—	1.18	.37	.33	.37	.40	.99
3	20	968	—	1.56	.40	.42	.32	.43	1.07

TABLE 2.9
Standard Error Estimates (SEE) Adjusted to Correspond to Eighty-Item Tests for Various Sample Sizes and Ability Levels with a Heterogeneous Item Pool

Sample size and replication	Actual		Ability level						
	Test length	Sample size	−3.0	−2.0	−1.0	0.0	1.0	2.0	3.0
50									
1	74	50	1.10	.35	.14	.14	.24	.24	.45
2	79	50	1.06	.48	.25	.17	.13	.32	.49
3	77	50	.93	.20	.19	.15	.17	.29	.48
200									
1	80	200	.89	.26	.22	.24	.19	.25	.44
2	80	200	.62	.29	.25	.19	.21	.25	.46
3	80	200	1.06	.35	.21	.19	.20	.25	.48
1000									
1	80	999	1.00	.35	.23	.21	.21	.24	.40
2	80	1000	.98	.32	.23	.22	.21	.23	.43
3	80	1000	1.08	.34	.20	.21	.20	.24	.46

EFFECT OF TEST LENGTH

Examination of the results reported in Tables 2.7–2.9 indicate that for samples of size 50, as test length increased variation in the SEE curves decreased at all ability levels. Results of the simulations for sample sizes of 200 and 1000 clearly show the following trends:

1. The most stable SEE curves were obtained for the longest test length.
2. For all ability levels, variation in the SEE curves decreased as test length increased.

Figure 2.1 illustrates the effect of test length and examinee sample size on the stability of the SEE curves at five ability levels for Item Pool 1. Each graph represents a plot of the values of the SEE curves obtained when sample size was held constant and test length was varied. It is clear from examination of these graphs that sample size has little effect on the stability of SEE curves of the 10-item tests. The effect of sample size on the stability of the SEEs was most apparent for the 20-item tests. For the 80-item tests, sample size showed the most pronounced effect when there was an increase from 50 examinees (Figure 2.1a) to 200 examinees (Figure 2.1c). An effect was also noticed when sample size was increased from 200 examinees (Figure 2.1a) to 1000 examinees (Figure 2.1c); the improvements in precision, however, were more modest in size.

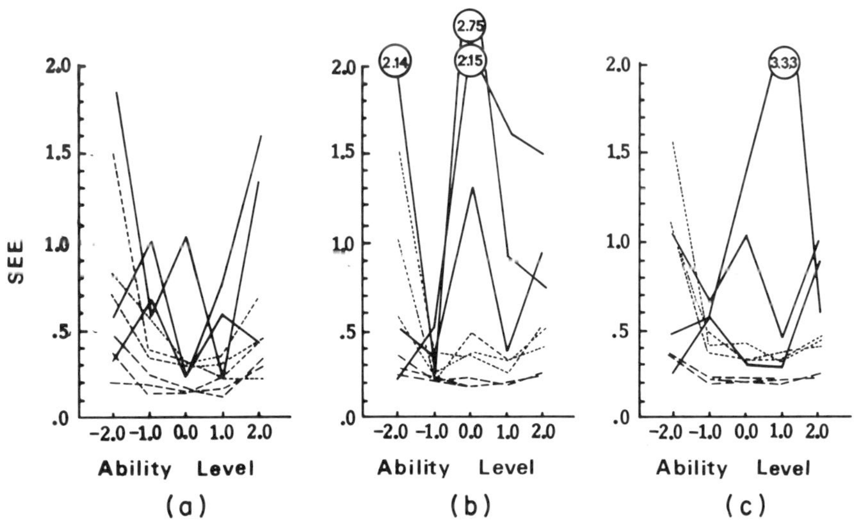

FIGURE 2.1. Standard error of estimate (SEE) for three test lengths (10, 20, and 80 test items), five ability levels, and three sample sizes (50, 200, and 1000 examinees). (Each combination of conditions was replicated three times.) (Solid line, 10-item test; dotted line, 20-item test; dashed line, 30-item test.)

Table 2.10 summarizes the data reported in Tables 2.7–2.9 and includes summary data for Item Pool 2. Entries in this table are the standard deviations of the SEEs obtained across the three replications of the various studies. Standard deviations are reported for each combination of test length and sample size across five ability levels. Also included in Table 2.10 is the average of the standard deviations across ability levels for each combination of test length and sample size.

TABLE 2.10
Standard Deviations of Standard Errors of Estimates across Three Replications at Several Ability Levels for Different Test Lengths and Examinee Sample Sizes, and for the Heterogeneous Item Pool (Pool 1) and the Homogeneous Item Pool (Pool 2)

Test length	Sample size and item pool	Ability level					Average variation across ability levels
		−2.0	−1.0	0.0	1.0	2.0	
10	50						
	Pool 1	.68	.20	.39	.23	.50	.40
	Pool 2		.17	.11	.41	.28	.24
	200						
	Pool 1	.85	.10	.60	.50	.31	.47
	Pool 2		.03	.07	.03	.22	.09
	1000						
	Pool 1	.32	.04	.47	1.40	.19	.60
	Pool 2		.07	.03	.04	.03	.04
20	50						
	Pool 1	.36	.10	.01	.05	.11	.16
	Pool 2	.78	.07	.10	.05	.08	.22
	200						
	Pool 1	.38	.06	.06	.04	.06	.12
	Pool 2	.37	.00	.02	.04	.00	.09
	1000						
	Pool 1	.22	.05	.04	.02	.02	.09
	Pool 2	.50	.03	.01	.00	.02	.11
80	50						
	Pool 1	.11	.04	.01	.05	.03	.06
	Pool 2	.16	.04	.01	.02	.04	.05
	200						
	Pool 1	.04	.02	.02	.01	.00	.02
	Pool 2	.03	.01	.01	.01	.01	.01
	1000						
	Pool 1	.01	.01	.00	.00	.00	.00
	Pool 2	.02	.00	.00	.00	.01	.01

Several trends are apparent from examination of the average variation of the SEEs for Item Pool 1: (*a*) the variation decreased as test length increased for all sample sizes; (*b*) when test length was fixed at 10 items, sample size had little or no effect on the stability of the SEE curves; and (*c*) examinee sample size, generally, had a noticeable effect on the stability of the SEE curves.

Examination of the average variation across ability levels for Item Pool 2 indicated that for all test lengths, sample size has a noticeable effect on the stability of SEE curves. In comparison to the results reported for Item Pool 1, the effect of test length on the average variation across ability levels was not so apparent. The reason for this is the smaller variation observed for short tests with this item pool.

EFFECTS OF STATISTICAL CHARACTERISTICS OF AN ITEM POOL ON PRECISION OF SEE CURVES

The results in Table 2.10 indicate that for tests of 20 and 80 items, the variation in the SEE curves, averaged across ability levels, was very similar for both item pools. For test lengths of 10, the situation was quite different. In order to make the average variation across ability levels at this test length comparable for both item pools, these values were recomputed for Item Pool 2 excluding the values obtained for ability level of -2. The recomputed average variation values were .33, .38, and .52 for sample sizes of 50, 200, and 1000, respectively. It is clear that for short tests, the homogeneous item pool (Item Pool 1) resulted in smaller average variations than did the heterogeneous item pool. A second point worth noting is that the heterogeneous item pool (Item Pool 2) provided more stable SEEs at an ability of -2 for test lengths of 10 or 20 items than did the homogeneous item pool. For test lengths of 80, the results appear to be about the same for both item pools.

Conclusions

Study 2 yielded results that were not unexpected. It is well known that the size of an examinee sample, the length of a test, and the statistical characteristics of an item pool will have an important influence on the shape and stability of SEE curves. The importance of this study is that it provided new results concerning the size of improvements in SEE curves relative to the three factors under investigation: (*a*) examinee sample size; (*b*) test length; and (*c*) item pool characteristics. Several conclusions appear to be warranted:

1. Both test length and examinee sample size are extremely important factors in the precision of SEE curves. The small number of reversals in the results was no doubt due to sampling fluctuations.

2. At the extremes of an ability continuum, precision of SEE curves is very poor, even with large examinee sample sizes. The results are substantially better when tests are lengthened, even if the sample size is small ($N = 50$). Also, more stability of SEE curves is obtained with heterogeneous item pools, especially when the tests of interest are short.
3. The precision of SEE curves will be acceptable in most instances if the curves are based on 200 or more examinees with test lengths of at least 20 items. This recommendation holds if primary concern is with values of the curves in the middle regions of the ability continuum $[-1, +1]$.
4. Increases in examinee sample sizes from 50 to 200 produce sizable improvements in the precision of SEE curves; however, gains in precision from increasing a sample size from 200 to 1000 produce only modest gains in precision of the SEE curves.
5. Similarly for test lengths, improvements in precision were substantially better when the change was from 10 to 20 items than from 20 to 80 items.

The results of this study suggest that if the items in a pool have typical statistical characteristics, the stability of SEE curves across readministrations of the test to similar groups of examinees will be quite good if the test includes at least 20 items and if 200 or more examinees are used in deriving the item statistics.

ACKNOWLEDGMENTS

This research was performed pursuant to a contract from the United States Air Force Office of Scientific Research. However, the opinions expressed here do not necessarily reflect the Air Force's position or policy, and no official endorsement by the Air Force should be inferred. A complete report of Study 2 is contained in Cook and Hambleton (1979). The authors are indebted to Janice Gifford for her assistance in the collection and analysis of data reported in Study 2.

REFERENCES

Cook, L. L., & Hambleton, R. K. *Effects of test length and sample size on the estimates of precision of latent ability scores* (Report No. 87). Amherst: University of Massachusetts, School of Education, Laboratory of Psychometric and Evaluative Research, 1979.

Dinero, T. E., & Haertel, E. Applicability of the Rasch model with varying item discriminations. *Applied Psychological Measurement*, 1977, *1*, 581–592.

Gulliksen, H. *Theory of mental tests*. New York: Wiley, 1950.

Hambleton, R. K. *An empirical investigation of the Rasch test theory model*. Unpublished doctoral dissertation, University of Toronto, 1969.

Hambleton, R. K., & Murray, L. Some goodness of fit investigations for item response models. In R. K. Hambleton (Ed.), *Applications of item response theory*. Vancouver, BC: Educational Research Institute of British Columbia, 1983.

Hambleton, R. K., & Rovinelli, R. A FORTRAN IV program for generating examinee response data from logistic tests models. *Behavioral Science*, 1973, *18*, 74.

Hambleton, R. K., & Swaminathan, H. *Item response theory: Principles and applications*. Boston: Kluwer-Nijhoff, 1983.

Hambleton, R. K., Swaminathan, H., Cook, L. L., Eignor, D., & Gifford, J. A. Developments in latent trait theory: Models, technical issues, and applications. *Review of Educational Research*, 1978, *48*, 467–510.

Hambleton, R. K., & Traub, R. E. Analysis of empirical data using two logistic latent trait models. *British Journal of Mathematical and Statistical Psychology*, 1973, *26*, 195–211.

Hambleton, R. K., & Traub, R. E. *The robustness of the Rasch test model* (Report No. 42). Amherst: University of Massachusetts, School of Education, Laboratory of Psychometric and Evaluative Research, 1976.

Lord, F. M. An analysis of the Verbal Scholastic Aptitude Test using Birnbaum's three-parameter logistic model. *Educational and Psychological Measurement*, 1968, *28*, 989–1020.

Lord, F. M. *Evaluation with artificial data of a procedure for estimating ability and item characteristic curve parameters* (Research Bulletin 75-33). Princeton, New Jersey: Educational Testing Service, 1975.

Lord, F. M. *Applications of item response theory to practical testing problems*. Hillsdale, NJ: Lawrence Erlbaum Associates, 1980.

Panchapakesan, N. *The simple logistic model and mental measurement*. Unpublished doctoral dissertation, University of Chicago, 1969.

Tinsley, H. E. A., & Dawis, R. Test-free person measurement with the Rasch simple logistic model. *Applied Psychological Measurement*, 1977, *1*, 483–487.

Urry, V. W. Approximations to item parameters of mental test models and their uses. *Educational and Psychological Measurement*, 1974, *34*, 253–269.

Wood, R. L., Wingersky, M. S., & Lord, F. M. *LOGIST: A computer program for estimating examinee ability and item characteristic curve parameters* (Research Memorandum 76-6). Princeton, New Jersey: Educational Testing Service, 1976.

Wright, B. D. Sample-free test calibration and person measurement. In *Proceedings of the 1967 invitational conference on testing problems*. Princeton, New Jersey: Educational Testing Service, 1968.

3

Small N Justifies Rasch Model

FREDERIC M. LORD

The usual Birnbaum item response function requires the determination of three parameters for each item; the Rasch model requires only one. If there is only a small group of examinees, the a parameter (the discriminating power) cannot be determined accurately for some of the items. The c parameters are even more of a problem. For small samples, is it perhaps better to use the Rasch model, estimating only one parameter per item, even though the Rasch model is incorrect?

For a better perception of the problem, consider a common prediction problem not related to item response theory: Suppose it is desired to predict variable y from measurements on five predictors. An available sample has been used to estimate the linear regression of y on the predictors. This regression equation may now be applied to estimate y for new individuals drawn from the same population. If the sample used to estimate the regression equation was large, the procedure is a good one; but if this sample was small, the procedure may be worse than simply using the sample mean of y as the predicted value of y for each new individual. Suppose, for example, that the true multiple correlation for predicting y was .40. If the sample had only 60 cases, the predictions from the sample regression equation would typically be no more accurate than a prediction that each new value of y will fall at the sample mean of y.

NEW HORIZONS IN TESTING
Latent Trait Test Theory and
Computerized Adaptive Testing

ISBN 0-12-742780-5

It would be useful to know how large the sample of examinees must be before it is worthwhile to use a two- or three-parameter item response model in preference to the Rasch model. The answer to this question will, of course, depend on the purpose to be served. This chapter is a modest beginning: it answers this question only for the two-parameter logistic model and only for one very limited situation. The purpose of this chapter, then, is to point out the problem, to indicate a method of solution, and to provide some numerical results, indicating the sample size required when there is no guessing.

METHOD

Under the Rasch model, ability must be estimated by some function of the number-correct score x, since this is a sufficient statistic under this model. Under the two-parameter logistic model, ability must be estimated by a function of the weighted sum $\Sigma_{i=1}^{n} a_i u_i$ of item responses u_i, the weight for each item being the item discriminating power a_i; under this model, this weighted sum is a sufficient statistic for estimating ability.

Given the a_i, the information function for number-correct score x and the information function for the weighted sum $\Sigma\, a_i u_i$ can be readily calculated and compared. The weighted sum always provides more information than the number-correct score except in the limiting case where the two scores are identical or proportional. In practice, the number-correct score perhaps provides up to 95% as much information as the weighted sum.

Now suppose that the a_i are *not* known but are only estimated. If the estimates $\hat{a}_i$ are sufficiently inaccurate, the weighted sum $\Sigma\, \hat{a}_i u_i$ will be less informative than the number-correct score x. The problem is to make a precise statement showing how the usefulness of the weighted sum $\Sigma\, \hat{a}_i u_i$ depends on the number of cases used to determine the estimated weights $\hat{a}_i$.

It is desired to compare $x \equiv \Sigma_i\, u_i$ and $\Sigma_i\, \hat{a}_i u_i$ as estimators of ability. Note, however, that expectations over the u_i for fixed $\hat{a}_i$ give

$$\mathscr{E}x = \sum_i P_i(\theta) \tag{3.1}$$

and

$$\mathscr{E} \sum_i \hat{a}_i u_i = \sum_i \hat{a}_i P_i(\theta), \tag{3.2}$$

where $P_i(\theta)$ is the two-parameter logistic item response function, the probability of answering the item correctly. This result shows that each scoring method provides an unbiased estimator of a function of ability, but the functions estimated are not the same.

A comparison must be made between a function of x and a function of $\Sigma_i \hat{a}_i u_i$ that estimate the same ability parameter. Moreover, the function of x should be independent of the a_i, since the estimation of a_i is not a part of any Rasch procedure. This will be done as follows:

The ability parameter to be estimated will be considered to be the individual's true number-correct score

$$\xi \equiv \sum_i P_i(\theta). \tag{3.3}$$

Note that this is simply a specified monotonic transformation of ability θ. Since x is an unbiased estimator of ξ, x is clearly the ideal statistic under the Rasch model for this purpose.

The optimal estimator of ξ under the two-parameter model is not x. If there is no prior distribution for ξ, an optimal estimator is the function of the sufficient statistic $\Sigma_i a_i u_i$ that is unbiased for ξ. This function is uniquely determined by the Rao–Blackwell theorem (Kendall & Stuart, 1973, sec. 17.35), but it is too complicated for practical use here.

If the a_i and the item difficulties b_i are known, an asymptotically optimal estimator of ξ under the two-parameter model is the maximum likelihood estimator (MLE)

$$\tilde{\xi} \equiv \sum_i P_i(\tilde{\theta}), \tag{3.4}$$

where $\tilde{\theta}$ is the MLE of θ when the a_i and b_i are known. This follows from the fact that the MLE of a given function of a parameter is, under regularity conditions, the same function of the MLE of the parameter. Moreover, this estimator $\Sigma_i P_i(\tilde{\theta})$ is actually a function of the weighted sum $\Sigma_i a_i u_i$, since the MLE is always a function of the sufficient statistic if such exists. In fact, $\tilde{\theta}$ is the solution of the likelihood equation

$$\sum_i a_i P_i(\theta) = \sum_i a_i u_i. \tag{3.5}$$

Since the a_i and b_i are not known, let $\hat{a}_i$ and $\hat{b}_i$, estimated from some previously available sample of examinees, be substituted. Thus the two-parameter estimator to be compared with x is

$$\hat{\xi} \equiv \sum_{i=1}^{n} \hat{P}_i(\hat{\theta}), \tag{3.6}$$

where $\hat{P}_i(\hat{\theta})$ is the item response function, with $\hat{a}_i$ and $\hat{b}_i$ substituted for the unknown true item parameters and $\hat{\theta}$ is the solution of

$$\sum_i \hat{P}_i(\hat{\theta}) - \sum_i \hat{a}_i u_i = 0. \tag{3.7}$$

If N is large enough, $\hat{\xi}$ will necessarily show the same advantage over x as does the weighted sum $\Sigma_i\, a_i u_i$ when the a_i are known. But what if N is small, so that the $\hat{a}_i$ and $\hat{b}_i$ are erroneous estimates?

Since x and $\hat{\xi}$ are both consistent estimators of the same ability parameter ξ, they are properly compared by their mean square errors (MSE). The exact sampling variance of x is

$$\text{Var}(x) = \sum_{i=1}^{n} P_i(\theta)Q_i(\theta). \tag{3.8}$$

Since x is unbiased for ξ, this is also the exact MSE.

The sampling variance of $\hat{\xi}$ depends on ξ, the true score of the examinee whose true score is to be estimated. Given ξ, the variance of $\hat{\xi}$ arises from two sources:

1. Sampling fluctuations in the data on N examinees used to estimate the a_i and the b_i
2. Sampling fluctuations in the responses (u_{ia}) of examinees at the given true-score level ξ

The examinee whose true score is to be estimated is *not* included in the sample of N examinees; thus the second source of error is independent of the first.

It does not seem feasible to obtain the exact MSE of $\hat{\xi}$ when $N < \infty$; consequently, this study deals only with its asymptotic variance, which is equal to its asymptotic MSE. Formulas for calculating the asymptotic sampling variance are given in the Appendix.

TEST STUDIED

Numerical results can be obtained only for particular numerical values of the item parameters a_i and b_i. The following procedure was used in the hope of obtaining realistic numerical values.

The responses of 3000 sixth-grade students to a 50-item Metropolitan (MAT) vocabulary test were analyzed by LOGIST. Since (for simplicity) this study was limited to the two-parameter model, all c parameters were

TABLE 3.1
Item Serial Numbers and Item Parameters for All Tests Studied

Item serial number	Item parameters a	Item parameters b	Item serial number	Item parameters a	Item parameters b
3	1.6	−1.9	28	1.8	−0.9
4	1.7	−1.5	29	0.9	−0.8
5	0.8	−1.7	30	0.5	0.3
8	1.3	−1.7	33	0.7	−0.8
9	0.4	0.5	34	0.7	−0.4
10	1.1	−1.3	35	1.0	−0.2
13	1.4	−1.2	38	0.8	0.0
14	0.9	−1.1	39	1.1	−0.4
15	0.6	−1.9	40	0.8	0.1
18	1.2	−1.0	43	0.5	0.8
19	1.6	−0.9	44	1.1	−0.3
20	0.6	−0.4	45	0.7	0.3
23	0.6	−1.3	48	0.6	0.8
24	0.5	−1.4	49	0.4	0.3
25	0.9	−0.9	50	0.7	0.9

held at 0. The $\hat{a}_i$ and $\hat{b}_i$ obtained were used as true item parameters for the tests to be studied here. These item parameters are listed in Table 3.1.

Table 3.2 shows how four different 10-item tests are defined in terms of the items listed in Table 3.1. Tests 3, 4, and 5 are nonoverlapping spaced samples of items. Since the items in Table 3.1 are arranged roughly in order of difficulty, in Table 3.2 test difficulty tends to increase from top to bottom. Table 3.2 also shows for each test the true test score ξ that corresponds to specified values of θ. Remember that for any given test, ξ and θ are equivalent measures of the same ability, differing only in scale.

TABLE 3.2
True Score ξ Equivalent to Specified Ability Levels θ for Four Ten-Item Tests

Test	Items in test	Specified values of θ: −2	−1	0	1	2
3	3, 8, 13, 18, . . . , 48	1.8	4.8	7.4	8.7	9.4
4	4, 9, 14, 19, . . . , 49	1.5	4.1	7.1	8.7	9.3
5	5, 10, 15, 20, . . . , 50	1.7	3.8	6.3	8.2	9.3
1B	10, 10, 20, 20, . . . , 50, 50	1.2	3.1	5.4	7.4	8.8

RESULTS

The number-correct score x is an unbiased estimator of ξ. On the other hand, $\hat{\xi}$ is only asymptotically unbiased. The exact small-sample bias of $\hat{\xi}$ was calculated for 10-item Test 1B and for 5-item Test 1A, parallel to Test 1B except for length. (The method used for computing $\mathscr{E}(\hat{\xi} - \xi | \xi)$ is entirely parallel to the method for computing $\text{Var}(\hat{\xi} | \xi)$ described in the Appendix.) Test 1B consisted of 2 items exactly like Item 10, 2 like Item 20, and so forth, for a total of 10 items. Test 1A consisted simply of Items 10, 20, 30, 40, and 50.

Table 3.3 compares the bias of these two tests that differed only in length. The bias was small, even for five-item tests. The true score ξ of Test 1B was exactly double the true score of Test 1A, but the bias in $\hat{\xi}$ increased more modestly, if at all, as the test length was doubled.

Table 3.4 shows the exact small-sample variance of $\hat{\xi}$ when the item parameters were known, determined from an infinitely large sample of examinees. In this table Tests 1A, 1B, and 1C, which were parallel except for length, are compared. Test 1C contained 3 items like Item 10, 3 like Item 20, and so forth, for a total of 15 items. As might be expected, the sampling variance increased almost exactly as test length.

As noted previously, the optimal estimator of ξ is the function of $\Sigma_i\, a_i u_i$ that is unbiased for ξ. Since the desired function (which can be found by the Rao–Blackwell theorem) is impractical to use, the consistent estimator $\hat{\xi}$ was used. The MSE is equal to the variance plus the square of the bias. For Tests 1A and 1B, it can be seen from Tables 3.3 and 3.4 that the MSE differed from the variance of $\hat{\xi}$ only in the third decimal place. Table 3.5 compares the variance of x with the variance of $\hat{\xi}$. Replacing variance by MSE would not change the picture. The relative efficiency of two consistent estimators is asymptotically proportional to the ratio of their sampling variances. A comparison of the last two columns of Table 3.5 shows that the efficiency of x ranged from .85 at $\theta = 0$ and $\theta = -1$ to .93 at $\theta = -2$.

Interpolating in Table 3.5, it can be seen that for $\theta = -2$, x was better than $\hat{\xi}$ when the item parameters were estimated from a sample with $N < 200$, to a rough approximation; $\hat{\xi}$ was better than x when $N > 200$. It can be said, therefore, that $N = 200$ is the *critical* sample size. For the other tabulated θ values, the critical sample size is in each case less than 100.

The critical Ns for Test 3 are listed in Table 3.6 along with similar values for Tests 4, 5, 1A, 1B, and 1C. Because of the heavy cost in computer time, no runs were made for 15-item tests other than Test 1C. It appears

TABLE 3.3
Bias ($\mathscr{E}\hat{\xi} - \xi$) in True Score Estimate $\hat{\xi}$ for Tests 1A and 1B, Which Were Parallel Except for Length, When Item Parameters Were Known ($N = \infty$)

θ	Test 1A ($n = 5$)	Test 1B ($n = 10$)
−2	−.028	−.035
−1	.029	.037
0	.045	.047
1	.029	.026
2	.020	.020

TABLE 3.4
Variance [Eq. (3.A2)] of True Score Estimate $\hat{\xi}$ When Item Parameters Were Known ($N = \infty$) for Tests 1A, 1B, 1C, Which Were Parallel Except for Length

θ	Test 1A ($n = 5$)	Test 1B ($n = 10$)	Test 1C ($n = 15$)
−2	.52	1.01	1.50
−1	.92	1.75	2.58
0	.93	1.87	2.82
1	.79	1.60	2.40
2	.47	.95	1.44

TABLE 3.5
Variance [Eq. (3.A2)] of True Score Estimate $\hat{\xi}$ for Test 3, as a Function of the Sample Size N Used to Estimate the Item Parameters; Also Variance [Eq. (3.8)] of Number-Correct Score x

		$\mathrm{Var}(\hat{\xi} \mid \xi)$				
θ	ξ	$N = 100$	$N = 300$	$N = 1000$	$N = \infty$	$\mathrm{Var}(x \mid \xi)$
−2	1.8	1.32	1.23	1.20	1.19	1.28
−1	4.8	1.84	1.76	1.73	1.72	1.90
0	7.4	1.18	1.13	1.12	1.11	1.30
1	8.7	.78	.74	.73	.72	.85
2	9.4	.47	.45	.44	.44	.50

TABLE 3.6
Approximate Number of Cases (N) Required for $\hat{\xi}$ to Have a Smaller Sampling Variance Than Number-Correct Score x

	Test					
θ	1A	3	4	5	1B	1C
−2	700	200	<100	300	250	200
−1	3000	<100	<100	250	200	100
0	<100	<100	<100	200	<100	<100
1	100	<100	<100	150	150	200
2	100	<100	<100	100	250	250

that for the 10- and 15-item tests, the Rasch estimator x may be slightly superior to the two-parameter estimator $\hat{\xi}$ when the number of cases available for estimating the item parameters is less than 100 or 200. This is the main conclusion of the study.

CONCLUSIONS

This study has been limited to a comparison of one-parameter (Rasch) and two-parameter estimators of the examinee's true score. Similar studies should be made for the three-parameter model. Estimators of other quantities, such as item difficulty, should also be compared. The same approach can, in principle, be applied to determine the relative effectiveness of the Rasch and other models for test equating and other practical purposes; however, the computational burden of doing this may prove to be excessive.

APPENDIX: ASYMPTOTIC SAMPLING VARIANCE OF $\hat{\xi}$

By a standard formula from analysis of variance, the error variance of $\hat{\xi}$ for fixed ξ can be written

$$\mathrm{Var}(\hat{\xi}\,|\,\xi) = \mathscr{E}_{\mathbf{u}}[\mathrm{Var}(\hat{\xi}\,|\,\xi,\mathbf{u})\,|\,\xi] + \mathrm{Var}_{\mathbf{u}}[\mathscr{E}(\hat{\xi}\,|\,\xi,\mathbf{u})\,|\,\xi]. \qquad (3.\text{A}1)$$

where $\mathscr{E}_{\mathbf{u}}$ and $\mathrm{Var}_{\mathbf{u}}$ are taken across all possible response vectors $\mathbf{u} \equiv \{u_i\}$. To understand the terms in brackets, note that when $\mathbf{u}$ is fixed, the only other source of variability is sampling error in the estimation of the a_i and the b_i. Remember that the $\hat{a}_i$ and $\hat{b}_i$ are obtained from a sample of N examinees and that $\mathbf{u}$ belongs to an examinee who is not part of that sample.

For large N, the last term in Eq. (3.A1) is adequately approximated by replacing the estimates $\hat{a}_i$ and $\hat{b}_i$ by their true values a_i and b_i; in other words, $\mathscr{E}(\hat{\xi}|\xi,\mathbf{u})$ can be replaced by $\mathscr{E}(\tilde{\xi}|\xi,\mathbf{u})$. By Eqs. (3.4) and (3.5), fixing $\mathbf{u}$ also fixes ξ, so now $\mathscr{E}(\tilde{\xi}|\xi,\mathbf{u}) \doteq \xi$. Thus the last term in Eq. (3.A1) becomes $\text{Var}_{\mathbf{u}}(\tilde{\xi}|\xi)$. By Eq. (3.3), whenever ξ is fixed, θ is fixed also; so Eq. (3.A1) can be written approximately

$$\text{Var}(\hat{\xi}|\xi) = \mathscr{E}_{\mathbf{u}}[\text{Var}(\hat{\xi}|\theta,\mathbf{u})|\theta] + \text{Var}_{\mathbf{u}}(\tilde{\xi}|\theta). \tag{3.A2}$$

The first variance on the right-hand side arises from sampling fluctuations in the $\hat{a}_i$ and the $\hat{b}_i$; the second variance is independent of these fluctuations. The second variance can be evaluated as follows:

1. For each possible item response pattern $\mathbf{u}$, determine $\tilde{\theta}$ by solving Eq. (3.5) numerically.
2. For each $\tilde{\theta}$ from Step 1, compute $\tilde{\xi} \equiv \Sigma_i P_i(\tilde{\theta})$.
3. For each $\mathbf{u}$, compute

$$\text{Prob}(\mathbf{u}|\theta) \equiv \prod_{i=1}^{n} P_i(\theta)^{u_i} Q_i(\theta)^{1-u_i} \tag{3.A3}$$

   for the given values of θ (not $\tilde{\theta}$). This gives the frequency distribution of $\mathbf{u}$.
4. Compute the variance for given θ of the $\tilde{\xi}$ obtained in Step 2, taken over the frequency distribution of $\mathbf{u}$ found in Step 3. The result is $\text{Var}_{\mathbf{u}}(\tilde{\xi}|\theta)$, as required.
5. Repeat the foregoing for different given values of θ, as desired.

Although the notation does not make it explicit, the results obtained depend, of course, on the a_i and b_i of the items in the test being studied. A separate study must be carried out for each different test. Because of the number of calculations required, practical considerations limit investigation to tests not much longer than 15 items.

It remains to evaluate the first term on the right-hand side of Eq. (3.A2). The quantity $\text{Var}(\hat{\xi}|\theta,\mathbf{u})$ will be evaluated by the delta method. Let $\xi'_{\hat{a}i}$ denote the partial derivative of $\hat{\xi}$ with respect to $\hat{a}_i$, and similarly for $\hat{b}_i$ and $\hat{\theta}$. The total derivative of $\hat{\xi}$ is

$$d\hat{\xi} = \xi'_{\hat{\theta}}\, d\hat{\theta} + \sum_i \xi'_{\hat{a}i}\, d\hat{a}_i + \sum_i \xi'_{\hat{b}i}\, d\hat{b}_i. \tag{3.A4}$$

But now, however, by Eq. (3.7), $\hat{\theta}$ is itself a function of the $\hat{a}_i$ and the $\hat{b}_i$. Denoting the left-hand side of Eq. (3.7) by ℓ, the total derivative of Eq. (3.7) is

$$\ell'_{\hat{\theta}}\, d\hat{\theta} + \sum_i \ell'_{\hat{a}i}\, d\hat{a}_i + \sum_i \ell'_{\hat{b}i}\, d\hat{b}_i = 0, \tag{3.A5}$$

where $\ell'_{\hat{a}i}$ is the partial derivative of $\hat{\ell}$ with respect to $\hat{a}_i$, and similarly for $\hat{b}_i$ and $\hat{\theta}$. Eliminating $d\hat{\theta}$ from Eqs. (3.A4) and (3.A5) gives

$$\ell'_{\hat{\theta}}\, d\hat{\xi} = \sum_i (\ell'_{\hat{\theta}}\xi'_{\hat{a}i} - \xi'_{\hat{\theta}}\ell'_{\hat{a}i})\, d\hat{a}_i + \sum_i (\ell'_{\hat{\theta}}\xi'_{\hat{b}i} - \xi'_{\hat{\theta}}\ell'_{\hat{b}i})\, d\hat{b}_i. \quad (3.A6)$$

When the delta method is applied to Eq. (3.A6), it is found that

$$\begin{aligned}\mathrm{Var}(\hat{\xi} \mid \theta,\mathbf{u}) = \frac{1}{l'^2_\theta}\Bigg[& \sum_i \{(\ell'_\theta\xi'_{ai} - \xi'_\theta\ell'_{ai})^2\, \mathrm{Var}(\hat{a}_i|\theta,\mathbf{u})\} \\ & + \sum_i \{(\ell'_\theta\xi'_{bi} - \xi'_\theta\ell'_{bi})^2\, \mathrm{Var}(\hat{b}_i|\theta,\mathbf{u})\} \\ & + 2\sum_i \{(\ell'_\theta\xi'_{ai} - \xi'_\theta\ell'_{ai})(\ell'_\theta\xi'_{bi} - \xi'_\theta\ell'_{bi})\,\mathrm{Cov}(\hat{a}_i,\hat{b}_i|\theta,\mathbf{u})\}\Bigg].\end{aligned} \quad (3.A7)$$

For given θ and $\mathbf{u}$, the variance needed for the first term on the right-hand side of Eq. (3.A2) can be computed from Eq. (3.A7). The necessary derivatives are

$$\xi'_\theta = \sum_i a_i\pi'_i, \quad (3.A8)$$

$$\xi'_{ai} = (\theta - b_i)\pi'_i, \quad (3.A9)$$

$$\xi'_{bi} = -a_i\pi'_i, \quad (3.A10)$$

$$\ell'_\theta = -\sum_i a_i^2\pi'_i, \quad (3.A11)$$

$$\ell'_{ai} = u_i - P_i - a_i(\theta - b_i)\pi'_i, \quad (3.A12)$$

$$\ell'_{bi} = +a_i^2\pi'_i, \quad (3.A13)$$

where $\pi'_i \equiv DP_iQ_i$ denotes the derivative of P_i with respect to $L_i \equiv a_i(\theta - b_i)$, and $D = 1.7$. The variance–covariance matrix of $\hat{a}_i$ and $\hat{b}_i$, needed in Eq. (3.A7), is found by inverting the Fisher information matrix:

$$\begin{bmatrix} D^2\sum_{a=1}^{N}(\theta_a - b_i)^2P_{ia}Q_{iq} & -D^2a_i\sum_{a=1}^{N}(\theta_a - b_i)P_{ia}Q_{ia} \\ -D^2a_i\sum_{a=1}^{N}(\theta_a - b_i)P_{ia}Q_{ia} & D^2a_i^2\sum_{a=1}^{N}P_{ia}Q_{ia}\end{bmatrix}, \quad (3.A14)$$

where $P_{ia} \equiv P_i(\theta_a)$.

To evaluate Eq. (3.A2) for fixed θ, compute $\mathrm{Var}(\hat{\xi}|\theta,\mathbf{u})$ by Eq. (3.A7) separately for each $\mathbf{u}$. Then take the average of these values across $\mathbf{u}$ [weighting each value by $\mathrm{Prob}(\mathbf{u}|\theta)$ given by Eq. (3.A3)] to find the first term on the right-hand side. Add on the second term, $\mathrm{Var}_{\mathbf{u}}(\tilde{\xi}|\theta)$, computed as described earlier. The resulting $\mathrm{Var}(\hat{\xi} \mid \xi)$ must be computed separately for each ξ or θ of interest. [Note again that fixing ξ is equivalent to fixing θ, because of Eq. (3.3).].

ACKNOWLEDGMENT

This work was supported in part by contract N00014-80-C-0402, project designation NR 150-453, between the Office of Naval Research and Educational Testing Service.

REFERENCE

Kendall, M. G., & Stuart, A. *The advanced theory of statistics* (Vol. 2, 3rd ed.). New York: Hafner, 1973.

4

Are We Correcting for Guessing in the Wrong Direction?

HOWARD WAINER

The development of multiple-choice tests had as a concomitant event the increased likelihood of correct guessing. Traditional corrections for guessing typically examine the incorrect responses and assume that, for, say, a k-choice item, for every $k-1$ incorrect responses observed there was one correct response that was arrived at through guessing. Thus, a so-called "formula score" on a k-possibility multiple-choice test was [number correct] − [number incorrect/($k-1$)]. There are many variants on this general scheme (e.g., DeFinetti, 1965), but all assume that the aim is to obtain some measure that has incorrect responses as a viable carrier of information; they all, given their assumptions about respondents' behavior, yield an unbiased estimate of number correct.

Use of these simple techniques for the correction for guessing is unsatisfactory for a variety of reasons. Lord and Novick (1968) wrote,

> examinees who have *partial information* about an item do not respond at random, nor do examinees with *misinformation* about that item. In these situations, wrong answers cannot be equally attractive to the examinee. No simple correction formula is appropriate in these cases [p. 303].

NEW HORIZONS IN TESTING
Latent Trait Test Theory and
Computerized Adaptive Testing

ISBN 0-12-742780-5

They later stated,

> The simple knowledge or random guessing model is used extensively despite its several weaknesses. One such weakness is that it ignores the possible day-to-day variations in examinee performance . . . which experience has taught us cannot be neglected. An equally serious weakness is that the model assumes that if the examinee is unable to pinpoint the correct response, then he is completely ignorant in this situation and has no basis for choosing among the possible responses. This second assumption can seldom be seriously entertained [p. 309].

Lord and Novick (1968, p. 196) have presented a convincing argument against the use of incorrect responses as the sole carrier of information about guessing. The basis of much of their argument is that simple guessing cannot be considered the cause of much of what is observed in a test response vector. Since most of these correction methods were working from a background of true-score theory, there was little else that could be done.

With the development of latent trait models [what is now being called *item response theory* (IRT)] two approaches became possible. The introduction of a nonzero lower asymptote to the item characteristic curve implies that a person can correctly answer an item without knowing the answer; it corrects for this in a way that is identical to formula scoring (see Appendix A for a proof). The consequence of this is thus identical to formula scoring—a reduction in ability estimates. In addition, this method for dealing with guessing can cause substantial increases in the standard errors of the estimates of item difficulty and should be avoided if possible [see Appendix B and Thissen & Wainer (1982)].

In using IRT, however, a new aspect of the test response vector became available to aid in the estimation of the testee's ability—the pattern of the response vector. Through the use of IRT the items could be ordered by their likelihood of being answered correctly, and unusual responses could be noted. Thus, an incorrect response to a very difficult item could be seen as a predictable event, tending to confirm the fit of the model. A correct response to an easy item was also anticipated and further corroborated the model. Unusual events were difficult items answered correctly (evidence of "guessing") or easy items answered incorrectly (evidence of "sleeping").[1] In practice, guessing occurred much more frequently than sleeping. A strategy to correct for guessing was either to eliminate items with

[1] Since events that are being called "guessing" or "sleeping" could have a wide variety of other causes, such as unusual patterns of knowledge, this may be a poor choice of terminology. Nevertheless, these terms will be used as a convenient notation to describe these two kinds of unusual events.

large residuals from the IRT model or to change their response to conform to expectation. The former was the more frequent choice, but both have the same outcome—the reduction of the ability estimate.

The principal difference between the traditional corrections for guessing (including IRT models with nonzero lower asymptotes) and the second kind of IRT correction is that the former focuses on the number of incorrect responses, the latter principally on the unlikely correct ones. In either case, the result is in general the same—the lowering of the score of the testee. It is clear that guessing inflates the testee's observed score, but it also adds variance. The question is, Is lowering the score the best way to go about correcting it?

Obviously, if there is certainty that the unusual response pattern was generated by guessing, the correct fix would be to reduce the ability estimate appropriately. The problem is that there can never be any certainty that it was guessing that yielded the observed unusual response pattern. Alternative explanations are possible, and sometimes even likely. One possible alternative explanation is that the trait being measured, although unidimensional for most of the tested population, is multidimensional for some small but identifiable subpopulation. For example, in a test of drug knowledge, one student indicated that he knew very little about drugs, but he correctly answered three rather difficult questions. The first impulse was to ascribe these correct responses to guessing. On closer examination, it was discovered that the questions answered correctly dealt with mescaline and peyote, and that the student was a Native American for whom these kinds of drugs were commonly used in religious ceremonies. Clearly, if this student's ability estimate were reduced based on the low probability of his knowing the correct answers to these questions (the probabilities coming from the test model), his drug knowledge would have been seriously underestimated. The problems associated with correcting for guessing when the responses observed were not generated by the sort of Bernoulli trials posited by most correction schemes are the subject of this chapter.

THE START OF AN ALTERNATIVE

Before alternative schemes for dealing with unlikely responses are discussed, what is desired from a test can be restated. Usually, what is desired is the best estimate of the testee's ability that can be obtained and an honest measure of that estimate's accuracy. This is usually concretized by an ability parameter and the standard error of that parameter. True-score theory yields standard errors through various reliability coefficients (Lord & Novick, 1968, p. 154ff). Typically, the shortening of the effective test length that is the result of guessing is not fully represented in the error estimate. Using the three-parameter IRT model (see, e.g., Birnbaum, 1968, p. 453ff), the standard error of the ability parameter is affected by the guessing parameter (see Jones, 1981a, for details of how to measure the size of this effect). There have been few applications in which the guessing parameter of the item has been successfully estimated; this is partially due to the problem of attempting to estimate a parameter in a region where there is little or no information.

Consider the following situations in which a test, scored from, say, 0 to 100, is administered and the mean ability is 50 with a standard deviation of 10. Suppose further that it is necessary to provide an estimate of each individual's ability.

Situation 1. Student *A* does not show up for the test.

Situation 2. Student *B* shows up but hands in a blank answer sheet.

Situation 3. Student *C* hands in an answer sheet on which he/she answered 20% of the items correctly, and these were all among the 25% easiest items.

Situation 4. Student *D* also answered 20% of the items correctly, and they were scattered randomly throughout the exam.

Situation 5. Student *E* answered the 20% most difficult items correctly and all others incorrectly.

How should these various situations be treated? Student *A*, most would agree, should be recorded as "missing data," and if he/she must be assigned a score, one should somehow be imputed based on the rest of the sample. Certainly, a zero ability estimate should not be assigned, for his/her absence from the test could have been caused by a wide variety of circumstances unrelated to his/her ability (e.g., illness, transportation failure, family circumstances). Thus a sensible way to treat such an individual (aside from saying "did not show up—no score assigned") would be to assign the mean and to use the standard deviation of the tested sample as the standard error of this estimate. This is in keeping with such correction schemes as Kelley's (1947, p. 409) equation

$$\tau = \rho x + (1 - \rho)\mu \qquad (4.1)$$

in which estimated true score τ is obtained by regressing observed score x toward the mean μ; the extent of the regression effect is the reliability ρ of the test. In this situation the test is very unreliable, and so assigning a score of 50 with a standard error of 10 would not be inappropriate.

The second situation is not very different from the first. Once again, the answer sheet is blank, the only difference being that the testee showed up at the testing center. Nevertheless, there can be many plausible reasons the test was handed in blank that are unrelated to ability. Once again, the testee's ability is probably being underestimated if a zero score is assigned; some sort of estimate based on an imputation scheme would surely serve better.

Note that in both these situations there is a sense in which the correction that is being used *raised* rather than lowered the examinee's observed score. Alternatively, the correction simply replaces the "nothing" at no-score-at-all with the population mean. This contrasts sharply with traditional corrections.

In the third situation, it is found that the response pattern fits the expectations of how a person ought to take the test; thus, confidence in the traditionally estimated ability estimate is well described by the estimated standard error. Note that in this situation there is a pattern that is quite likely; so even though the total score is about what would be expected by chance, the pattern of responses argues against it.

The fourth situation shows a pattern that might have been generated by guessing, but it might also have been generated by a pattern of knowledge different from what was anticipated. Thus the ability estimate might be lowered following traditional practice or, instead, it might be concluded that this person has not been measured very well (his/her pattern does not fit the test); his/her ability estimate is therefore regressed in the direction of the mean and the standard error is expanded. When the standard error is 10, it can be concluded that what is known about this person after viewing his/her test response vector is as much as would have been known had he/she not taken the test. Of course, experience has taught that individuals who hand in answer sheets like this come from a population whose mean is considerably lower than that of the general population; thus, the score might be regressed downward.

The last situation is the most difficult to explain. Most explanations that suggest themselves indicate a much higher ability, for example, the test results of Galois in his two entrance examinations for the École Polytechnique. Perhaps such a pattern was generated by a very bright student who was bored by the exam and only started to answer questions when they became interesting. Certainly, it cannot be concluded that this student was guessing; it would be far better to conclude that he/she was not measured very well and to impute a more middle score and a large standard error.

The preceding five situations represent in extreme what all test patterns show to some extent or other, a pattern that does not conform exactly to what the test makers expected. A more honest approach to scoring, when a test pattern is found that is unlikely to have been generated by someone behaving according to a psychometric model, should reflect uncertainty in their ability by regressing their score toward the middle and by suitably increasing the standard error of that estimate. Additional measures (e.g., retesting) can then be used to obtain ability estimates for those for whom the standard errors of estimate are unacceptably large. This process can be thought of in a Bayesian way by having a prior ability distribution generate a posterior score, the observed test score *with an associated fit statistic* being the data that pull the prior toward the posterior. If the fit is poor, the data have little effect. If the fit is good, the data have a more profound effect. The fit statistic can also be used to inflate the standard error of the estimate (see Finney, 1952).

A MORE SPECIFIC PLAN

The foregoing discussion was more epistemological than operational. It was pointed out that the problems associated with assuming the wrong answers or unusual response patterns were the result of guessing and not, say, idiosyncratic knowledge. The schemes proposed for correcting this problem were general, and few details were elucidated, although some may see some similarities between the use of Kelley's (1947) equation and the more general James–Stein estimators (Efron & Morris, 1975).

This discussion has purposely gone too far. Certainly, cogent arguments can be made for regressing the ability estimates obtained from unusual response patterns inward from their observed values. There has been an attempt to elucidate some of them here. Yet, experience has taught that guessing is an activity that occurs frequently and accounts for many (although by no means all) of the anomalous responses observed. Therefore a middle way will be proposed. It would seem that a proper scheme to correct for unusual response patterns would be in between the traditional correction (that assumes that *all* anomalies are from guessing) and one that professes complete ignorance of the cause of such unusual response patterns. The method that is preferred regresses the corrected ability estimates inward, although not usually as far as using an uncorrected raw score.

Wainer and Wright (1980) proposed an amalgam of the Jackknife and the Sine M-Estimator (the so-called AMJACK estimation scheme) to robustly estimate ability in the one-parameter logistic (1-PL) model. The aim of this procedure was to obtain an estimate of ability from each item responded to, and to robustly estimate the middle of this distribution. The notion was that items that were responded to in a nonstereotypic manner would look like outliers and would be largely discounted in their contribution toward the ability estimate. The result of this was that ability estimates tended to be regressed inward toward the center of the ability continuum from their position after the traditional correction. The extent of this regression was dependent on the fit of the response vector to the 1-PL model: the better the fit, the less the regression effect. Using this scheme, it was found that the estimates obtained were better in many ways than the maximum likelihood estimators (MLEs) of ability. In particular, even when the data fit the model, the AMJACK estimator was more efficient than the MLE for short- to modest-length tests (test lengths of less than 40 or so items); as tests grew longer, the asymptotic optimality of MLE began to manifest itself. When the data became noisier, the superiority of the AMJACK estimator over MLE grew larger.

In addition to the increase in efficiency and accuracy of the AMJACK estimator, it was found that the standard error of the estimator, calculated directly from the Jackknifed pseudovalues, increased as the deviation from the 1-PL model increased. This, once again, is in line with the general guidelines described above for a sensible test-scoring scheme, although the actual estimates of standard error tended to be too conservative.

AN AMJACK EXAMPLE

To understand better how the AMJACK estimator works, consider a simple numerical example. Suppose a 10-item test whose difficulties uniformly span the range from −2 to +2 logits is given; among the many patterns of responses, the several shown in Table 4.1 are observed.

The estimators shown reflect different philosophies of correction. The first column, uncorrected θ, is the standard MLE of ability obtained from the 1-PL model. As is obvious, the pattern is unrelated either to the ability estimate or to its standard error. This reflects the well-known property of the 1-PL model that the raw score is a sufficient statistic for estimating ability. Ordinarily, unusual patterns in this model are detected by examining the goodness of fit of the model to each person rather than the standard error.

The two columns in Table 4.1 labeled "corrected" are two different methods of correcting for guessing that are representative of two classes of corrections. θ_{c1} examines the probability (under the model) of each

TABLE 4.1
Ability Estimates on a Short Test (with Some Noise) by Five Methods[a]

	MLE							
	Uncorrected		Corrected		AMJACK		Biweight	
Response patterns	θ	SE	θ_{C_1}	θ_{C_2}	θ	SE	θ	SE
Some with guessing								
1. 1110000000	−1.15	.79	−1.15	−1.15	−1.05	.64	−1.07	.78
2. 1110000001	−.56	.76	−1.12	−1.15	−.86	.79	−.91	.77
3. 1110000010	−.56	.76	−1.11	−1.15	−.82	.78	−.81	.77
4. 1110000100	−.56	.76	−1.09	−1.15	−.70	.76	−.69	.76
5. 1110001000	−.56	.76	−1.07	−1.15	−.60	.74	−.61	.76
6. 1110010000	−.56	.76	−1.01	−1.15	−.54	.70	−.55	.75
7. 1110100000	−.56	.76	−.56	−.56	−.49	.64	−.52	.75
8. 1111000000	−.56	.76	−.56	−.56	−.52	.56	−.53	.75
9. 1111110000	.56	.76	.56	.56	.52	.56	.49	.75
10. 1111110001	1.15	.79	.68	.56	.66	.83	.95	.78
11. 1111110010	1.15	.79	.73	.56	.80	79	1.00	.78
12. 1111110100	1.15	.79	1.15	1.15	1.02	.71	1.02	.78
13. 1111111000	1.15	.79	1.15	1.15	1.05	.64	1.01	.78
Some with sleeping								
14. 0110000000	−1.84	.89	−1.37	−1.15	−1.20	1.04	−1.66	.85
15. 0111000000	−1.15	.79	−.68	−.56	−.66	.83	−.96	.78
16. 0111110000	0.00	.74	.51	.56	.34	.76	.27	.75
17. 0111111000	.56	.76	1.12	1.15	.86	.79	.90	.77
18. 0000011111	0.00	.74	?	?	0.00	1.36	0.00	.74

[a] Item difficulties are −2.0, −1.6, −1.1, −0.7, −0.2, 0.2, 0.7, 1.1, 1.6 and 2.0.

response. If a response is too unlikely to have occurred under the model, that item is omitted from consideration in the determination of the person's ability estimate. This reduces the test length and so increases the standard error. The gradual increase in ability that is evident in Response Patterns 1–7 is due to the average difficulty of the test getting progressively greater as less and less difficult items are omitted. The column marked θ_{c2} uses the same determination rule as θ_{c1} except that instead of omitting the item it *changes* the response so that the pattern fits the model. Thus, it makes Response Patterns 2–6 identical to Response Pattern 1, and similarly changes Response Pattern 14 to be like Response Pattern 1.

The last two estimators are the AMJACK and Mislevy and Bock's (1982) biweight estimator. This latter scheme is a robust measure that uses the adaptive biweight in the likelihood equation. Its behavior has not been thoroughly tested, but illustrative examples, like that in Table 4.1, indicate that it (or methods that develop from it) may be worth pursuing.

A careful examination of Table 4.1 will illustrate how these corrections work and indicate that the two robust methods shown offer a middle ground. They are less extreme than either of the two corrections, yet not so forgiving as the uncorrected estimator. The AMJACK estimator seems to react to unusual response patterns in a way that closely mirrors what a thoughtful test scorer would do (i.e., discount an unlikely event as bizarre and provide a more or less identical ability estimate with an increased standard error). The biweight estimator seems to follow this behavior very closely as well, except that the standard error does not seem to expand apace. This latter effect is most visible in the last case (Case 18) in which only AMJACK indicates the unusual nature of this response pattern (by giving an almost doubled standard error).

Note particularly Cases 1 and 13, where the robustification schemes shrink the ability estimate inward *even though the data fit the model.* In these instances, the Jackknife part of AMJACK is doing what the Jackknife was traditionally designed to do—reduce bias. As is well known, the bias of IRT ability estimates increases at the extremes (Lord, 1980, 1981), and an extra added benefit of the use of these robustifications is a correction for it.

The AMJACK and the biweight estimators are good first choices to correct for unusual response patterns in that they move in the correct direction. The AMJACK is preferred because it expands the standard error in a way that mirrors anticipation. Yet even the AMJACK is just a first step. It may be that a complete Bayesian approach that determines the amount of regression effect from more direct evidence of guessing and sleeping will be better still. Such data, however, are hard to find; thus for many applications, AMJACK may do as well as can be expected in the absence of extensive empirical prior distributions.

Under What Circumstances Can This Be Used?

Whether or not a robustification scheme like AMJACK (or some improved Bayesian method) can be successfully applied in paper-and-pencil tests is uncertain. For political reasons, having a complicated scoring method sitting behind a nonlinear test model may not be easy to implement widely at the current time. There is, however, one avenue of application that is particularly appropriate—Computerized Adaptive Testing (CAT).

To understand the problem area better, and to understand how AMJACK-like techniques can be helpful, consider the task in CAT. As an individual answers questions, the computer algorithm must determine new items to ask. There are several item-selection procedures that have been suggested (i.e., choose the available item that maximizes information or one that reduces the posterior variance). All these are (or at least should be) very similar to one another in practice. The reason that this is the case has to do with the structure of a well-developed item pool. More specifically, the slopes of item characteristic curves of items in the pool ought to be nearly equal. This is because items whose slopes are low or negative are poor items and will have been eliminated from consideration in initial pretesting and item screening. Items whose slopes are very steep are rare; indeed, finding one almost always signals an artifact. Thus, even though AMJACK can be run with an unequal slope model, equal slopes tend to be the rule.

A further argument supporting the use of an equal slope model in adaptive testing is that even though modest differences in slopes have virtually no effect on the fit of the model, they have a profound effect on the estimate of information or posterior variance. This is a serious problem in adaptive testing since items are chosen on the basis of their contribution to overall information.

Since this problem does not appear to be either often mentioned or well understood, some discussion is in order. Adaptive testing is at its greatest advantage over traditional paper-and-pencil tests at the extremes of the ability distribution. The standard errors of the slope parameter, however, are largest at these points (Thissen & Wainer, 1982), easily reaching 2 logit units (see Appendix B for details). Information is a function of the *square of the slope,* thus compounding the error. Moreover, it cannot be assumed that errors in the estimation of slope will balance themselves out in the computation of information, since the choice algorithm will be guided toward those items that have errors in a positive direction. The resulting bias will cause vast overestimation of the accuracy of ability estimates. By way of example, consider that the average "good" item has a slope a of between 1 and 1.5; squared this is between 1 and 2.25. If the adaptive test is working perfectly, each item presented will have a probability of a correct response of .5; thus the information obtained for each item will range [based on $a^2(PQ)$] from .25 to .56. For purposes of illustration, let it be said that items will average information of .5. To obtain an ability estimate that is accurate to one decimal place (i.e., has a standard error of .05), information will need to equal 400, so that $1/\sqrt{\text{information}}$ will be .05. Obviously, for information to total 400 when each item contributes .5

implies that 800 items will be needed for this degree of accuracy. How, it may be asked, do existing adaptive tests achieve acceptable accuracy with 20–30 items?

Obviously, the item-selection algorithm will choose those items for which the sampling variance (the standard error) has gone in the positive direction. Thus, while the slope may have a true value of 1.5, a standard error of 2 would mean that it might easily be mistaken for an a of 4 or so. Squaring this yields a contribution to information 8 times the correct amount. This bias in choosing items is difficult to correct for within the context of a model that allows differential slopes. Using a complex parameterization for the initial item screening is suggested; but after the items are chosen, the selected items should be recalibrated with an equal slope model. This will avoid the biased (overoptimistic) estimate of the accuracy of the ability estimate. Perhaps the conclusion from this is that there ought not to be any a in CAT.

Second, if CAT is working properly, data on inappropriately administered items will be observed even more rarely than is currently the case with paper-and-pencil tests. Thus the estimation of a nonzero lower asymptote will be almost impossible (there is no c in CAT). This implies that with information gathered in any reasonably efficient CAT system, it will not be possible to tell the difference between an item with a nonzero lower asymptote from one with a zero asymptote (see Appendix B for further details).

These two characteristics lead to the conclusion that the single characteristic of an item that should determine its selection within a CAT mode is its difficulty. Thus, regardless of the item-selection method used, the item difficulty will be what is of most interest. To choose a reasonable next item, it will merely be necessary to match, as closely as possible, the difficulty to the current estimate of the examinee's ability. It is here that the AMJACK ability estimator comes into its own: After each item, it gives a good estimate of ability and an honest estimate of its standard error. The former aids in item selection; the latter, in determining appropriate stopping.

AMJACK has only been tested thus far within the context of the 1-PL model. Certainly, there are circumstances where this model will not fit; yet, within CAT it would appear that the chances of the 1-PL model fitting are very good indeed. Furthermore, the robust character of the ability estimation scheme gives protection from even substantial departures from fit as well as characterizing poor fit with large standard errors. Obviously, AMJACK is not the final solution; it merely moves a little in the appropriate direction. More work is needed.

CONCLUSIONS

This chapter has been concerned with the problem of how to treat unusual responses in an individual test pattern. These are most often caused by what has been referred to previously as guessing. Traditional corrections for these unusual responses yield reduced ability estimates. It has been argued that there are enough alternative explanations for unusual responses to indicate that a useful strategy might be to allow that the unusual response pattern indicates that the person taking the test did not respond in the way expected. The cause of this could lie with the examinee (he/she guessed) or the test constructor (the test was not unidimensional). It is certain that not so much is known about the testee as would have been, had he/she responded in a manner described by the test model. This increase in the tester's ignorance is better admitted by regressing inward the ability estimate that has been corrected in a traditional way and by increasing the standard error.

It has been suggested that one way to use such scores would be to determine the accuracy required for the purpose at hand, setting aside all individuals who were not measured with sufficient accuracy (scanning the standard errors) and using the ability estimates of the rest. Those who were not measured with sufficient accuracy would then be subjected to different treatment (e.g., another test, a different kind of test). It is clear that far less damage is done to individuals for whom the test is inappropriate by recognizing that fact than by reducing their score. This is particularly true now with the necessity of making public test vectors and scoring keys. The explanation that an ability estimate of a student was reduced because the items responded to correctly were difficult is an explanation that will cause more problems than it solves. The alternative offered is just a first approximation but one that moves the correction for guessing in what can be considered a more appropriate direction. It is, in prospect, a fruitful approach to use within the context of CAT.

The original intent of this chapter was to point out how a robust estimator of ability can aid the testing enterprise in two ways. The first advantage is that of obtaining an estimate of ability less biased and less strongly affected by unusual responses than the MLE. The second advantage is that its standard error is calculated directly, thus providing a more honest estimate of its variability. This tends to be somewhat larger than the asymptotic estimates obtained from maximum likelihood theory because it takes into account the goodness of fit of the model. In the course of accomplishing these laudable ends, however, the estimator shrank the estimate toward the center of the ability distribution in much the same fashion as a ridge estimator, or an estimator following Kelley's (1947) equation, or any Bayesian scheme.

This chapter was originally meant as an epistemological discussion of the shrinking, since it seems, on initial examination, to be going in the wrong direction. A number of points arose that expanded these initial modest aims, however. First was the question of an appropriate model to use in estimating ability in CAT. After carefully studying the behavior of the standard errors of estimate of complex parameterizations and the realities of CAT, it was concluded that the estimation of a lower asymptote within the confines of CAT is generally impractical (i.e., to obtain a sufficient accuracy so as not to compromise the integrity of the test, the sample size would have to be very large, and items would have to be administered to many individuals for whom they were inappropriately difficult). The conclusion was drawn that "there is no c in CAT." Next, when the standard errors of slope for extreme items were considered, it was noted that these could be large. When this was coupled with the selection bias in CAT for choosing items whose slopes are largest (i.e., items for which the standard error has pushed them upward), it was concluded that the inclusion of slopes in the estimation model will result in a very optimistic estimate of the accuracy of the ability estimate. To avoid this was difficult, and a simple but effective solution was opted for—an equal slope model when the CAT is operational. The second conclusion was then that "there is no a in CAT."

It is not suggested that the one-parameter model will fit in all situations. Rather, it will be difficult to show that in most operational situations it does not. Furthermore, any bias in the model caused by ignoring differences in lower asymptotes or slopes can be better reduced through robust measures than by increasing the complexity of the parameterization. It must be emphasized that the advocacy of the simple one-parameter model is only in the operational phase of CAT. In pilot testing, in which items are to be selected or not for inclusion in item banks, a more complex parameterization may be helpful.

Finally, the AMJACK method is not the only robust method available. It is merely the first. Mislevy and Bock's (1982) biweight scheme seems promising, and perhaps the best so far is Jones's (1981b) H estimate. In this method the maximum likelihood estimator of ability is modified by the "slopeless information" raised to the H power, i.e., the value of θ that solves

$$0 = \sum a(x - P(\theta))(PQ)^H. \tag{4.2}$$

Note that when $H = 0$, this reverts to the standard MLE; as H gets larger, this tends to weight observations near the person's ability more and those in the extremes less. Theoretical investigations have indicated that this estimator is very efficient when the MLE is the estimator of choice and robust when bias is introduced. Of course, when all *a*s are equal, it still applies.

The area of robustification of parameter estimation within the confines of IRT is still developing. There are at least three promising methods available. Each is meant to provide a more accurate estimate than maximum likelihood when the model is violated. The approach presented in this chapter is to use a simpler parameterization and to improve the estimate with a robustified estimator. This is done because of the consequences of accuracy loss involved in increasing the complexity of the parameterization. In some situations this decision may usefully go the other way, but in a CAT circumstance a simple parameterization with robustification is the approach with the most potential application.

AFTERTHOUGHT

The problems of unusual responses discussed in this chapter can be diminished in other nonstatistical ways. Most obviously, the standard format multiple-choice test could not be used. Gulliksen (1981) has often suggested that instead of having several choices next to each item, there could just be a long list of choices at the beginning of each test section. With a sufficiently long list the probability of choosing the correct response by chance is small enough not to yield serious difficulties. Gulliksen also mentioned that this would make the task of item writing much easier, since it becomes simply a matter of creating questions with their answers, not performing the very difficult task of developing reasonable wrong answers.

Another suggestion (Wright & Stone, 1979, p. 188) is not to present to individuals items that are far away from their ability levels. In particular, if the probability of a person responding correctly to an item is less than chance, that item should not be presented. This does not cause much of a loss in the accuracy of ability estimation, since items that far in difficulty from an ability level carry very little information. The problem with implementing this suggestion in general conditions for wide-range tests is that an examinee's ability is not a known a priori. With adaptive testing this becomes more feasible. A second problem with this approach is that there is no check on the fit of the model if no items are offered that would allow an unusual response. Of course, a middle path is to ask enough items that are extreme so that model fit can be estimated, but not so many to compromise the efficiency and integrity of the test. The trade-off between efficiency–redundancy and the testing of model fit cannot be avoided entirely, but only minimized. The extent of redundancy used should reflect the match between the test being used and the model that scores it. Viewed as several parts of a single solution, it seems that each of these types of suggestions can be implemented to better accomplish the specific aims of more precise estimation of ability and more honest measurement of the accuracy of that estimate.

APPENDIX A

The inclusion of a lower asymptote c on the item characteristic curve implies that the maximum likelihood estimator of ability θ will be a function of a formula score.

Proof: The MLE for θ is the solution of

$$\sum_{j=1}^{n} (x_j - P(\theta)) = 0 \qquad (4.\text{A1})$$

where

$$x_j = \begin{cases} 0 & \text{if item } j \text{ is incorrectly answered} \\ 1 & \text{otherwise.} \end{cases}$$

Let $P(\theta) = c + (1 - c)\,\Phi(\theta)$, where $\Phi(\theta)$ is the response model. From Eq. (4.A1) is obtained

$$\sum x = r = \sum P(\theta) = cn + (1 - c) \sum \Phi(\theta) \qquad (4.\text{A2})$$

where r is the number correct and n the number of items. Therefore

$$\sum \Phi(0) - \frac{r - cn}{1 - c} = r - \left(\frac{c}{1 - c}\right)(n - r). \qquad (4.\text{A3})$$

Note that if random guessing on a k-choice item (i.e., $c = 1/k$) is assumed, it can be seen that

$$c/(1 - c) = 1/(k - 1) \qquad (4.\text{A4})$$

or

$$\sum \Phi(\theta) = r - [1/(k - 1)]\,(n - r) = \text{formula score.} \qquad (4.\text{A5})$$

If all item characteristic curves have identical slope and intercept,

$$\sum \Phi(\theta) = n\Phi(\theta) \qquad (4.\text{A6})$$

or

$$\Phi(\theta) = \frac{1}{n}\left[r - \left(\frac{1}{k-1}\right)(n-r)\right] \tag{4.A7}$$

or

$$\hat{\theta} = \Phi^{-1}\ [(1/n)\ (\text{formula score})]. \tag{4.A8}$$

If an obiter dictum is permitted, note further that the familiar delta (the traditional measure of item difficulty) derives in exactly the same way, assuming a response model with equal slopes and equal lower asymptotes. This result implies that the use of delta is based on the validity of the 1-PL IRT model.

APPENDIX B

Shown in Table 4.2 are the asymptotic errors for the three-parameter logistic model when it fits. These errors are for the slope parameter for a sample of size 1. To determine what it will be for any other sample size, divide the tabulated values by $\sqrt{n}$. Thus, for a sample of 1000, the standard errors for steep items at the extremes of the difficulty dimension range from 1 to 2. Note that these values are lower bounds on what is likely to be achieved in practice, since in a real testing situation it is unlikely that these asymptotic values will be reached, that the idealized fit hypothesized will be achieved, or that ability (which was assumed known in the derivation of these values) will typically be known.

TABLE 4.2
Minimal Asymptotic Standard Errors for Slopes for the Three-Parameter Model, with Lower Asymptote at .1

	Item difficulty						
Slope	−3	−2	−1	0	1	2	3
.25	14	13	13	13	13	14	15
.50	9	8	8	8	8	9	10
.75	8	7	6	6	7	8	11
1.00	8	7	6	6	6	8	13
1.50	11	7	6	6	7	10	19
2.00	16	9	7	7	8	13	29
2.50	30	15	10	9	11	21	56

TABLE 4.3
Minimal Asymptotic Standard Errors for Item Difficulties for the Three-Parameter Model, with Lower Asymptote at .1

	Item difficulty						
Slope	−3	−2	−1	0	1	2	3
.25	806	643	514	411	324	248	178
.50	194	124	82	55	36	21	19
.75	102	54	31	18	11	9	18
1.00	70	31	16	9	6	7	17
1.50	43	16	7	4	3	5	15
2.00	31	10	4	3	2	4	14
3.00	20	6	3	2	2	3	12

If the slopes are poorly determined at the extremes of the ability–difficulty distribution, the difficulties are even more poorly estimated, as shown in Table 4.3. This is due to the very large covariance between the lower asymptote and difficulty in the three-parameter model. This near singularity of the Hessian (required in the MLE of the parameters) implies that for items whose difficulty is near the ability of the examinee sample, the lower asymptote will be very poorly estimated indeed. This uncertainty of estimation is passed through to the estimation of item difficulty.

ACKNOWLEDGMENTS

The research reported here was supported by ETS Program Statistics Research project and the National Institute of Justice (Grant No. 80-IJ-CX-0075) in a grant to the Bureau of Social Science Research, Howard Wainer, Principal Investigator. The author is indebted to Gary Marco, David Thissen, and David J. Weiss for their comments on earlier drafts of this manuscript.

REFERENCES

Birnbaum, A. Some latent trait models and their use in inferring an examinee's ability. In F. M. Lord & M. R. Novick, *Statistical theories of mental test scores*. Reading, Massachusetts: Addison-Wesley, 1968.

DeFinetti, B. Methods for discriminating levels of partial knowledge concerning a test item. *British Journal of Mathematical and Statistical Psychology*, 1965, *18*, 87–123.

Efron, B., & Morris, C. Data analysis using Stein's estimator and its generalizations. *Journal of the American Statistical Association*, 1975, *70*, 311–319.

Finney, D. J. *Probit analysis*. London and New York: Cambridge University Press, 1952.

Gulliksen, H. Personal communication, 1981.

Jones, D. H. *Precision of the ability estimator in the absence of exact item parameters.* Unpublished manuscript, Princeton, New Jersey, 1981. (a)

Jones, D. H. *Foundations for the mathematical notion of information in item response theory and robust ability estimation.* Paper presented at the annual meeting of the Military Testing Association, Arlington, Virginia, October 1981. (b)

Kelley, T. L. *Fundamentals of statistics.* Cambridge, Massachusetts: Harvard University Press, 1947.

Lord, F. M. *Applications of item response theory to practical testing problems.* Hillsdale, New Jersey: Erlbaum, 1980.

Lord, F. M. *Statistical bias in IRT estimates of ability parameters.* Unpublished manuscript, Princeton, New Jersey, 1981.

Lord, F. M., & Novick, M. R. *Statistical theories of mental test scores.* Reading, Massachusetts: Addison-Wesley, 1968.

Mislevy, R., & Bock, R. D. Biweight estimates of latent ability. *Educational and Psychological Measurement,* 1982, *42,* 725–737.

Thissen, D., & Wainer, H. Some standard errors in item response theory. *Psychometrika,* 1982, *47,* 397–412.

Wainer, H., & Wright, B. D. Robust estimation of ability in the Rasch model. *Psychometrika,* 1980, *45,* 373–391.

Wright, B. D., & Stone, M. H. *Best test design.* Chicago, Illinois: MESA, 1979.

II

BEYOND THE ABILITY ESTIMATE

In Chapter 4, Wainer described how ability estimates can be improved under certain circumstances by not scoring item responses that are not predictable from the IRT model being used. That idea of examining how an individual's item responses fit a specified IRT model can be extended to examining individual response patterns to a set of items to determine whether the individual is responding to the items in accordance with the model. This is an important problem in the use of IRT models, since these models make strong assumptions about the functional relation between the observed item responses and the unobserved trait levels. If individuals are responding in accordance with the model, the full power of the models can be used to estimate ability levels and the errors associated with them. On the other hand, if the individual is not responding in accordance with the model, the ability estimates have little utility.

In Chapter 5, Trabin and Weiss examine the data from a group of college students on a vocabulary test to determine whether these individuals are responding in accordance with the IRT model. These data are essential if the model is to be used for individual measurement. Prior to the data in Chapter 5 there were no data available to show that individuals answered test items in accordance with a three-parameter IRT model, although the fit of persons to IRT models had been examined within the assumptions of the one-parameter model. Chapter 5 provides some first data on person fit across items and some indications of some of the potential kinds of information in addition to ability estimates that might derive from an examination of person fit.

In Chapter 6, Levine and Drasgow extend this reasoning to investigating a specific set of indices for person fit, or as they call it, appropriateness measurement. Appropriateness measurement is designed to index the appropriateness for an individual of a particular set of items with specified parameters and a specified underlying model, based on the responses of the individual to those items. The focus of appropriateness indices is to yield values that can be used to identify aberrant response patterns and to classify individuals into appropriate and inappropriate test score groups. Levine and Drasgow investigate the behavior of their appropriateness indices under a variety of conditions.

5

The Person Response Curve: Fit of Individuals to Item Response Theory Models

TOM E. TRABIN DAVID J. WEISS

The development of group ability tests more than 50 years ago has enabled the comparison of the total test score of an individual with the scores of a population norm group, thus allowing for more meaningful interpretation of ability estimates than can be done with the use of simple number-correct scores. For example, the statement, "On the XYZ aptitude test John scored at the seventy-third percentile of college students," gives more information than the statement, "On the ABC ability test Mary correctly answered 64 questions out of 90, whereas Sam correctly answered 33 questions." Both examples have in common the report of a person's test performance on a specific dimension given in terms of an overall test score; but this single summary score, while more parsimonious than a description of a testee's entire response pattern, may not reveal the operation of other factors on test-taking behavior, such as guessing, anxiety, cultural bias, or lack of motivation. Therefore total scores on a test do not indicate whether that test is inappropriate for a certain individual or group of individuals.

The emergence of item response theory (IRT) brings with it the promise of better tests conveying more accurate information about testee ability levels. This is partially accomplished by use of ability estimation procedures that take into account the testee's total response pattern in estimat-

NEW HORIZONS IN TESTING
Latent Trait Test Theory and
Computerized Adaptive Testing

ISBN 0-12-742780-5

ing ability levels (Bejar & Weiss, 1979; Kingsbury & Weiss, 1979a). These scoring methods can provide individualized error bands around a testee's ability level estimates, which indicate the precision of those ability estimates (e.g., Kingsbury & Weiss, 1979b). Thus, in addition to providing methods designed to permit more adequate test construction by the use of test information curves, IRT also allows for the development of powerful methods of adaptive testing for the solution of many practical measurement problems (e.g., Brown & Weiss, 1977; Kingsbury & Weiss, 1979b; McBride & Weiss, 1976; Vale & Weiss, 1977; Weiss, 1973, 1975).

In contrast to classical test theory, IRT makes strong assumptions in order to achieve its objectives. The major operational forms of IRT assume (*a*) local independence, (*b*) unidimensionality, and (*c*) a specified shape for the item characteristic curve. Although local independence cannot be directly demonstrated, data supporting the unidimensionality assumption in a variety of settings (e.g., Bejar, Weiss, & Kingsbury, 1977; Church, Pine, & Weiss, 1978; Martin, Pine, & Weiss, 1978; McBride & Weiss, 1974; Reckase, 1978) lend indirect support to the assumption of local independence. Lord (1968) has presented data showing that the assumption of a normal ogive item characteristic curve is tenable and, given the minor differences between a logistic ogive and a normal ogive, has indirectly supported the use of the logistic item response function in IRT.

There has been very little research, however, to demonstrate that *individuals* behave in accordance with the IRT model, although a growing concern has been exhibited in the testing literature for the development of methods to extract more information from test response data than simply a total score. Use of IRT models with individuals must rest on a demonstration that the test responses of individuals are in accordance with the testing model hypothesized. If this can be demonstrated for most individuals on a number of tests, IRT models can be used with confidence to their full power. On the other hand, if a majority of individuals respond in ways contrary to the assumptions of IRT, the utility of the theory for individual measurement can be seriously questioned.

A major advantage of the assumptions of IRT for individual measurement is that the question of an individual's fit or nonfit to the model can be investigated on an individual basis. The practical implications of identifying nonfitting persons were realized by Levine and Rubin (1979) in their study of methods to identify response patterns of the type of student who "may be so atypical and unlike other students that his or her aptitude test score fails to be a completely appropriate measure of his or her relative ability [p. 269]." Examples of such students are low-ability examinees who copy answers to several difficult items from a much more able neighbor and very high-ability examinees fluent in another language but not fluent in English who misunderstand the wording of several relatively easy questions. Levine and Rubin (1979) recommended the development of indices to identify such test item response patterns as a "rich and fertile area for future research [p. 287]."

The appropriateness of a certain test or of certain items for specific individuals has also been an important concern for test developers working with the one-parameter logistic IRT (Rasch) model. Wright and his associates (Mead, 1980; Wright, 1977; Wright & Stone, 1979; Wainer & Wright, 1980) have proposed identification of such factors as guessing, carelessness, and bias, using the Rasch model. According to Lumsden (1977), a bright but careless student may have the same overall ability score as a careful and consistent average student, but there are differential instructional implications for teaching these two types of students or differential counseling implications if the two students are seeking vocational counseling.

Thus, the question of fit or nonfit of individuals to IRT testing models has important practical and theoretical importance. Fit of individuals must be demonstrated in order to realize the full potential of the model for practical use. At the same time, the development of reliable and valid methods for quantifying and identifying aberrant response patterns would provide a potentially useful source of additional information on test-taking behavior of individuals.

RELATED RESEARCH

The question of fit of individuals to IRT models can be conceptualized as investigating the variability of a single individual in a single testing situation. Wright (1977), in suggesting that to postulate and to study such a phenomenon would be to "wreak havoc with the logic and practice of measurement [p. 103]," exemplifies an attitude that may, in part, account for the meager literature on the topic. It is more likely, however, that the development of sufficiently refined measurement techniques to handle such a difficult problem has not occurred until very recently. The development of computerized testing together with the development of latent trait test theory was necessary to bring about the possibility of measuring individual variability with a single test.

Mosier's Psychophysical Approach

The first reference in the testing literature to an individual's variability within a single ability testing situation was in a two-part study by Mosier (1940, 1942). The emphasis in this study was on the fundamental relationships between the field of mental test theory and the methods of measuring psychophysical processes. This comparison included relating the constant method of psychophysical measurement with scoring by the number-correct method in mental testing. Mosier asserted that a composite score is an imperfect representation of an individual's test score and depends on the individual's variability, just as an individual's threshold in psychophysics depends on the ambiguity of the stimulus; as a stimulus is variable with respect to a group of judges, so an individual is variable with respect to a group of items.

Mosier likened the ambiguity (discriminal dispersion) of a stimulus in psychophysics to individual variability in mental test performance. He postulated the distribution of the proportion of correct answers for one individual across items of differing difficulty as the integral of the normal probability curve, and the variability of that individual as the standard deviation of the probability function whose integral is the proportion of correct answers as a function of difficulty. Mosier applied the constant process of psychophysics to a set of test data (of unspecified characteristics) and estimated the difficulty of median error for individuals (ability level) and its dispersion. He found odd–even reliability of ability level estimated by this method to be .88. The reliability of the person variability index was .55, a value significantly different from zero. It was perhaps this apparent low reliability estimate that was responsible for a complete lack of research on person variability for the next 30 years.

Weiss's Stradaptive "Trace Line"

The idea of person variability within one test was independently developed by Weiss (1973) in the context of computerized adaptive testing. In the design of the stratified adaptive (stradaptive) test, he ordered ability test items by difficulty levels into strata. In examining testee performance on stradaptive tests, Weiss noted that individuals who correctly answered items of the same average difficulty level differed in terms of the proportion of items they answered correctly at different difficulty levels.

To examine differences in individual variability, Weiss proposed the concept of a "trace line" for a testee's item responses, with items divided into strata of increasing difficulty on the x axis and proportion correct for an individual on each stratum on the y axis, independently duplicating the suggestion of Mosier 30 years earlier. Weiss hypothesized, as did Mosier, that proportion correct would decrease as stratum difficulty increased. Also echoing Mosier, he proposed that the steepness of the slope be interpreted as an index of the consistency of an individual's item responses and the capability of the item pool to discriminate an individual's ability level. The point of inflection of the curve, where 50% of the items were answered correctly (for free-response items) was proposed as an indicator of the difficulty of the item pool for an individual or the position of that individual on the trait continuum. To operationalize the concept of person variability, or what Weiss called "consistency," he suggested calculating several indices, including the standard deviation of item difficulties answered correctly and the standard deviation of item difficulties encountered.

Vale and Weiss (1975) empirically studied some aspects of individual consistency as part of a larger study of computer-administered adaptive testing. Included in this study was a test of the hypothesis that more consistent individuals—those with smaller errors of measurement in Mosier's (1940, 1942) formulation—would have more stable ability estimates. The five operationalizations of consistency originally proposed by Weiss (1973) were studied as moderators in the prediction of test-retest reliability of ability estimates. The standard deviation of item difficulties encountered significantly moderated the stability of ability estimates in the expected direction as, to a lesser extent, did the standard deviation of item difficulties answered correctly.

In addition, Vale and Weiss (1975) studied the test–retest reliability of the trace line plots for individuals and introduced the new term *subject characteristic curve* for these trace lines. They used canonical redundancy analysis (Weiss, 1972) on the proportion correct within strata data (i.e., the subject characteristic curves) in a retest situation. The results indicated a high degree of predictability of subject characteristic curves on one test from the test scores on the other; redundancies indicated from 47 to 67% common variance across the two testing times. These results indicated a good degree of stability in the proportion of correct responses within the strata of the stradaptive test as indexed by the subject characteristic curves.

Lumsden's Person Characteristic Curve

The person characteristic curve was again independently proposed by Lumsden (1977, 1978) as a derivation from Thurstone's law of categorical judgment. Lumsden proposed an attribute-based model of test performance in which a person's ability fluctuates in trends (long-term developmental changes), swells (short-term mood swings), and tremors (moment-to-moment shifts). He assumed tremors to be rapid, random, and normally distributed shifts of ability occurring from moment to moment within a single test situation: The discriminal dispersion of item difficulties stays at zero, and it is only person ability that fluctuates. If the momentary location of a person's ability level is higher than the point location of the item's difficulty, the person will answer an item correctly. If ability is lower at any moment than the item difficulty location, the person will answer that item incorrectly. Lumsden then extended the idea to the plot, for a single person, of item responses at different difficulty levels, which he called the "person characteristic curve." He suggested that the person characteristic curve is "perfectly analogous to the item characteristic curve [p. 478]." Lumsden's basic assumptions, however, are different from the IRT assumptions underlying item characteristic curves; IRT assumes that ability level is constant, not fluctuating, but that the response to a given test item includes a random error component causing observed item responses to fluctuate around the true ability level.

Levine and Rubin's Aberrancy Indices

Other approaches to the study of intraindividual variability within a test have concentrated on the use of intraindividual variability for test validation rather than on individual ability assessment. Levine (1980) and Levine and Rubin (1979) initiated several studies concerned with individuals or groups of individuals for whom a given test might be invalid, inappropriate, or both. Among the populations of concern were those who obtain higher scores because of cheating and those who obtain lower scores because of lack of proficiency in English. Levine and Rubin developed several types of "aberrance indices" to determine at greater than chance level, without reference to demographic data, examinees for whom a given test would be inappropriate. Their basic assumption was that an aberrant examinee's response pattern to items of varying difficulty should have a low marginal probability, since it is unlikely that a high-ability examinee would incorrectly answer an easy item or a low-ability examinee correctly answer a difficult item. Marginal probability was operationally defined as the average of the conditional probabilities of a correct response on each item of difficulty level b for an individual of ability level θ. If n is the number of items, there are 2^n marginal probabilities. These were ranked, with all probabilities below an arbitrary cutoff point considered to represent aberrant response patterns.

Using a Monte Carlo simulation with 3000 hypothetical examinees, 200 of whom were aberrant responders, Levine and Rubin (1976) conducted several studies at different cutoff points on the marginal probabilities to determine if aberrant examinees could be identified at a rate significantly greater than chance. Receiver operator curves (ROC) from signal detection theory were used to evaluate the performance of their experimental methods of identifying aberrance. The best method identified 50% of the spuriously low and 80% of the spuriously high examinees, while only mistaking 10% of the normal examinees as aberrant.

When compared to the chance level predictions of only 10% of the spuriously high or low examinees identified, while mistaking 10% of the normal examinees, this study seemed to have yielded results that merit further study. A closer look, however, reveals the impracticality of Levine and Rubin's best method. Even if the aberrance indices identified 80% of the aberrant examinees (160 out of 200) and only misclassified 10% of the nonaberrant examinees (280 out of 2800), this would still result in eliminating as invalid the test results of 280 nonaberrant examinees. Levine and Rubin seemed to ignore this problem completely in their paper.

Wright's Residual Analysis

Wright's (1977; Wright & Mead, 1977; Wright & Stone, 1979) concern with intraindividual variability in a single testing situation focuses on the interaction of a person with specific test items, within the context of the one-parameter logistic IRT model. Wright has developed methods for identifying items that may be invalid for a certain person or group of persons and that can then be excluded from consideration when calculating ability estimates from those items. Wright (1977; Wright & Stone, 1979, pp. 165–180) cited tendencies such as guessing, cheating, "sleeping" (getting bored with a test and answering later items in a more haphazard fashion), "fumbling" (e.g., answering earlier items with difficulty because of confusion with test format), and cultural bias. Wright's method (Mead, 1980, Wright & Stone, 1979) uses standardized squares of the residuals between an item's difficulty level and a person's ability level after fitting the one-parameter logistic model to the test data. If these residuals indicate a significantly low probability of responding correctly or incorrectly and the person responded in that way, the tester then has reason to suspect that the item or item set may be invalid for that person.

This approach is consistent with Wright's use of the one-parameter Rasch model, which recognizes only a difficulty level of items but not a discrimination parameter or a guessing parameter. Following the assumptions of that model, Wright maintained that the probability of success on more difficult items should always be less than on easier items no matter who attempts the items, so the test developer must prevent variation in item discrimination sufficient to produce item characteristic curves that cross. Also, following this logic, a higher ability person should have a better chance for success no matter what the difficulty of the item attempted, so the test developer must prevent variation in person sensitivity; the result is that person characteristic curves must not cross each other. Wright claimed that the practical problem of variations in item discrimination and person sensitivity can be treated through supervision rather than estimation, using residuals and deleting inappropriate items from a person's responses without interfering with estimates of a person's ability. Wright's method seems to oversimplify response data by ignoring the effects of item discrimination and guessing, as well as precluding the possibility of more subtle diagnoses of added dimensions acting as moderator variables in the testing situation.

Summary and Objectives

The limited literature on person variability within a test thus seems to have three major trends:

1. The direct analysis of person variability as originally suggested by Mosier, later called the testee's trace line by Weiss, the subject characteristic curve by Vale and Weiss, and the person characteristic curve by Lumsden
2. The designation of highly variable persons as aberrant by Levine and Rubin
3. The elimination of aberrant person–item interactions by Wright

Careful analysis of these three approaches indicates that the first approach (that of the person characteristic curve) is the most general of the three, subsuming the other two as special cases: If the entire pattern of a testee's responses is studied as a function of the difficulty levels of the items, the identification of aberrant response patterns or person–item interactions follows directly. In addition, postulating a person characteristic curve in conjunction with IRT provides a means of testing whether or not the response patterns of single individuals fit the theory, regardless of the number of parameters assumed.

The purpose of this study was to explore further the Mosier–Weiss–Lumsden idea of the person characteristic curve, to determine its utility as a means of describing testee response variability, and to study the fit of individuals to the IRT model. To emphasize that the curve is derived from the responses of an individual to a set of test items, it was renamed the "person response curve." The focus of this research is on the investigation of the reliability and other psychometric characteristics of the person response curve.

THE PERSON RESPONSE CURVE

Observed Person Response Curves

Figure 5.1 is a plot of person response curves (PRCs) for each of three hypothetical testees. To obtain these plots, a number of items of different difficulty levels are administered to a testee. For each difficulty level, the proportion of items answered correctly is plotted as a function of difficulty level. The resulting PRC is representative of one person's performance on one test.

Figure 5.1 shows the PRC plots of three different persons (A, B, and C) who have all obtained the same score on the test by answering 50% of the total test questions correctly. Therefore all the curves cross at the .50 point on the vertical axis, and their average proportion correct across all item difficulty levels is .50. The center point of the curve can then be projected downward to the horizontal axis to obtain an ability level estimate ($\hat{\theta}$) of 0.0, which in standard score terms is at the mean of a population. Yet Figure 5.1 illustrates that although these three persons all achieved the same total score on this test, they obtained that score in substantially different ways.

As shown in Figure 5.1, the three testees differ in a number of variables. Note that the curve for Person A has a substantially steeper slope

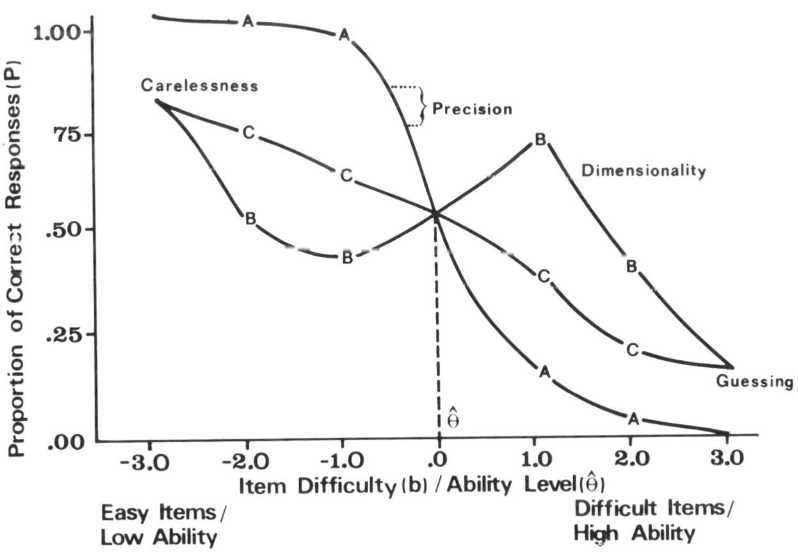

FIGURE 5.1. Observed person response curves for three hypothetical persons with the same ability level ($\hat{\theta} = 0.0$).

around its center point than do those for Persons B and C. This shows that with this item pool, Person A was measured more precisely than either Person B or C or [in Mosier's (1942) terms] that the error of measurement for Person A was smaller. Thus, in addition to ability level scores, information on individual precision of measurement is derivable from the PRC.

The third type of information derivable from the study of PRCs is a person's guessing behavior. This is shown in Figure 5.1 as the lower right-hand portion of the curve for each testee. Note that Persons B and C correctly answered very difficult items at a nonzero level. It may therefore be hypothesized that they were guessing. Person A, however, answered none of the difficult items correctly. It may be hypothesized that this testee, unlike the other two, was not guessing.

A fourth type of information possibly derivable from the PRC is a carelessness index, shown in the upper left-hand corner of Figure 5.1. Persons B and C answered only about 80% of a set of very easy items correctly, even though their ability levels were considerably higher. On the other hand, Person B answered the same items all correctly, as would be expected for a person with a relatively high ability level. Therefore it could be hypothesized that Persons B and C were more careless than Person A.

Finally, the fifth kind of potential information derivable from a study of PRCs is shown for Person B and is a deviation from a unidimensional response pattern, as suggested by Mosier (1940, p. 364). That is, the test performance of Person B shows that he/she was answering correctly beyond the chance level some difficult items which were beyond his/her ability level. Since such test response behavior is inconsistent with a unidimensional hypothesis, there may be, for this individual, some dimension accounting for test performance other than the one being measured by the test for other persons.

Thus the PRC provides the potential for considerable additional information from an individual's test response record. All that is required to obtain an observed PRC is (*a*) to administer to an individual a number of items of varying difficulty levels; (*b*) to determine the proportion of items answered correctly at each difficulty level; and (*c*) to plot those proportions as a function of item difficulty level.

Expected Person Response Curves

Although the observed PRCs are useful in describing a person's test behavior, by themselves they provide no means of determining whether observed fluctuations in the curve represent important characteristics of the individual or merely chance deviations. IRT theory, however, permits the derivation of expected PRCs, which can then be used to evaluate whether aspects of the observed PRCs are real or chance fluctuations. In addition, these expected PRCs permit testing the fit of individual persons to the IRT model for a given set of test item responses.

Expected PRCs are derivable from the one-, two-, or three-parameter IRT models. Derivation of the expected PRC requires an ability estimate $\hat{\theta}$ and the item parameters for all the items administered. Generally, the parameters of the item characteristic curves (ICCs) of the items administered will have been estimated in advance by a method such as LOGIST (Wood, Wingersky, & Lord, 1978) or one of Urry's (e.g., Schmidt & Urry, 1976) estimation procedures; the difficulty (b) parameters will have been used to order the items by difficulty level to obtain the observed PRC. Estimates of ability level ($\hat{\theta}$) may be obtained using programs such as described by Bejar and Weiss (1979).

In the case of the three-parameter logistic IRT model, the expected probability of a correct response for any given test item P_g is given as a function of $\hat{\theta}$, a, b, and c by the three-parameter logistic equation

$$P_g(\hat{\theta}) = \frac{c_g + \exp[Da_g(\hat{\theta} - b_g)]}{1 + \exp[Da_g(\hat{\theta} - b_g)]} \tag{5.1}$$

where $\hat{\theta}$ is the person's estimated ability score, g an item, a_g the item discrimination parameter, b_g the item difficulty parameter, c_g the item lower asymptote ("guessing") parameter, and $D = 1.7$. If a two-parameter IRT model is used, the terms in Eq. (5.1) with c are deleted; if the one-parameter (Rasch) model is used, the a values are set to 1.0.

The expected PRC can be constructed from the estimated probability of a correct response for each item resulting from Eq. (5.1) and from averaging the probabilities for items close in difficulty level. Since the observed PRC uses the proportion of correct responses to a set of items of similar difficulty, averaging of the probabilities of correct responses in the expected PRC will facilitate the direct comparison of observed and expected PRCs. Lord (1980) has referred to

$$\zeta = \sum_{g=1}^{k} P_g(\theta) \tag{5.2}$$

as the expected true score on a set of test items, where k is the number of items for which the expected probability of a correct response has been computed from Eq. (5.1) and ζ is the expected number of correct responses in k items. An estimate of the proportion of correct responses on a subset of items is

$$\hat{p}_s = \frac{\zeta}{k} = \sum_{g=1}^{k} \frac{P_g(\theta)}{k} \tag{5.3}$$

or the average number correct on the k-item subset. Connecting these values with a curve gives the expected PRC based on the three-parameter logistic IRT model, which for any individual is directly comparable to his/her observed PRC.

The expected PRC is therefore simply a function of $\hat{\theta}$ and the item parameters. Thus, for a given $\hat{\theta}$ and a given set of items, the expected values of the PRC will be constant. The observed PRC, on the other hand, results from the interaction of an individual with the items. If an individual answers the set of test items strictly in accordance with the IRT model, the observed PRC should conform to the expected PRC. If an individual's test item responses are determined by factors other than a single unidimensional trait, deviations of the observed PRC from the expected PRC will appear.

Observed versus Expected PRCs

Figure 5.2 shows hypothetical observed and expected PRCs for an individual with $\hat{\theta} = 0.0$. The observed PRC (solid line) is plotted from data on test items grouped at seven points on the item difficulty continuum: $b = \pm 3, \pm 2, \pm 1$, and 0. The expected PRC data points (dashed line) were derived from Eqs. (5.1) and (5.3) for the test items adminis-

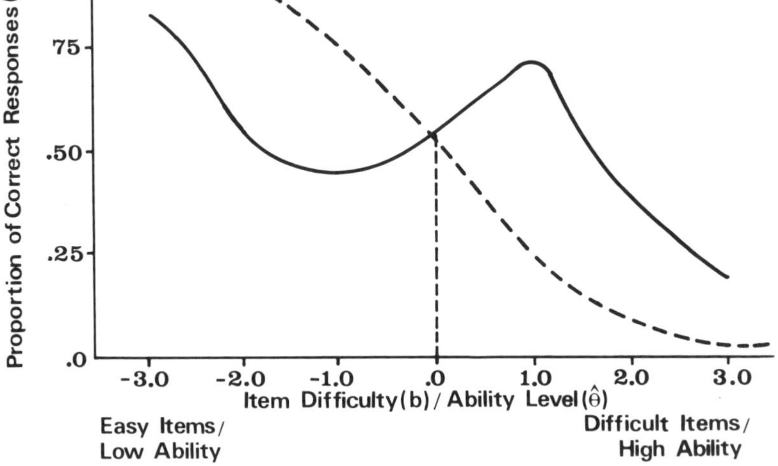

FIGURE 5.2. Observed and expected person response curves for a person with $\hat{\theta} = 0.0$. (Solid line, observed PRC; dashed line, expected PRC.)

tered, using the same item difficulty groupings. To determine whether a person's carelessness, guessing, dimensionality, or precision are significantly different from those predicted by the model, an expected PRC may be determined for any person on any set of test items with estimated ICC parameters, and the observed PRC may be compared to it. If the observed PRC differs from the expected model-based prediction in any respect, the observed PRC describes a significant aspect of the person's testing behavior. Once quantified, these person-fit variables might then be usable in prediction situations to increase the accuracy of predictions made from test scores. This could be done by including additional information on guessing, carelessness, precision, and dimensionality and on other aspects of a person's test performance as reflected in the relationship of observed and expected PRCs.

METHOD

The following data analyses constitute a first examination of observed PRCs and their relationships with expected PRCs for a group of individuals on a test designed to permit study of the characteristics of PRCs. The analyses were directed at establishing the reliability of observed PRCs and the fit of observed and expected PRCs.

Subjects

Subjects were 151 undergraduate students in the introductory psychology course at the University of Minnesota. These students volunteered for the study in return for bonus points that would count toward their final grade. Students were given a posttest debriefing, which consisted of a brief explanation of the purpose of the study. No test results were given, because of the lengthy procedures for keypunching and scoring the data.

Test Instrument

The test consisted of 216 five-option multiple-choice vocabulary items. The items were chosen from a preexisting item pool of over 500 items with IRT difficulty and discrimination parameters that had been developed on a similar population of undergraduates in the introductory psychology course in previous years (McBride & Weiss, 1974). The 216 items were selected for high discriminations and for spread of difficulty (c parameters were set at .20 for all items).

The test was given as a paper-and-pencil test without time limits. Items were randomly ordered for administration so that easy and difficult items were spread throughout the test. In addition, to control for any effects of item order, the pages of test questions were ordered in six different ways so that only one-sixth of the students took the test in the same page order.

Observed PRCs

STRATIFYING THE TEST

In order to transform student response data into observed PRCs, test items were divided into strata containing an equal number of items, with each stratum representing a different level of difficulty. This was done by reordering the items by difficulty level (b parameter), then dividing them into nine separate groups (or strata) of 24 items each. In this way, Stratum 1 contained the 24 easiest items and Stratum 9 contained the 24 most difficult items.

Items were then ordered within each stratum by discrimination (a) level, with the most discriminating item the first item in the stratum and the least discriminating item the twenty-fourth item in the stratum. To investigate the parallel forms reliability of observed PRCs, each stratum was then randomly split into two parallel substrata of items with similar difficulty and discrimination.

Items were scored as either correct (1) or incorrect (0), with omitted items scored as incorrect. The correct–incorrect response vectors were then reordered by item difficulty level for each student. The proportion of correct responses was then computed on each of the nine strata and on each of the 18 substrata for each student, providing information for observed PRCs based on all 216 items (i.e., nine 24-item subtests of differing difficulty levels) and split-half parallel observed PRCs, each based on nine 12-item subtests.

To examine the characteristics of the items constituting the strata, internal consistency reliability of each of the nine strata was determined using Cronbach's alpha. Parallel forms reliability of the nine pairs of parallel substrata was determined by the product-moment correlation coefficient between proportion-correct scores on each of the nine pairs of substrata.

Expected PRCs

Using Program Lindsco (Bejar & Weiss, 1979), Owen's Bayesian ability estimation procedure was used to compute ability estimates $\hat{\theta}$ for each student based on his/her responses to all 216 items in the test. This $\hat{\theta}$ was then used in Eqs. (5.1) and (5.3), in conjunction with the item parameters for the 24 items in each stratum, to obtain the expected proportion-correct score in each of the nine strata $\hat{p}_s$. The $\hat{p}_s$ values then constituted the expected PRC for each student, assuming the three-parameter IRT model. This process was repeated for each of the parallel substrata, yielding expected PRCs for each student from each of the two 108-item parallel pools.

Reliability of Observed PRCs

A χ^2 test of independence was used to determine the reliability of the observed PRCs. For each student the 2×9 contingency table included the number of correct responses on each of the parallel substrata in each of the rows of the nine-column table. χ^2 tests of independence were computed separately for each student. If the paired substrata were parallel, a nonsignificant value of χ^2 would be supportive of the reliability of observed PRC data. Although this χ^2 test violated the usual assumption of independence because the cell frequencies were based on the same student's responses to all the questions, it may be argued that the students' test item responses are locally independent (i.e., are independent for a given student who has a fixed value of θ) and, therefore, that the test is not entirely inappropriate. Further study of this problem is necessary, however, in future applications of this index.

PRCs and Person-Fit

OBSERVED VERSUS EXPECTED PRCs

To determine if students' responses to these ability test items were consistent with the three-parameter IRT model, a χ^2 goodness-of-fit statistic was computed between each student's observed and expected PRC data across the nine strata. If the PRC is an adequate index of model fit, the mean χ^2 for the group would be nonsignificant. On an individual level, at an .05 level of significance, χ^2 goodness-of-fit values should be statistically significant for 7.55 of the 151 students by chance alone, assuming the null hypothesis of no significant deviations from person-fit. More significant χ^2 values would indicate a tendency for lack of fit in these data.

When the overall level of fit in the data is substantially different from the chance expectation, it is still difficult to conclude from the overall goodness-of-fit tests that a specific individual exhibited reliable and meaningful lack of fit to the IRT model, since a certain number of such deviations from fit will occur by chance alone. To identify such individuals, two separate goodness-of-fit tests were conducted for each student using their observed and expected PRC data on each of the parallel substrata. This yielded two χ^2 model fit statistics for each student, one for each of the two sets of substrata. Assuming that the two χ^2 values were independent, reliable person-nonfit would be indicated by identifying persons with significant ($p < .05$) χ^2 values for each of the substrata tests of independence; the probability of observing such a result by chance alone would be $.05 \times .05$, or .0025.

PRCs AND ABILITY LEVEL

If the responses of most persons fit the IRT model, the observed PRC should be a function of ability level θ, just as the expected PRC is a function of ability level. To investigate this possibility, a D^2 analysis was used. Based on observed PRC data within substrata, students were first matched on ability level $\hat{\theta}$ before between-persons D^2 measures were computed. These mean $\hat{\theta}$-matched between-persons D^2 values were then compared to a within-persons D^2 value computed between the parallel substrata observed PRCs, on the hypothesis that there should be little difference between these means—and considerably less difference than when persons were matched without regard to $\hat{\theta}$—if observed PRCs were primarily a function of ability level.

RESULTS

Test Characteristics

Table 5.1 shows the means, standard deviations, and range of item difficulties (b) and proportion-correct scores (p) in each of the nine strata, and the values of Cronbach's alpha internal consistency coefficient for each of the 24-item strata. The strata contained items of increasing difficulty: Stratum 1 contained the easiest items and Stratum 9 contained the most difficult items. This distribution of items was mirrored in the proportion-correct data for each stratum. The average proportion correct decreased as difficulty level of items increased. An exception to this tendency occurred for Strata 8 and 9, in which average proportions correct were very similar. Although average proportion correct was related to the item difficulties in accordance with expectations, the data on the range of individual proportion-correct scores show considerable variability in proportion correct within each of the nine strata. The largest range of propor-

TABLE 5.1
Mean, Standard Deviation, and Range of Item Difficulties (*b*) and Proportion-Correct Scores (*p*), and Cronbach's Alpha Coefficient for Each of the Nine Strata

	Item difficulties (*b*)				Proportion correct (*p*)				
			Range				Range		
Stratum	Mean	*SD*	Min	Max	Mean	*SD*	Min	Max	Alpha
1	−2.40	.31	−2.97	−1.97	.866	.146	.130	1.000	.82
2	−1.55	.20	−1.93	−1.25	.794	.186	.210	1.000	.85
3	−1.01	.14	−1.24	−.77	.713	.216	.170	1.000	.86
4	−.56	.13	−.76	−.37	.615	.202	.040	1.000	.80
5	−.15	.11	−.36	.01	.545	.209	.080	1.000	.81
6	.26	.13	.06	.47	.481	.210	.040	.960	.80
7	.75	.19	.51	1.12	.416	.197	.080	.960	.78
8	1.32	.12	1.13	1.52	.330	.135	.040	.880	.54
9	1.98	.37	1.52	2.67	.334	.124	.040	.790	.44

tion correct was in Stratum 4, where at least one student answered only .04 of the items correctly and the maximum observed proportion correct was 1.0. The smallest range of observed proportion correct was for Stratum 9, in which the minimum proportion-correct score was .04 and the maximum was .79. These data suggest a wide range of individual differences in the proportion-correct scores for each stratum and consequently the potential for individual differences in observed PRCs.

Table 5.1 shows that the alpha internal consistency coefficients for Strata 1–7 were fairly high and quite similar, ranging from .78 to .86. Alpha coefficients for Strata 8 and 9 were lower—.54 and .44, respectively. The low alphas for Strata 8 and 9 were likely due to large amounts of random guessing for most students as the average proportion of correct responses of .33 for the two strata approached the theoretical expectation of .20 for the five-alternative multiple choice items.

Table 5.2 shows the estimated alpha coefficients for the 12-item substrata and the parallel forms correlations obtained by correlating proportion-correct scores for the 151 students on each of the nine pairs of substrata. The estimated 12-item alphas were obtained using the Spearman–Brown formula from the 24-item alphas for the strata shown in Table 5.1. As Table 5.2 shows, the uncorrected parallel forms correlations between pairs of substrata ranged from .63 to .74 for the first seven strata; for the two most difficult strata the correlations were .42 and .28. These correlations were fairly substantial, considering the low internal consis-

TABLE 5.2
Estimated Alpha Coefficients for Twelve-Item Substrata, and Parallel Forms Correlation of Proportion-Correct Scores—Uncorrected, Corrected Spearman–Brown Formula, and Corrected for Attenuation—on Each of the Nine Pairs of Parallel Substrata

		Parallel forms correlations		
Stratum	Estimated 12-item alpha	Uncorrected	Spearman–Brown corrected	Attenuation corrected
1	.69	.64	.78	.93
2	.74	.74	.85	1.00
3	.75	.73	.84	.97
4	.67	.70	.82	1.04
5	.68	.64	.78	.94
6	.67	.66	.80	.99
7	.64	.63	.77	.98
8	.40	.42	.59	1.00
9	.28	.28	.44	1.00

tency reliabilities for the two most difficult strata. Using the Spearman–Brown formula to correct the parallel forms correlations based on two 12-item tests to the 24-item length of the strata, the average corrected correlation between the pairs of Substrata 1–7 was slightly above .80. For the two most difficult strata, the corrected correlations were .59 and .44.

To determine whether scores on the paired substrata correlated as highly as possible, given the reliabilities of the substrata, the estimated 12-item alphas for the substrata were used along with the uncorrected parallel forms correlations to estimate the correlation between proportion-correct scores on the paired substrata, assuming that the substrata had been perfectly reliable. These attenuation-corrected correlations are shown as the last column in Table 5.2. As the data show, attenuation-corrected correlations were .97 or above for seven of the nine strata; for Strata 1 and 5, these correlations were .93 and .94, respectively. These data indicate that the paired substrata scores were as parallel as possible, given their estimated internal consistencies.

Reliability of Observed PRCs

The χ^2 test of independence, based on a 2×9 contingency table with number-correct scores on each of the nine pairs of parallel substrata for each student, resulted in a minimum value of .14 and a maximum of 12.94; mean of the distribution was 3.67, with a standard deviation of 2.34. A χ^2

value of 15.51 is statistically significant at the $p = .05$ level with eight degrees of freedom (from the 2×9 contingency table). Since all the individual χ^2 values were less than 15.51, the data indicated that the two split-pool observed PRCs for all students were not significantly different from each other. These results indicated that the observed PRCs obtained from these data were reliable.

PRCs and Person Fit

The frequency distribution of individual χ^2 values reflecting the fit of observed PRCs to the PRCs expected from the three-parameter IRT model is shown in Figure 5.3. The lowest χ^2 value obtained was 1.88 and the highest was 23.17. Mean χ^2 was 8.76, with a standard deviation of 4.14; modal value was about 6.0.

Since ability estimates used in calculating the theoretically expected proportion-correct scores were taken from the data being analyzed, an extra degree of freedom was subtracted to determine the significance of the χ^2 values. Thus, with seven degrees of freedom, a χ^2 value of 14.07 is significant at the .05 level. The group mean χ^2 was well below this value, which would suggest that the three-parameter logistic IRT model served as a fairly good predictor of test response behavior for the majority of this group of students. Of 151 students, 8 would be expected to have signifi-

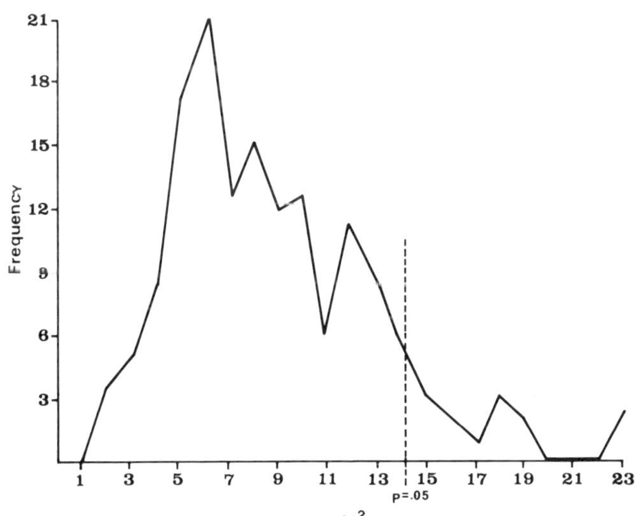

FIGURE 5.3. Frequency distribution of intraindividual χ^2 values for goodness of fit between observed and expected person response curves (PRCs).

cant χ^2 values by chance alone at the .05 level; in this group, 15 students had χ^2 values greater than 14.07.

To identify persons reliably deviating from the model, the χ^2 person-fit statistics were recomputed for each student separately on the two sets of substrata. Six students out of 151 had significant χ^2 values for both pairs of substrata; of these six, four were also significantly nonfitting on the overall χ^2 goodness-of-fit test. The observed and expected total PRCs for two of these students are shown in Figures 5.4a and b. The ways in which these

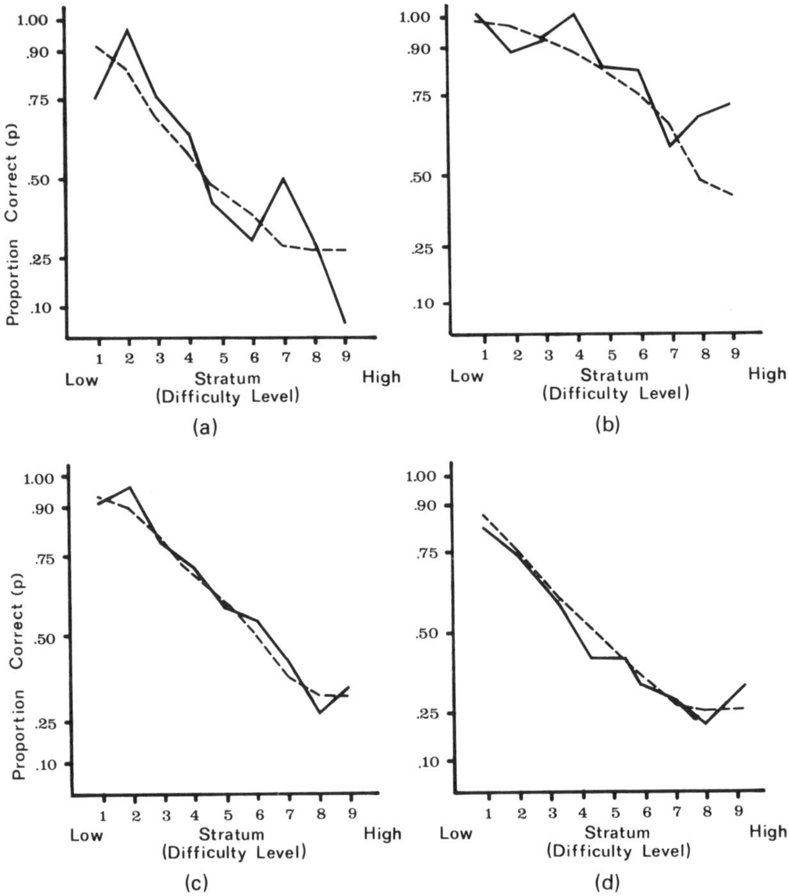

FIGURE 5.4. Observed and expected person response curves (PRCs) for two persons whose responses did not reliably fit and two persons whose responses reliably fit the three-parameter IRT model. (a) Person 111 (χ^2 = 23.17). (b) Person 117 (χ^2 = 22.75). (c) Person 128 (χ^2 = 1.88). (d) Person 135 (χ^2 = 1.92). (Solid line, observed PRC; dashed line, expected PRC.)

students' response curves deviated from their expected curves differed. Person 111 (Figure 5.4a) seems to have been careless with easier items, as indicated by a proportion correct of .75 on items in Stratum 1, and then to have been fortunate in guessing on some of the more difficult ones ($p = .50$ on Stratum 7). On the other hand, this may be the type of profile to be expected from a person with an unusual educational history, such as an international student with a specialized knowledge of English. Person 117 (Figure 5.4b) seems to have done much better on difficult (Strata 8 and 9) items than was predicted by the model; this student might be sophisticated at guessing or high in "testwiseness."

By contrast, Figures 5.4c and d show observed and expected PRCs for two students with low overall χ^2 person-fit values. Person 128 (Figure 5.4c) obtained the lowest χ^2 value among the 151 students tested. As Figure 5.4c shows, the observed PRC for Person 128 (solid line) was quite close to the expected PRC (dashed line) for each of the nine strata, as was also the case for Person 135 (Figure 5.4d).

PRCs and Ability Level

Additional data supporting the overall fit of persons to the three-parameter IRT model are shown in Table 5.3, which summarizes the distribution of between-persons D^2 data on parallel substrata when students were matched as closely as possible for $\hat{\theta}$ values before the substrata D^2 indices were calculated. As Table 5.3 shows, none of the mean between-persons D^2 indices was significantly different from the within-persons D^2; in three of the four cases the mean between-persons D^2 was

TABLE 5.3
Mean, Standard Deviation, and Range of Within-Persons D^2 and $\hat{\theta}$-Matched Between-Persons D^2, and Results of t Tests Comparing the Mean Within-Persons D^2 with Each Between-Persons D^2 Index

D^2 Index	N	Mean	SD	Range		t	p^a
				Min	Max		
Within-Person	150	.28	.15	.02	.86		
Between-Person							
D(AA)	75	.25	.11	.03	.48	1.50	<.20
D(BB)	75	.26	.13	.05	.74	1.00	<.50
D(AB)	75	.29	.14	.05	.79	0.50	<.80
D(BA)	75	.28	.12	.07	.64	0.00	1.00

[a] Probability of error in rejecting null hypothesis of no difference in group means (two-tailed test).

slightly lower than the mean within-persons D^2. In addition, the standard deviations and ranges of the two kinds of D^2 indices were very similar. Thus the data in Table 5.3 show that observed PRCs for this group of students were highly dependent upon their ability levels, further supporting the fit of these individuals to the three-parameter IRT model.

CONCLUSIONS AND DIRECTIONS FOR FUTURE RESEARCH

The feasibility of the person response curve (PRC) approach to investigating the fit of persons to the three-parameter IRT model was explored in this study. To operationalize the PRC it was necessary to subdivide ability test items into separate strata of varying difficulty levels. For the vocabulary test used in this study, strata possessed sufficient internal consistency and parallel forms reliability to justify their use, although the more difficult strata were much less reliable than the easier strata.

Conclusions

The PRCs proved to be highly reliable. χ^2 tests for independence between profiles of parallel forms for each individual were nonsignificant for all 151 students. The results of the D^2 statistics between individuals matched on ability level suggest that for the majority of this sample, observed PRCs were predictable as a function of ability level.

At the .05 level, eight students were expected to have significant χ^2 goodness-of-fit values for observed and expected PRCs. Fifteen students had significant χ^2 values, leaving somewhat in question whether these students deviated from the model because of chance or because of interaction with another dimension. One method of investigating this question was to calculate separately the goodness of fit of each student's observed and expected PRCs on the odd-numbered substrata and on the even-numbered substrata. Of the 15 students with significant χ^2 values on the overall nine-strata goodness-of-fit test, four had significant χ^2 values on both substrata goodness-of-fit tests. These four students were identified as reliably deviating from the IRT model predictions. The nature of this lack of fit, however, would best be investigated in a future study with an experimental design that included interactions with additional dimensions other than the ability being measured.

The results of this study demonstrate that the PRC can be useful in studying the fit of individuals to IRT models by testing the fit of the observed PRC to the theoretically expected PRC. Although the three-parameter IRT model was used here, the method can be used with the two-parameter or one-parameter logistic (Rasch) model or with any of the normal ogive IRT models. The data also demonstrated that the three-parameter IRT model adequately accounted for the test response behavior of the vast majority of the students studied. More research is, of course, necessary to explore further the use of the PRC in examining model-fit in test behavior.

Directions for Future Research

Guessing and "testwiseness" are variables unrelated to abilities, but they may affect ability test scores. Future research should be designed to determine whether these variables can be detected by PRCs or PRC fit to theoretical predictions. Cultural bias is another dimension that may differentially affect ability test performance (e.g., Church *et al.*, 1978; Martin *et al.*, 1978; Pine & Weiss, 1978). One approach to testing for the existence of such bias by use of PRCs would be to compare the goodness of fit of observed and expected PRCs for a control group of white middle-class testees and a group of testees who would be hypothesized to have uneven educational development by white middle-class American standards. This latter group might involve international students with a specialized knowledge of English or some American minority group persons. It would be expected that the PRCs would show greater deviation from the model predictions for the latter group, particularly in terms of deviations from the unidimensionality required by the IRT model. Carelessness and nervousness are two other dimensions that may contribute to unexpected performance on ability tests and that may be detected by PRC analysis.

Further investigation of the measurement properties of observed versus expected χ^2 goodness-of-fit statistics for assessing nonfit of persons is also of importance. Monte Carlo simulations should be run in order to determine the null distribution of the χ^2 values. These should be repeated at a number of θ levels to determine whether goodness-of-fit distributions differ as a function of ability level.

Finally, since the research literature on methods for assessing nonfitting profiles has begun to branch in several directions, it would be informative and useful to compare the efficacy of several different methods using the same data base. The one-, two-, and three-parameter IRT models could each be used in computing ability estimates so that nonfit measures based on these different models could be used. This would best be done in simulation, with nonfitting data experimentally induced so that the different methods of evaluating model fit could be compared on their degree of "hits" and "misses."

These are only a few of many research possibilities in investigating the properties and diagnostic utility of PRCs. A closer look at these properties of the PRC test performance profiles and their use in determining person fit may provide important information on selected individuals and improve the validity of ability tests for individual prediction and diagnosis.

ACKNOWLEDGMENTS

The assistance of Isaac Bejar, Austin T. Church, J. Stephen Prestwood, and James B. Sympson in various aspects of the design of this study and the analysis of these data is gratefully acknowledged.

This research was supported by funds from the Navy Personnel Research and Development Center and the Office of Naval Research under Contract N00014-76-C-0243, NR 150-382.

REFERENCES

Bejar, I. I., & Weiss, D. J. *Computer programs for scoring test data with item characteristic curve models* (Research Report 79-1). Minneapolis: University of Minnesota, Department of Psychology, Psychometric Methods Program, February 1979.

Bejar, I. I., Weiss, D. J., & Kingsbury, G. G. *Calibration of an item pool for the adaptive measurement of achievement* (Research Report 77-5). Minneapolis: University of Minnesota, Department of Psychology, Psychometric Methods Program; September 1977.

Brown, J. M., & Weiss, D. J. *An adaptive testing strategy for achievement test batteries* (Research Report 77-6). Minneapolis: University of Minnesota, Department of Psychology, Psychometric Methods Program, October 1977.

Church, A. T., Pine, S. M., & Weiss, D. J. *A comparison of levels and dimensions of performance in black and white groups on tests of vocabulary, mathematics, and spatial ability* (Research Report 78-3). Minneapolis: University of Minnesota, Department of Psychology, Psychometric Methods Program, October 1978.

Kingsbury, G. G., & Weiss, D. J. *Relationships among achievement level estimates from three item characteristic curve scoring methods* (Research Report 79-3). Minneapolis: University of Minnesota, Department of Psychology, Psychometric Methods Program, April 1979. (a)

Kingsbury, G. G., & Weiss, D. J. *An adaptive testing strategy for mastery decisions* (Research Report 79-5). Minneapolis: University of Minnesota, Department of Psychology, Psychometric Methods Program, September 1979. (b)

Levine, M. V. Psychometric models and appropriateness measurement. In D. J. Weiss (Ed.), *Proceedings of the 1979 Computerized Adaptive Testing Conference*. Minneapolis: University of Minnesota, Department of Psychology, Psychometric Methods Program, Computerized Adaptive Testing Laboratory, 1980.

Levine, M. V., & Rubin, D. Measuring the appropriateness of multiple-choice test scores. *Journal of Educational Statistics*, 1979, *4*, 269–290.

Lord, F. M. An analysis of the verbal scholastic aptitude test using Birnbaum's three-parameter logistic model. *Educational and Psychological Measurement*, 1968, *28*, 989–1020.

Lord, F. M. *Applications of item response theory to practical testing problems*. Hillsdale, N.J.: Erlbaum, 1980.

Lumsden, J. Person reliability. *Applied Psychological Measurement*, 1977, *1*, 477–482.

Lumsden, J. Tests are perfectly reliable. *British Journal of Mathematical and Statistical Psychology*, 1978, *31*, 19–26.

Martin, J. T., Pine, S. M., & Weiss, D. J. *An item bias investigation of a standardized aptitude test* (Research Report 78-5). Minneapolis: University of Minnesota, Department of Psychology, Psychometric Methods Program, December 1978.

McBride, J. R., & Weiss, D. J. *A word knowledge item pool for adaptive ability measurement* (Research Report 74-2). Minneapolis: University of Minnesota, Department of Psychology, Psychometric Methods Program, June 1974.

McBride, J. R., & Weiss, D. J. *Some properties of a Bayesian adaptive ability testing strategy* (Research Report 76-1). Minneapolis: University of Minnesota, Department of Psychology, Psychometric Methods Program, March 1976.

Mead, R. J. Using the Rasch model to identify person-based measurement disturbances. In D. J. Weiss (Ed.), *Proceedings of the 1979 Computerized Adaptive Testing Conference*. Minneapolis: University of Minnesota, Department of Psychology, Psychometric Methods Program, Computerized Adaptive Testing Laboratory, 1980.

Mosier, C. I. Psychophysics and mental test theory: Fundamental postulates and elementary theorems. *Psychological Review*, 1940, *47*, 355–366.

Mosier, C. I. Psychophysics and mental test theory II: The constant process. *Psychological Review*, 1942, *48*, 235–249.

Pine, S. M., & Weiss, D. J. *A comparison of the fairness of adaptive and conventional testing strategies* (Research Report 78-1). Minneapolis: University of Minnesota, Department of Psychology, Psychometric Methods Program, August 1978.

Reckase, M. D. Unifactor latent trait models applied to multifactor tests: Results and implications. In D. J. Weiss (Ed.), *Proceedings of the 1977 computerized adaptive testing conference*. Minneapolis: University of Minnesota, Department of Psychology, Psychometric Methods Program, 1978.

Schmidt, F., & Urry, V. W. Item parameterization procedures for the future. In C. L. Clark (Ed.), *Proceedings of the first conference on computerized adaptive testing* (U.S. Civil Service Commission, Personnel Research and Development Center, PS-75-6). Washington, D.C.: U.S. Government Printing Office, 1976. (Superintendent of Documents Stock No. 006-00940-9)

Vale, C. D., & Weiss, D. J. *A study of computer-administered stradaptive ability testing* (Research Report 75-4). Minneapolis: University of Minnesota, Department of Psychology, Psychometric Methods Program, October 1975.

Vale, C. D., & Weiss, D. J. *A comparison of information functions of multiple-choice and free-response vocabulary items* (Research Report 77-2). Minneapolis: University of Minnesota, Department of Psychology, Psychometric Methods Program, April 1977.

Wainer, H., & Wright, B. D. Robust estimation of ability in the Rasch model. In D. J. Weiss (Ed.), *Proceedings of the 1979 Computerized Adaptive Testing Conference*. Minneapolis: University of Minnesota, Department of Psychology, Psychometric Methods Program, Computerized Adaptive Testing Laboratory, 1980.

Weiss, D. J. Canonical correlation analysis in counseling psychology research. *Journal of Counseling Psychology*, 1972, *19*, 241–252.

Weiss, D. J. *The stratified adaptive computerized ability test* (Research Report 73-3). Minneapolis: University of Minnesota, Department of Psychology, Psychometric Methods Program, September 1973.

Weiss, D. J. (Ed.). *Computerized adaptive trait measurement: Problems and prospects* (Research Report 75-5). Minneapolis: University of Minnesota, Department of Psychology, Psychometric Methods Program, November 1975.

Wood, R. L., Wingersky, M. S., & Lord, F. M. *LOGIST: A computer program for estimating examinee ability and item characteristic curve parameters* (Research Memorandum 76-6) (revised) Princeton, New Jersey: Educational Testing Service, 1978.

Wright, B. Solving measurement problems with the Rasch model. *Journal of Educational Measurement,* 1977, *14,* 97–115.

Wright, B. D., & Mead, R. J. *The use of measurement models in the definition and application of social science variables* (Technical Report DAHC 19-76-G-011). Arlington, Virginia: U.S. Army Research Institute for the Behavioral Sciences, 1977.

Wright, B. D., & Stone, M. H. *Best test design.* Chicago, Illinois: MESA, 1979.

6

Appropriateness Measurement: Validating Studies and Variable Ability Models[1]

MICHAEL V. LEVINE FRITZ DRASGOW

In a large test administration a few examinees may be so unlike other examinees that their multiple-choice aptitude test scores have limited value as ability measures. A particularly transparent example is provided by a hypothetical low-ability copier who copies half of his/her answers from a much more able neighbor. Other test anomalies include

1. Improperly coached examinees who are shown the answers to some items before the exam begins
2. Examinees with high ability but atypical schooling or low English fluency
3. Exceptionally creative examinees who discover novel interpretations for some items
4. Examinees who are very conservative in their use of partial information
5. Examinees who make alignment errors over a block of items, answering the ninth item in the tenth place, the tenth item in the eleventh place, and so forth

[1] This chapter is an expanded version of a paper by Levine and Drasgow (1982), which appeared in the *British Journal of Mathematical and Statistical Psychology*. The chapter contains a more detailed treatment of variable ability models than the article. The authors thank the *Journal* editor for permission to reproduce portions appearing in both papers.

NEW HORIZONS IN TESTING
Latent Trait Test Theory and
Computerized Adaptive Testing

ISBN 0-12-742780-5

In each of these cases it can be argued (*a*) that the test score is not an appropriate measure of ability and (*b*) that the item-by-item pattern of answers may be recognizably unusual. For example, the hypothetical low-ability copier is likely to have relatively many easy items incorrect and many hard items correct. A very bright examinee with an alignment error is likely to have an unusual pattern in which most of his/her wrong answers are on consecutive items.

Thus a multiple-choice aptitude test may be a dubious measure of ability because of any one of many (although possibly rare) causes. In at least some cases the item-by-item answer pattern may contain evidence of this fact. Appropriateness measurement attempts to recognize inappropriate test scores by identifying unusual answer patterns.

OBJECTIVES AND LIMITATIONS OF APPROPRIATENESS MEASUREMENT

Appropriateness measurement is a general approach to the problem of inappropriate test scores. Unlike most alternative approaches, its purpose is simply to identify inappropriate test scores and not to correct the causes of inappropriateness. It is limited to cases like those noted, in which inappropriate test scores and unusual answer patterns tend to occur together.

Appropriateness measurement is implemented by statistics, called *appropriateness indices,* that measure the degree to which an examinee's answer pattern is "unusual," that is, unlike the pattern expected from typical examinees. In appropriateness measurement studies, examinees are sorted into two groups: (*a*) examinees with very unusual answer patterns, as indicated by very extreme index values; and (*b*) all other examinees. Perfect classification is impossible. Appropriateness measurement is successful to the extent that the group of examinees with extreme index values has a larger proportion of examinees with inappropriate scores than the group with more typical index values.

The goal of appropriateness measurement is to recognize inappropriate test scores. Appropriateness measurement is only incidentally concerned with questions of "person fit," the goodness of fit of a person's data to a test model. We take the view that all tractable models are essentially incomplete and inaccurate as descriptions of test-taking behavior. A model and a measure of model fit cannot be considered useful for appropriateness measurement until they have been shown to effectively classify appropriate and inappropriate test scores.

Three Problems in Appropriateness Measurement

The first large-scale systematic appropriateness measurement study was reported by Levine and Rubin (1979). That study, reviewed in the following section, provides the background and context for the theoretical developments and empirical results reported in this chapter.[2]

Levine and Rubin (1979) identified three types of appropriateness indices and reported positive empirical findings. Their indices adequately recognized aberrant examinees in their validation studies, but, unfortunately, they used simulation data exclusively and thereby bypassed several important questions that arise in applications. The research reported in this chapter uses both actual and simulated data to attack three problems left unanswered by the Levine and Rubin study, namely:

1. *Estimated versus known item parameters:* With simulated data, item parameters (e.g., item difficulties) are known and need not be estimated prior to computing appropriateness indices. With actual data, parameters must be estimated. How seriously will appropriateness measurement be affected by estimation errors?
2. *Unidentified aberrants:* In a simulation study atypical examinees are unequivocally identified and a sample of truly normal examinees is available for item parameter estimation. With actual data, an unknown proportion of unidentified aberrants will be included in each large sample of nominally normal examinees. How will the presence of these aberrants affect parameter estimation and, consequently, appropriateness measurement?
3. *Model validity:* Simulated data conform precisely to the psychometric model used to generate data and to formulate appropriateness indices, but there will be reliable contradictions to the assumptions of any tractable, nontrivial psychometric model in a large sample of actual data. Are currently available psychometric models sufficiently valid to support appropriateness measurement with actual data?

Results will be presented to show (*a*) that appropriateness indices are not seriously degraded by the use of estimated parameters, (*b*) that appropriateness indices are not seriously degraded even when relatively many aberrant examinees are initially (and improperly) treated as normal, and (*c*) that detection rates with actual test data are comparable to detection rates with simulated data.

[2] For independent, contemporary work that appears similar in conception, see van der Flier (1977). For possibly related research based on classical test theory, see Ghiselli (1956, 1960a, 1960b) and Donlon and Fischer (1968).

REVIEW OF TEST THEORIES AND BASIC APPROPRIATENESS MEASUREMENT

Appropriateness measurement involves a two-stage process: (*a*) test norming or item parameter estimation and (*b*) person measurement or index computation. This distinction parallels the separation of parameters in latent trait or item response theory models into (*a*) item or test characterizing parameters (such as item difficulties) and (*b*) individual difference or person characterizing parameters (such as abilities). A discussion of the two stages is included in the following review of basic appropriateness measurement theory. A discussion of estimation with "variable ability" models is also included.

Test Norming and the Standard Model

The studies to be reported here use what will be called the *standard model* of item response theory. (Levine and Rubin, being concerned with apparent changes in ability during testing, called this model the constant ability model.) The defining equation of the standard model is

$$\text{Prob}\{\mathbf{U} | \text{ability} = \theta\} = \prod_{i=1}^{n} P_i(\theta)^{u_i}[1 - P_i(\theta)]^{1-u_i}, \tag{6.1}$$

where $\mathbf{U} = \langle u_1, u_2, \ldots, u_n \rangle$ is the vector of item scores, coded 1 for correct and 0 for incorrect, and $P_i(\theta)$ is the conditional probability of an examinee, randomly selected from the population of all examinees with ability θ, correctly answering the ith item.

Equation (6.1) reflects a conceptualization of test data as the outcome of a two-stage experiment. First, an examinee with (unobserved) ability θ is sampled; then, the examinee's item responses are generated as a series of independent Bernoulli trials with success probabilities $P_i(\theta)$. By specifying a functional form for the P_i, the most commonly used models of item response theory are obtained as special cases, namely, the Rasch model and the three-parameter logistic model. The standard model differs from alternative models in that each examinee's ability is constant during testing, responses are scored dichotomously, and the range of possible θ values is unidimensional.

In this research the three-parameter logistic formula (Birnbaum, 1968) was used for the P_i. Thus P_i is represented as

$$P_i(\theta) = c_i + (1 - c_i)\{1 + \exp[-a_i(\theta - b_i)]\}^{-1} \tag{6.2}$$

for parameters a_i, b_i, c_i. This particular functional form is conventional; it is widely used; and it is supported by empirical studies in which the form of P_i was not assumed (Levine & Saxe, 1976; Lord, 1970). The findings in this chapter constitute an additional demonstration of its value.

"Test norming" consists of estimating the numerical values for a_i, b_i, and c_i. This is done by selecting a large sample of apparently normal examinees and by fitting Eq. (6.1) to their data. Lord's LOGIST[3] algorithm (Wood & Lord, 1976; Wood, Wingersky, & Lord, 1976) was used to approximate maximum likelihood estimates of the item parameters. (The θs are also estimated by Lord's algorithm, but they are not used in the test norming stage of appropriateness measurement.)

The standard model, the logistic formula for the P_i, and Lord's algorithm have been criticized for various reasons. The relevance of the results of the present studies to these criticisms is summarized in the last section of this chapter.

Person Measurement and L_0

The test norming stage can be considered as specifying a continuum of submodels for test-taking behavior. There is one model for each value of θ. Equation (6.1) gives an explicit formula from which the probability of each answer pattern can be calculated. There is one formula for each choice of θ. The problem in the person measurement stage is to decide whether any of these formulas adequately fits the person's data.

An intuition that has guided much of this research and that will be used to introduce a simple but useful appropriateness index follows:

> Suppose that for all values of θ, a pattern $\mathbf{U}^*$ appears improbable in the sense that $\text{Prob}(\mathbf{U}^*|\theta)$ is very small. Then, $\mathbf{U}^*$ is badly fit by *all* submodels for individual data developed in the test norming stage.

Thus a relatively low maximum of the function of θ, $\text{Prob}(\mathbf{U}^*|\theta)$, is expected to be indicative of aberrance. This heuristic reasoning leads to the appropriateness index L_0

$$L_0(\mathbf{U}^*) = \max_{\theta} \text{Prob}(\mathbf{U}^*|\theta). \tag{6.3}$$

L_0 is one of the simplest aberrance indices yet proposed. Nonetheless, it detects some forms of aberrance surprisingly well. In a recent study in

[3] LOGIST version 2b was used in the present research. LOGIST has since been modified.

which an optimal index could be specified and computed, L_0 had roughly 80% of the statistical power of the optimal index. Improvements on L_0 may be achieved by modeling the aberrance-producing process and by testing for specific departures from the standard model. An example of this approach is reviewed in the next section.

Variable Ability Models and Appropriateness Indices

L_0, like the familiar χ^2 statistic, can be considered an omnibus measure, indiscriminately testing for any kind of departure from the standard model. One strategy for increasing statistical power is to test for a specific type of abnormality. To implement this strategy, Levine and Rubin (1979) formulated models of aberrance and attempted to detect inappropriate scores by noting the degree of improved fit of the examinee's data obtained with the more general model. In the process they obtained what might be considered a measure of the "amount" or "seriousness" of inappropriateness.

In some of the most important types of inappropriateness, the aberrant examinee behaves as if his/her ability were fluctuating from item to item. Thus the low-ability copier appears to have a much higher ability for those items on which he/she has been coached. The high-ability, low-English-fluency candidate behaves as if he/she had lower ability on linguistically demanding items.

In the standard model the examinee's ability θ is constant across items. In the *variable ability* models introduced by Levine and Rubin (1979), the examinee's ability is conceptualized as varying from item to item. For example, in the *Gaussian model,* each examinee is characterized by a pair of individual difference parameters: *central ability* θ_0 and *ability variance* σ^2. According to the Gaussian model, the examinee has ability θ_1 on Item 1, θ_2 on Item 2, . . . , and θ_n on Item n. The model asserts that for given θ_0 and σ^2, the θ_i are independent identically distributed normal or Gaussian[4] random variables with mean θ_0 and variance σ^2. (Comments on the choice of the normal distribution are in the next section.)

Note that the standard model is a limiting special case of the Gaussian model in which the ability variance is zero. To specify the model more precisely, the likelihood function for the Gaussian model $\text{Prob}(\mathbf{U}|\theta_0, \sigma^2)$ is written and simplified below. $\text{Prob}(\mathbf{U}|\theta_0, \sigma^2)$ is the conditional probability that an examinee, randomly sampled from all those with central ability θ_0 and ability variance σ^2, has vector of responses $\mathbf{U}$.

[4] The model is called "Gaussian" rather than "normal" to avoid having a normal model for abnormal behavior.

$$
\begin{aligned}
\text{Prob}(\mathbf{U} \mid \theta_0, \sigma^2) \\
&= \int_{-\infty}^{\infty} \cdots \int_{-\infty}^{\infty} \prod_{i=1}^{n} P_i(\theta_i)^{u_i}[1 - P_i(\theta_i)]^{1-u_i}\phi(\theta_i; \theta_0, \sigma^2)\, d\theta_1 \cdots d\theta_n \\
&= \prod_{i=1}^{n} \int_{-\infty}^{\infty} P_i(t)^{u_i}[1 - P_i(t)]^{1-u_i}\phi(t; \theta_0, \sigma^2)\, dt, \qquad (6.4)
\end{aligned}
$$

where $\phi(t; \theta_0, \sigma^2)$ is the Gaussian density

$$\phi(t; \theta_0, \sigma^2) = \frac{1}{\sigma\sqrt{2\pi}} \exp\left[-\frac{1}{2}\left(\frac{t - \theta_0}{\sigma}\right)^2 \right]. \qquad (6.5)$$

Equation (6.4) is analogous to Eq. (6.1) for the standard model.

Since the standard model can be regarded as a submodel of the Gaussian model, the usual maximum likelihood ratio test statistic can be used to test for the presence of the sort of aberrance described in the Gaussian model. It is defined as

$$LR = L_G/L_0, \qquad (6.6)$$

where

$$L_G = \max_{\theta_0, \sigma^2} \text{Prob}(\mathbf{U} \mid \theta_0, \sigma^2)$$

and the statistic L_0 is the maximum of the standard model likelihood function as before.

LR measures the degree of improvement of fit obtained by permitting abilities to vary during testing. It does not measure the amount of variation. If σ^2 is small but positive, a large value of LR may reflect strong evidence for a relatively unimportant amount of variation. The Gaussian model permits the formulation of an index which estimates the amount of variation, i.e., an estimate of the ability variance. This is the index $\hat{\sigma}^2$, the maximum likelihood estimate of ability variance obtained during the calculation of L_G.

COMPUTATION OF GAUSSIAN MODEL INDICES

Numerical considerations have influenced the development of variable ability models. A more plausible model for the item ability distribution would be

$$\theta_i = \begin{cases} \theta_{\text{High}} & \text{with probability } P \\ \theta_{\text{Low}} & \text{with probability } 1 - P. \end{cases} \qquad (6.7)$$

However, appropriateness measurement results with this model were poor, probably because a satisfactory procedure for estimating the person parameters θ_{High}, θ_{Low}, and P was never found. In any event, the examinee's fewer than 100 binary responses seemed too few to reject the assumption of normally distributed item abilities, and Gaussian model appropriateness measurement techniques classify examinees well.

If the item abilities are assumed to have a Gaussian distribution, then approximations can be used to expedite parameter estimation. These approximations were used in the Levine and Rubin (1979) study and the replication presented below. With them the likelihood ratio test seems to be at least as powerful as L_0. A thorough study of the accuracy of the approximations has not been attempted, and a more accurate numerical algorithm may give substantially better appropriateness measurement. The inclusion of the approximations here is not an endorsement of them; it is a response to several requests for a variable ability numerical algorithm from other researchers. With the approximations in this section, variable ability results can be replicated with off-the-shelf software.

The approximations use the following well-known mathematical facts:

1. The difference of independent normal variables is normal, with mean equal to the difference of means and variance equal to the sum of the variances.
2. The logistic function $\Psi(t) = 1/[1 + e^{-t}]$ has very nearly the same shape as the Gaussian cumulative

$$\Phi(t) = \frac{1}{\sqrt{2\pi}} \int_{-\infty}^{t} e^{-x^2/2}\, dx. \tag{6.8}$$

   Specifically for $D = 1.702$ and for all t, $|\Psi(t) - \Phi(t/D)| < .01$.
3. If X has cumulative distribution function H, if Y has density function f, and if X and Y are independent, then

$$\text{Prob}\{X - Y \leq 0\} = \int_{-\infty}^{\infty} H(y) f(y)\, dy. \tag{6.9}$$

The logarithm of the likelihood function for the variable ability model is the sum of terms of the same form:

$$\begin{aligned}
&\log \int_{-\infty}^{\infty} P_i(t)^{u_i}[1 - P_i(t)]^{1-u_i}\phi(t;\theta_0,\sigma^2)\,dt \\
&\quad = \log \int_{-\infty}^{\infty} [1 - u_i + (2u_i - 1)P_i(t)]\phi(t;\theta_0,\sigma^2)\,dt \\
&\quad = \log \Bigg[1 - u_i + (2u_i - 1)\bigg(c_i + (1 - c_i) \\
&\qquad \times \int_{-\infty}^{\infty} \Psi[a_i(t - b_i)]\phi(t;\theta_0,\sigma^2)\,dt\bigg)\Bigg] \\
&\quad = \log \bigg(1 - c_i + u_i(2c_i - 1) \\
&\qquad + (2u_i - 1)(1 - c_i)\int_{-\infty}^{\infty} \Psi[a_i(t - b_i)]\phi(t;\theta_0,\sigma^2)\,dt\bigg) \\
&\quad = \log\{A_i + B_i F(\theta_0, \sigma^2; a_i, b_i)\},
\end{aligned} \tag{6.10}$$

where A_i, B_i are independent of person parameters, and F is the function of person and item parameters defined by an integral.

To maximize the likelihood function and to estimate θ_0 and σ^2, it is sufficient to develop an approximation to each summand and to the partial derivatives of each summand. To do this, a logistic approximation of F is now developed.

Using the logistic approximation to the normal cumulative

$$\Psi[a_i(t - b_i)] \doteq \Phi[(a_i/D)(t - b_i)] \tag{6.11}$$

and

$$F \doteq \int \Phi\left[\frac{a_i}{D}(t - b_i)\right] \phi(t; \theta_0, \sigma^2)\, dt. \tag{6.12}$$

The first factor in the integrand is the cumulative distribution function of a normal variable with mean b_i and variance D^2/a_i^2. Thus the integral has the form given in Eq. (6.9), $\int Hf$, and is therefore equal to the probability that the difference between two independent variables is negative. Thus the integral is the probability that a normal variable with mean $b_i - \theta_0$ and variance $D^2/a_i^2 = \sigma^2$ is negative,

$$\Phi\left[\left(\frac{D^2}{a_i^2} + \sigma^2\right)^{-1/2} (\theta_0 - b_i)\right]. \tag{6.13}$$

Finally, using the logistic approximation of the normal cumulative distribution function

$$F(\theta_0, \sigma^2) \doteq \Psi\left[D\left(\frac{D^2}{a_i^2} + \sigma^2\right)^{-1/2} (\theta_0 - b_i)\right]. \tag{6.14}$$

The approximation is easily calculated. To maximize the likelihood function, the authors (and Levine & Rubin, 1979) used generally available Fletcher–Powell algorithm (Gruvaeus & Jöreskog, 1970). The algorithm begins with a gradient procedure requiring partial derivatives with respect to θ_0 and σ. The partial derivatives are especially easy to calculate using the approximation in Eq. (6.14) because the logistic function satisfies the equation $\Psi'(t) = \Psi(t)[1 - \Psi(t)]$.

Index Evaluation and Receiver Operating Characteristic Curves

The design and evaluation of an inappropriate score detection procedure will depend on the relative frequency of inappropriate scores, the rewards for correctly identifying an inappropriate score, and the penalties for incorrectly classifying a normal examinee as aberrant. The receiver operating characteristic (ROC) curve of statistical decision theory provides a convenient way to evaluate indices prior to the specification of frequencies, rewards, and penalties.

When an appropriateness index is used to classify examinees, a cutoff or criterion value of the index is specified. At each criterion value t, the proportion of aberrant examinees correctly identified as aberrant and the proportion of normal examinees improperly identified as aberrant can be denoted by using

> $x(t)$ for the proportion of normal examinees with index values less than t,
> $y(t)$ for the proportion of aberrant examinees with index values less than t.

An ROC curve is obtained by plotting the $\langle x(t), y(t) \rangle$ pairs obtained for various values of t. The proportion $x(t)$ is called the *false-alarm rate* and the proportion $y(t)$ is called the *hit rate*.

In most applications high false-alarm rates cannot be tolerated. Therefore, in this research the minimum requirement for an index was that its ROC curve have a high initial segment. In other words, it is required that $y(t)$ be large for values of t such that $x(t)$ is small (i.e., that high hit rates are achieved at low false-alarm rates).

SIMULATION PROCEDURES AND EARLIER RESULTS

Some of Levine and Rubin's (1979) simulation methods, results, and data files are reviewed and tabulated in this section. In following sections their data are reanalyzed and their methods tested with actual data.

To simulate a normal vector of item scores, Levine and Rubin first sampled an ability θ from a normally distributed population with zero mean and unit variance. The examinee's first item score was generated by sampling a number uniformly distributed in the unit interval. If the sampled number was less than or equal to $P_i(\theta)$ in Eq. (6.1), then the first item was scored as correct; otherwise, the item was scored as incorrect. The remaining item scores were obtained by independently drawing new uniformly distributed numbers and comparing them with $P_i(\theta)$ for $i = 2, 3, \ldots, n$. The parameters a_i, b_i, c_i used in Eq. (6.1) were obtained from Lord's (1968) fitting of a three-parameter logistic model to a large sample of Scholastic Aptitude Test, Verbal Section (SAT-V) data.

Aberrant examinees were simulated by modifying simulated normal answer sheets in various ways. This chapter is primarily concerned with their "20% spuriously low" modification. To create a *spuriously low* examinee, 20% of a normal examinee's responses were sampled without replacement. A random number generator was then used to simulate random guessing over the five multiple-choice alternatives on the sampled items. No matter what the original sampled item score had been, the sampled item was scored correct with probability $\frac{1}{5}$ and scored incorrect with probability $\frac{4}{5}$.

Levine and Rubin (1979) observed that the modification of a normal answer pattern to create a spuriously low answer pattern frequently resulted in little or no change in the actual response vector. Clearly, if an aberrance-producing process does not alter the objective response pattern, it is futile to attempt to detect the presence of the process with an appropriateness index. Furthermore, if there is no effect on the objective response pattern and test score, there is little motivation for detecting aberrance. Consequently, Levine and Rubin separately analyzed the data from spuriously low examinees who had at least 10% of their scores changed. Both correct responses changed to incorrect and incorrect responses changed to correct were counted toward the 10% figure.

Levine and Rubin observed good aberrance detection with the total sample of spuriously low examinees and excellent detection for the selected, large-score-change sample of spuriously low examinees. Their three indices L_0, LR, and $\hat{\sigma}$ yielded similar results.

Most of the results of the present research were obtained with their simplest index L_0. In one study their likelihood ratio test for aberrance LR was also used, which for theoretical reasons was expected to be somewhat more powerful than L_0.

Numerical details of the Levine and Rubin data files needed for this research are summarized below:

1. NORMAL 3200: 3200 simulated answer sheets with normally distributed abilities; item parameters from Lord's (1968) fitting of the SAT-V; items scored either as correct or incorrect.
2. NORMAL 2800: The first 2800 records from NORMAL 3200.
3. LOW 200: Records 3001, 3002, . . . , 3200 from NORMAL 3200 modified to simulate ability-unrelated responding on 20% of the test according to the 20% spuriously low modification.
4. LOW 102: The 102 records selected from LOW 200 that had at least 10% of the simulated examinee's original responses changed on the spuriously low modification.

EMPIRICAL STUDIES

Four studies are presented below to answer the three questions raised by the Levine and Rubin (1979) studies. Throughout these studies an attempt was made to determine the severity of an "overfitting" artifact, described in Study 1. Actual (not simulated) data from the Graduate Record Examination and the SAT-V were included in Studies 3 and 4 to guarantee that these findings were not limited to simulated data or to only one test.

Study 1: Estimated Parameters and Overfitting

The problem considered in this study was whether or not high detection rates could be achieved with estimated item parameters.

METHOD

Levine and Rubin's exact values of L_0 calculated with exact simulation item parameters were compared with index values computed by the following procedure. First, item parameters were estimated by applying LOGIST to NORMAL 2800 data. L_0 was then computed for each NORMAL 2800 examinee by evaluating the likelihood function in Eq. (6.1) at $\theta = \hat{\theta}$, the LOGIST maximum likelihood estimate of ability. The LOGIST estimated item parameters were then used to compute L_0 for the LOW 102 response vectors in two steps. First, maximum likelihood ability estimates were obtained by rerunning LOGIST for the LOW 102 data with all item parameters fixed at the values obtained from NORMAL 2800 data; L_0 was then calculated for each LOW 102 examinee by substituting $\theta = \hat{\theta}$ in Eq. (6.1).

RESULTS

A very close agreement between values of L_0 computed with simulation parameters and estimated parameters was observed. In Figure 6.1 a bivariate scatterplot demonstrates this agreement for response vectors in the critical LOW 102 sample. The x coordinate of the point is $\log(L_0)$ computed with exact item parameters; the y coordinate, with estimated item parameters. If there were perfect agreement between the two measures, then the points would fall on the diagonal line shown. There is a tendency for estimated parameter index values to be slightly smaller than the simulation parameter index values for the most aberrant examinees in LOW 102 ($\log L_0 < -60$). Even closer agreement was observed for the NORMAL 2800 sample. This is shown in Figure 6.2 for the first 100 simulated examinees in the NORMAL 2800 data.

Detection rates with estimated parameters were high—at least as high as with simulation parameters. This is evident from Figure 6.3a, which shows the initial segment of the L_0 ROC curve. Of the aberrant examinees, 12.7% can be correctly classified with an L_0 cutoff score that did not

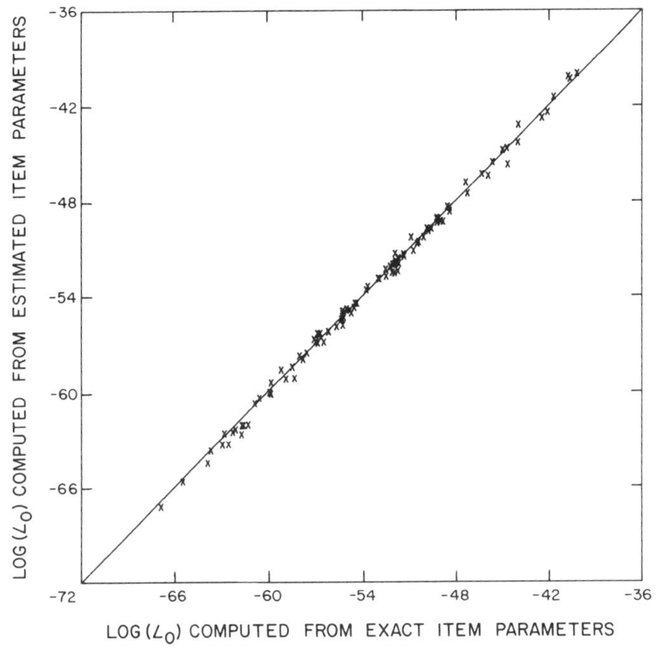

FIGURE 6.1. Bivariate scatterplot of $\log(L_0)$ computed from exact and estimated (from NORMAL 2800 data) item parameters for LOW 102 response vectors.

misclassify a single NORMAL 2800 examinee; 21.6% were detected at a false-alarm rate of .9%; and 47.1% were detected at a 4.4% false-alarm rate.

DISCUSSION

These results may initially seem surprising in view of Lord's (1975a) study of the disparity between LOGIST-estimated item parameters and simulation parameters. Lord found much more variation around the diagonal in his bivariate plots of simulation item parameters versus estimated item parameters than appears in the plots in this study of estimated parameter $\log(L_0)$ versus simulation parameter $\log(L_0)$. The discrepancy becomes less surprising when it is recalled that L_0 is the maximum value of a likelihood function, whereas a LOGIST parameter estimate gives the location of a point at which the maximum is obtained. If the likelihood function is relatively flat, then the maximizing values of the arguments of the function will be difficult to determine precisely because a small parameter change results in a very small likelihood change. Somewhat paradoxically, flatness of the likelihood function simplifies the problem of calculat-

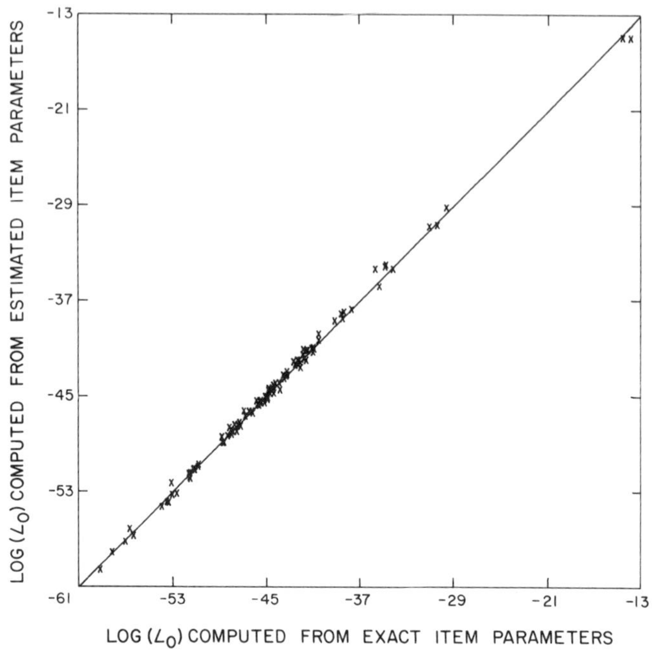

FIGURE 6.2. Bivariate scatterplot of log(L_0) computed from exact and estimated (from NORMAL 2800 data) item parameters for the first 100 simulated examinees in NORMAL 2800.

ing L_0 because the likelihood function is almost constant for parameter values in the neighborhood of the maximizing value.

Complicating the interpretation of these results is a possible artifact that will be called *overfitting*. The term *overfitting* will be used in the remainder of this chapter to refer to artifactual classification results caused by overlapping test norming and classification samples. In Study 1 the norming and classification sample were the same. Therefore, the high detection rates evident on the ROC curve could to some degree be due to the estimated parameters describing aspects of the norming sample that were not present in other samples of normal data. If another normal sample were used, much lower detection rates could have been observed. That estimated parameter L_0 scores for the most aberrant LOW 102 examinees were lower than exact parameter aberrance scores supports the suspicion of overfitting.

Significant overfitting would result in relatively high NORMAL 2800 scores (i.e., many points above the diagonal in Figure 6.2), and this was not found. Furthermore, the discrepancy between estimated and exact parameter index values was so small that the false-alarm rates in Figure

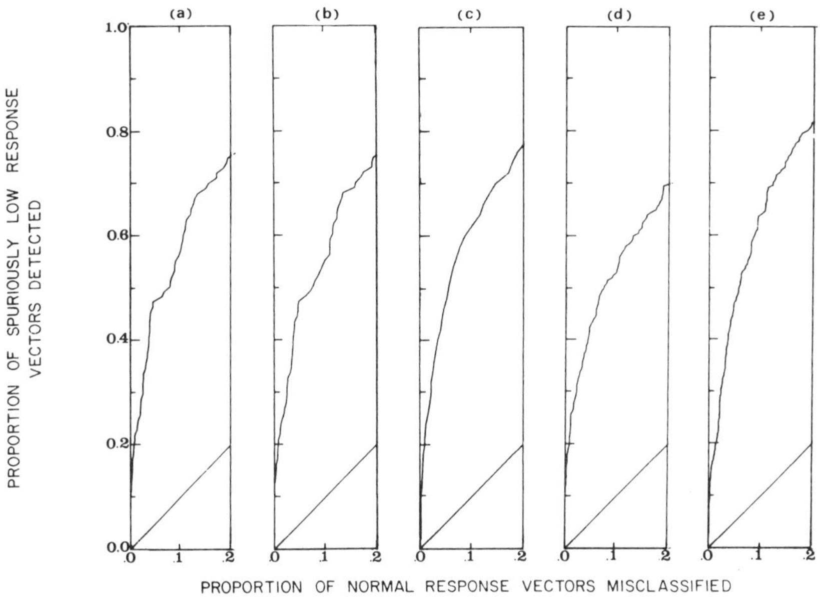

FIGURE 6.3. ROC curves for L_0 in (a) Study 1; (b) Study 2; and (c) Study 3; (d) *LR* in Study 3; and (e) L_0 in Study 4.

6.3a would be changed less than .0004 if the lowest LOW 102 index values fell exactly on the diagonal in Figure 6.1.

The other studies described below also support the conclusion that overfitting was not a serious problem in these studies. If overfitting were a significant artifact, then poor detection would be expected (*a*) if the normal group used to evaluate an index were distinct from the norming group used for item parameter estimation or (*b*) if the aberrant and normal groups were pooled to form the norming sample. In Study 4 the norming and normal groups were distinct; in Study 2 the aberrant and normal groups were pooled.

Study 2: Heterogeneous Norming Sample—Classification and Norming Sample Equal

The problems considered in Study 2 were

1. Will unidentified aberrants in the norming sample seriously degrade item parameter estimates and undermine the person measurement stage of appropriateness measurement?
2. To what extent has overfitting contributed to the high detection rates in Study 1?

METHOD

NORMAL 2800 and LOW 200 were merged to form a data file with a large proportion of aberrant examinees. Item parameters were estimated using all 3000 simulated examinees. As before, L_0 was computed for all examinees by evaluating the likelihood function at the LOGIST maximum likelihood estimate of ability. New index values for LOW 102 were compared with exact parameter index values, and detection rates were calculated.

RESULTS AND DISCUSSION

Figure 6.3b shows that estimating item parameters in a large sample with a large proportion of spuriously low examinees did not noticeably degrade appropriateness measurement. The simulation parameter L_0 ROC curve had hit rates of 11.7, 21.6, and 47.1% at false-alarm rates of .1, .9, and 4.8%. The corresponding heterogeneous norming sample false-alarm rates were .1, 1.2, and 4.9%. Clearly, the net effect on appropriateness measurement of estimating item parameters from this heterogeneous sample was small.

This study is important because in applications with actual data it will generally be necessary to evaluate appropriateness for examinees in the norming sample. The 255 parameters of the three-parameter logistic model for 85 items may fit the peculiarities of the aberrant examinees in the heterogeneous norming sample so well that they will be undetectable with L_0 or any other index of goodness-of-model fit.

Figure 6.4 contains the bivariate scatterplot of exact parameter $\log(L_0)$ values plotted against $\log(L_0)$ values computed from item parameters estimated in the heterogeneous sample. The relatively high frequency of points above the diagonal in Figure 6.4 indicates that there was indeed a tendency to overfit the aberrants in the norming sample; but the figure shows that the magnitude of the effect was small. The high detectability exhibited in the ROC curve (Figure 6.3b) also supports the contention that overfitting was small because a large overfitting effect would have reduced normal-aberrant group differences and thereby reduced detectability of aberrance.

The seriousness of overfitting should be reevaluated in each appropriateness measurement application. In this application, item parameters were estimated from a large sample ($N = 3000$), and only one type of aberrance was investigated. The effects of overfitting will be relatively small with large norming samples and, as explained below, for spuriously low scores.

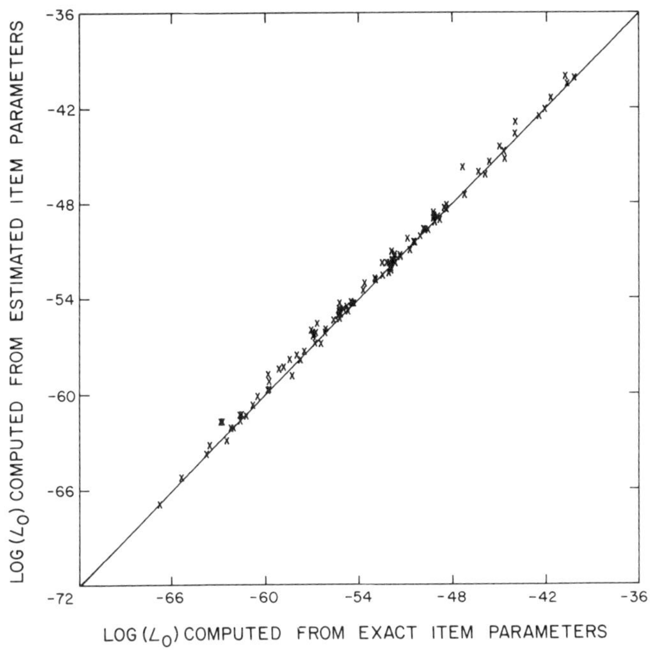

FIGURE 6.4. Bivariate scatterplot of log(L_0) computed from exact and estimated (from NORMAL 2800 and LOW 200 data) item parameters for LOW 102 response vectors.

The distinction between bias and sampling error is useful in understanding why overfitting effects are small in detecting spuriously low scores. The processes that are of concern in these studies—atypical schooling, exceptional creativity, low English fluency—are *essentially* unsystematic. Different aberrant examinees will tend to have different incorrect responses on different examination items. Thus the aberrant examinees will have competing effects on item parameter estimates. Consequently, the presence of many of these aberrant examinees in the norming sample may increase the sampling error of item parameter estimates without having much effect on the bias of the estimates.

Study 3: Actual Data—Overlapping Norming and Classification Samples, SAT-V Test

The problem considered in Study 3 was, Is the three-parameter logistic model sufficiently descriptive to detect spuriously low examinees in actual test data?

METHOD

Three thousand "low-omitting-rate" examinees were sampled from an administration of the SAT-V. All 3000 examinees responded to at least 90% of the test items. LOGIST was used to estimate item parameters from these 3000 nominally normal examinees. A file of 200 aberrant examinees was then created by applying the 20% spuriously low modification to answer sheets for examinees 2801, 2802, . . . , 3000.

L_0 was computed by maximizing the individual likelihood functions for the three-parameter logistic model. As in Study 1, the LOGIST-estimated item parameters were held constant and the likelihood of the individual's response vector was maximized by selecting an optimal ability estimate. The ability estimates were slightly different from the LOGIST ability estimates because the programs used in this study ignored both omitted and unreached items. (The LOGIST procedure ignores items not reached but attempts to use the omitted items by a technique that can be thought of as giving partial credit for omitted items.) Levine and Rubin's (1979) methods were used to calculate their likelihood ratio index *LR*.

RESULTS

Figures 6.3c and d present detection rates for the L_0 and *LR* appropriateness indices. High detection rates were obtained at low false-alarm rates. In particular, there were .2, .8, and 3.3% false alarms at hit rates of 10, 20, and 40% for the L_0 index; and the *LR* index yielded .04, .8, and 4.3% false alarms at the same hit rates. These results are even more impressive when it is noted that the aberrant group consisted of all 200 spuriously low examinees; no spuriously low examinee was deleted from the analysis because of an insufficient change in item scoring. In fact, 42 examinees had seven or fewer item responses changed (from correct to incorrect and incorrect to correct) when subjected to the spuriously low modification.

The *LR* index seemed more powerful than the L_0 index at very low false-alarm rates. *LR* detected 19 aberrant examinees (out of 200) without misclassification, whereas L_0 detected only 6 without a misclassification. At a false-alarm rate of .5%, *LR* detected 37 aberrant examinees and L_0 detected 32. At high false-alarm rates, however, L_0 was more powerful than *LR*.

There probably was very little overfitting artifact in this study, since the effects of overfitting were small in Studies 1 and 2 and since most of the aberrant examinees' item responses were included in the norming stage. (At least 80% of every aberrant examinee's responses were unchanged.) However, it seemed wise to confirm this result with a study in which the test norming and index-evaluation classification samples were separate.

Study 4: Actual Data—Distinct Norming and Classification Samples, GRE-V Test

The problems in Study 4 were

1. Are the methods that Levine and Rubin developed for SAT-V data generalizable to other tests?
2. Can high detectability be achieved when norming and classification samples are distinct?

METHOD

The responses to the Verbal Section of the Graduate Record Examination (GRE-V) by 10,000 examinees were used in the following manner. First, a file of 3000 examinees (FILE 1) with a wide range of ability and unrestricted omitting was created by selecting examinees 1, 2, 3, 11, 12, 13, . . . , 9991, 9992, 9993. This data set was then analyzed by LOGIST to obtain item parameter estimates. A second file was created by examining the item responses of the remaining 7000 examinees and selecting examinees with a low omitting rate. A total of 2470 examinees who had answered at least 86 of the 95 GRE-V items was obtained. Two hundred of these examinees were subjected to the 20% spuriously low modification. These modified response vectors formed the aberrant group for Study 4, and the remaining 2270 response vectors formed the normal group. L_0 was computed for the 200 aberrant and 2270 normal response vectors, as in Study 3, using the FILE 1 item parameter estimates.

RESULTS

The detection rates presented in Figure 6.3e for L_0 are quite high and are not substantially different from those obtained in Study 3. Hit rates of 10, 20, and 40% were obtained at false alarm rates of .3, 1.4, and 3.7%.

DISCUSSION

The finding that appropriateness measurement is effective for GRE-V data is important because Levine and Rubin (1979) originally considered a variety of appropriateness indices. After examining the effectiveness of each index *using simulated SAT-V data,* they selected the most effective indices for further study. The extent to which the most effective indices were capitalizing on test characteristics unique to the SAT-V was unknown.

In this study there was no overlap between the test norming and classification samples. Therefore, the high detection rates cannot be attributed to overfitting.

SUMMARY AND CONCLUDING REMARKS

In these experimental studies it has been shown that some appropriateness measurement techniques are robust to errors in estimation of item parameters, to the inclusion of unidentified aberrants in the test norming sample, and to violations of the three-parameter logistic model, which surely must exist in actual data. The detection rate of spuriously low examinees was high in all the analyses undertaken and could not be attributed to an overfitting artifact. This chapter concludes with some loosely interrelated remarks on appropriateness methods and current research.

Model Validity and Variable Ability Models

The three-parameter logistic standard model with its local independence and unidimensionality assumptions is admittedly simplistic. Lord (1975b) has described a very likely violation of local independence on the SAT and GRE. Some paragraphs associated with several items on the reading comprehension part of the examinations may be misunderstood or, alternatively, relevant to an area in which the examinee has an unusually strong background. Thus it seems virtually certain that responses to items referring to the same passage will be more highly interrelated than the local independence assumption [Eq. (6.1)] predicts. However, in spite of its shortcomings, the standard model has been able to describe regularities in data well enough to support appropriateness measurement. Presumably, a more valid model could support even more powerful appropriateness indices.

Parenthetically, it can be noted that violations of local independence can be accommodated by variable ability generalizations of the standard latent trait model. In variable ability models the examinee's ability may change during testing. A variable ability model is now being considered for dealing with covarying blocks of items [such as those called to the authors' attention by Lord (1975b)]. In the *blocks model* a test is analyzed into interrelated blocks of items. The examinee's ability on an item in a particular block is his or her central ability θ_0 plus a (normally) distributed correction. This correction is constant throughout the block of items. Levine and Rubin's (1979) Gaussian model is a limiting special case in which each item forms a one-item block. The standard model is another limiting case in which the entire test forms one block. If an adequate item parameter estimation procedure were developed for the blocks model, a substantial improvement in appropriateness measurement may be achieved.

The variable ability models are related to the independent work of Lumsden (1978). Lumsden (and Trabin and Weiss in Chapter 5) suggested "person characteristic curves" to study individual differences between examinees at the same ability level in the way they perform on items of varying difficulty. The Gaussian model and its individual difference parameter σ^2 seem to be a quantitative step in the direction Lumsden recommends for this purpose.

Item Parameters versus Conditional Probabilities

In appropriateness measurement and in many other applications of latent trait models, even very large standard errors of item estimation can sometimes be tolerated. This is true because the probability measure determined by a latent trait model depends directly on the conditional probability functions or item characteristic curves P_i and only indirectly on the item parameters. The item parameters are simply a convenient way to encode the shape of item characteristic curves. In fact, some curves are adequately described by a broad range of parameters.

This point is developed in a recent study of item bias (Linn, Levine, Hastings, & Wardrop, 1981). In that study item parameters were estimated for a reading achievement test from two groups with widely different distributions of achievement scores. A remarkable degree of invariance was observed when the estimated curves from the two groups were compared. The two sets of estimated curves were generally very nearly the same, although a superficial comparison of parameters often showed large differences in the estimated item parameters.

Estimation of Item Characteristic Curves Having More Than One Parameter

In spite of several Monte Carlo studies and numerous successful applications of the three-parameter logistic model, there seems to be some doubt about whether three-parameter item characteristic curves can be estimated by the currently used programs or by any method whatsoever. Some psychometricians evidently believe that adequate parameter estimates can be obtained only with the one-parameter Rasch model (i.e., the specialization of the three-parameter model obtained by setting $a_i = 1$ and $c_i = 0$).

In fact, parameter estimation techniques are available for models with much more complex item characteristic curve shapes than those of the three-parameter logistic model. Lord (1970) and Samejima (1977) have formulated, programmed, and applied nonparametric curve estimation techniques suitable for estimating curves of arbitrary shape. Furthermore, Levine (1976) has proved a consistency result for an estimation technique for estimating points on curves of arbitrary shape. It seems that the estimation difficulties ensuing from departing from very simple curve shapes have been exaggerated.

Work in Progress

Our next step in the development of appropriateness measurement will be to develop techniques for conventional tests in which there is substantial omitting. The research in this chapter with actual data has been restricted to answer sheets with 90% or higher response rates. However, a substantial proportion of the examinees have nonresponse rates greater than 10% on the SAT-V and GRE-V.

We have found that there is an orderly relation between omitting and ability on many items on conventional tests, and latent trait models are being developed for polychotomously scored items to exploit this relationship. By treating "omitted" and "not reached" as option choices, each answer sheet can be analyzed as if every item were answered. The preliminary investigations indicate that in this way both the range of applicability and statistical power of appropriateness measurement can be significantly increased.

ACKNOWLEDGMENTS

This research was primarily supported by funds provided by the Educational Testing Service and grants from the Graduate Record Examination Board (No. 75-3) and the College Entrance Examination Board (PJ 82201) to Michael V. Levine. The points of view and opinions expressed here do not represent the position or policy of the Educational Testing Service, the Graduate Record Examination Board, or the College Entrance Examination Board. Final manuscript preparation was supported by ONR Contract N00014-79-C-0752, Michael V. Levine, Principal Investigator.

REFERENCES

Birnbaum, A. Some latent trait models and their use in inferring an examinee's ability. In F. M. Lord & M. R. Novick, *Statistical theories of mental test scores*. Reading, Massachusetts: Addison-Wesley, 1968.

Donlon, T. F., & Fischer, F. E. An index of an individual's agreement with group determined item difficulties. *Educational and Psychological Measurement,* 1968, *28,* 105–113.

Ghiselli, E. E. Differentiation of individuals in terms of their predictability. *Journal of Applied Psychology,* 1956, *40,* 374–377.

Ghiselli, E. E. Differentiation of tests in terms of the accuracy with which they predict for a given individual. *Educational and Psychological Measurement,* 1960, *20,* 675–684. (a)

Ghiselli, E. E. The prediction of predictability. *Educational and Psychological Measurement,* 1960, *20,* 3–8. (b)

Gruvaeus, G. T., & Jöreskog, K. G. *A computer program for minimizing a function of several variables* (Research Bulletin 70-14). Princeton, New Jersey: Educational Testing Service, 1970.

Levine, M. V. *Item-item curves and consistent mental test parameter estimates* (Research Bulletin 76-36). Princeton, New Jersey: Educational Testing Service, 1976.

Levine, M. V., & Drasgow, F. Appropriateness measurement: Review, critique and validating studies. *British Journal of Mathematical and Statistical Psychology,* 1982, *35,* 42–56.

Levine, M. V., & Rubin, D. B. Measuring the appropriateness of multiple-choice test scores. *Journal of Educational Statistics,* 1979, *4,* 269–290.

Levine, M. V., & Saxe, D. H. *The use of periodic functions to measure the difficulty of aptitude test items* (Research Bulletin 76-17). Princeton, New Jersey: Educational Testing Service, 1976.

Linn, R. L., Levine, M. V., Hastings, C. N., & Wardrop, J. L. Item bias in a test of reading comprehension. *Applied Psychological Measurement,* 1981, *5,* 159–173.

Lord, F. M. An analysis of the Verbal Scholastic Aptitude Test using Birnbaum's three-parameter logistic model. *Educational and Psychological Measurement,* 1968, *28,* 989–1020.

Lord, F. M. Item characteristic curves estimated without knowledge of their mathematical form—A confrontation of Birnbaum's logistic model. *Psychometrika,* 1970, *35,* 43–50.

Lord, F. M. *Evaluation with artificial data of a procedure for estimating ability and item characteristic curve parameters* (Research Bulletin 75-33). Princeton, New Jersey: Educational Testing Service, 1975. (a)

Lord, F. M. Personal communication, 1975. (b)

Lumsden, J. Tests are perfectly reliable. *British Journal of Mathematical and Statistical Psychology,* 1978, *31,* 19–26.

Samejima, F. A method of estimating item characteristic functions using the maximum likelihood estimate of ability. *Psychometrika,* 1977, *42,* 163–191.

van der Flier, H. Environmental factors and deviant response patterns. In Y. H. Portinga (Ed.), *Basic problems in cross-cultural psychology.* Amsterdam: Swets & Zeitlinger, 1977.

Wood, R. L., & Lord, F. M. *A user's guide to LOGIST* (Research Bulletin 76-4). Princeton, New Jersey: Educational Testing Service, 1976.

Wood, R. L., Wingersky, M. S., & Lord, F. M. *LOGIST: A computer program for estimating examinee ability and item characteristic curve parameters* (Research Bulletin 76-6). Princeton, New Jersey: Educational Testing Service, 1976.

III

APPLICATIONS TO TESTING PROBLEMS

IRT has applications to many testing problems. Some of these applications include problems of test design, the identification of test bias, and, as indicated in Section 2, the problem of person fit. One important problem in large-scale testing applications is constructing item pools, which, in principle, will have all items on the same metric. Indeed, one of the major promises of IRT is the capability to create such item pools, since the metric of measurements in IRT is assumed to be invariant, within a linear transformation, throughout the range of ability. This allows the creation of item pools that permit adaptive testing, in which each individual is administered a different subset of items, but ability estimates for those individuals are on the same numerical scale.

The problem of creating item pools from items administered to subsets of different individuals is the *linking* problem, or the problem of *equating* item parameters onto a common metric. In Chapter 7, Ree and Jensen investigate the factors affecting this process when an anchor test is used in different samples as a means of equating item parameters.

In addition to equating item parameters, large-scale testing programs need to ensure that the ability estimates obtained over different subsets of items, which comprise different tests of the same ability, will be on the same scale. IRT in principle accomplishes this, just as it facilitates item pool construction. In Chapter 8 Marco, Petersen, and Stewart compare IRT score equating with other approaches, and investigate factors that influence the adequacy of the equating.

Another problem of classical test theory that has not been adequately addressed is that of speeded tests. In classical test theory, testing time is confounded with ability scores on most tests, unless they are pure speed or power tests. In Chapter 9, Thissen approaches the problem of speeded tests from an IRT viewpoint, and provides a variation of the IRT model for using response time as an additional variable in addition to correctness of response. Thissen's model incorporates both speed and power as separable independent variables characterizing items and individuals, and uses these data in estimating ability levels of individuals.

The invariant metric properties of IRT, useful in linking and score equating, have other applications as well. In Chapter 10, Bock describes the use of IRT for the measurement of group change in "mental age." In this study, Bock clarifies findings that resulted from an inappropriate interpretation of an earlier metric used by Thurstone, by redefining the metric onto an IRT base.

7

Effects of Sample Size on Linear Equating of Item Characteristic Curve Parameters

MALCOLM JAMES REE HARALD E. JENSEN

The application of the technology of computer-driven adaptive testing requires the development of large banks of test items. Each bank may contain 250–400 items, and all must measure the same ability on the same metric or scale. It is unreasonable and impracticable to assemble a single group of 2000 subjects for 250–400 min in order to obtain data on all the items; therefore, a method for linking subsets of items administered to varying groups must be investigated. Item response theory (IRT) offers a unique method of linking subsets of test items due to the invariance property of the item parameters. This invariance property rests on the two major theoretical assumptions of latent trait theory: (*a*) unidimensionality and (*b*) local independence.

Unidimensionality means that only a single ability is being measured and is assumed to be the property of an item pool, even when assembled into subsets. Local independence means that testees' responses to an item are independent of their responses to another item. More simply stated, this means that an item response is a function of ability and no other factor. In effect, this is a restatement of the unidimensionality assumption. If an item pool is unidimensional, then any shift in score metric that is due to a linear transformation may be corrected or adjusted by application of the proper complementary linear transformation. This is

NEW HORIZONS IN TESTING
Latent Trait Test Theory and
Computerized Adaptive Testing

ISBN 0-12-742780-5

what is meant by the idea that IRT parameters are invariant to a linear transformation, and it is this theoretical property that allows item pools to be linked and transformed to a common metric.

In previous research efforts, item pools have been linked by the method of linear equating (see Lord, 1977; Ree, 1977; Sympson, Weiss, & Ree, 1983) with apparent success. To date there has been little research on the efficacy of these linking procedures and the effects of errors in item parameter estimation on their (linearly) transformed values.

ITEM CHARACTERISTIC CURVE PARAMETERS

The three-parameter logistic model of Birnbaum (1968) is the most frequently used for relating item responses to testee ability. The three parameters of the item characteristic curve (ICC)—a, b, and c—are item discrimination, item difficulty (or location), and probability of chance success (or lower asymptote), respectively.

The curve described by these parameters takes the shape of an (cumulative frequency) ogive or an S, with the upper asymptote approaching a probability of 1.0 and, usually, with a lower asymptote of a probability greater than 0.0. The ogive describes the probability of obtaining a correct answer to an item as a monotonically increasing function of ability.

The item discrimination parameter a is a function of the slope of the ICC and generally ranges from .5 to about 2.5. The value of a equal to about 1.0 is typical of many test items; a values below .5 are insufficiently discriminating for most testing purposes; and a values above 2.0 are infrequently found.

The item difficulty parameter b describes the point of inflection of the ICC and is usually scaled between -2.5 and $+2.5$, although the metric is arbitrary. The item guessing parameter c is the lower asymptote of the ICC and is generally interpreted as the probability of selecting the correct item option by chance alone. Most test items have c parameters greater than 0.0 and less than or equal to .30.

Figure 7.1 shows three ICCs. The horizontal axis is scaled in units of ability, θ, and the vertical axis is the probability of answering the item correctly $[P(1|\theta)]$. The solid curved line shows an ICC for an item of average difficulty with acceptable discrimination and the lower asymptote appropriate for a five-alternative multiple-choice item. The dashed line shows an item of identical difficulty, with a c value of .28, but a lower a value. Note how the slope of the curve is less steep. The third curve, the dot–dashed line, shows an item with a c value of .30, an a parameter of

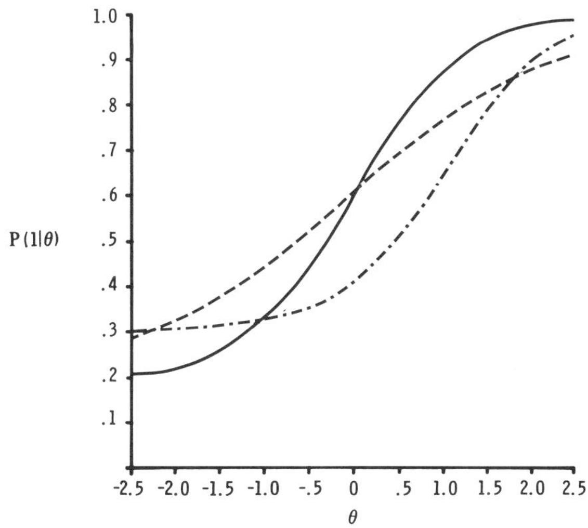

FIGURE 7.1. Item characteristic curves.

1.0, and a b parameter equal to 1.0. As the b parameter changes, the location of the inflection point of the curve is displaced along the horizontal axis. The mathematical function describing the curve is

$$P(\theta)_j = c_i + (1 - c_i)\{1 + \exp[-1.7a_i(\theta - b_i)]\}^{-1}. \qquad (7.1)$$

Previous research (Ree, 1978) indicates that the ICC parameters may be estimated with some reasonable degree of accuracy, provided a sufficient sample of examinees with an appropriate distribution of ability (θ) is available.

LINKING PARADIGMS

Two fundamental linking procedures may be defined and are known as the anchor items method (AIM) and the anchor subjects method (ASM). In AIM every subset of items is administered to a different sample of subjects, but embedded into the group of items to be analyzed is a common (or anchor) set of items. During analysis the anchor items are identified, and the following linear transformation is applied to the resultant ICC parameters:

$$b_t = (sb_t/sb_2)b_2 + [\bar{b}_t - (sb_t/sb_2)\bar{b}_2], \qquad (7.2)$$

where b_t is the item location parameter transformed to the desired scale and sb_t and sb_2 are standard deviations of the desired scale and observed scale, respectively. A similar procedure for the a parameter is defined by

$$a_t = a_2 sb_2/sb_t, \tag{7.3}$$

where a_t is the item discrimination parameter transformed to the desired scale; a_2 is the observed a parameter; and sb_t and sb_2 are as in Eq. (7.2). Because the c parameter is measured on the probability axis, it does not change and no transformation need be applied.

The ASM requires that the same group of subjects be available to take each subset of items. It is extremely unlikely that the same 2000 subjects could be assembled to take items over a long period of time, as would be requird to place tests on the same metric from year to year. For this reason, the ASM method seems less likely to find long-term practical application. Because of its potential for use, the AIM procedure was the subject of the present study.

METHOD

In order to have a known standard for reference, a simulation study was run using two groups of simulated testees, a single set of 20 anchor items, and two differing groups of 60 experimental, or nonanchor, items. These two groups of items were assembled into two tests. Both groups of simulees, designated $S1$ and $S2$, were specified to have about the same normal distribution of θ. Table 7.1 shows the mean, standard deviation, and minimum and maximum of θ for Groups $S1$ and $S2$. These two groups represent what might be expected if subjects for experimental testing were chosen from a larger pool, such as candidates for military enlistment. Response vectors for these simulees were generated on the two tests.

TABLE 7.1
Mean, Standard Deviation, and Minimum and Maximum of θ for Groups $S1$ and $S2$

	Group	
Statistic	$S1$	$S2$
Mean	−.014	.025
SD	.998	1.004
Minimum	−2.600	−2.600
Maximum	2.600	2.600

Generation of Item Responses

In order to generate a vector of item responses for each simulee, the θ values were used in Eq. (7.1) to compute the likelihood of correctly answering each item.

Because Eq. (7.1) yields a number $P(\theta)_j$ such that $0.0 < P(\theta)_j < 1.0$, a number X_j was drawn from a uniform (rectangular) distribution ranging from 0.0 to 1.0 and compared to $P(\theta)_j$. If X_j was larger than $P(\theta)_j$, an incorrect response was specified for the item; otherwise, a correct response was specified. Thus, a simulee with $P(\theta)_j = .90$ would answer an item correctly 9 times in 10, and a vector of item responses was developed for each simulee in each data set. These response vectors were then used to investigate the AIM linking procedures.

Table 7.2 shows the distribution of ICC parameters for the 80 items for Test 1 ($T1$) and Test 2 ($T2$), and Table 7.3 shows the ICC parameters for the 20 anchor items common to both tests.

Simulees from Group $S1$ were administered only the items in Test 1, and simulees from Group $S2$ only the items in Test 2. In order to study the effects of sample size, the ICC parameters were estimated on four samples drawn with replacement as follows: 250, 500, 1000, and 2000. The ICC parameters were estimated on these four sample sizes for both groups. Anchor ICC parameter values from the four samples administered Test 1 served as the input values for the anchor item parameters to the second test. This permitted the four sizes of the calibration sample (250, 500, 1000, 2000) to be varied and to be applied in the four samples used to estimate the anchor item ICC parameters.

TABLE 7.2
Mean and Standard Deviation of the Generated Item Parameters for Tests 1 and 2

Item parameter and statistic	Test 1	Test 2
a		
Mean	1.056	1.045
SD	.279	.239
b		
Mean	.085	−.056
SD	.844	.858
c		
Mean	.188	.202
SD	.054	.047

TABLE 7.3
ICC Item Parameters of the Twenty Anchor Items Common to Both Tests

Anchor item number	ICC item parameter *a*	*b*	*c*	Anchor item number	ICC item parameter *a*	*b*	*c*
1	.80	−1.50	.10	12	1.30	.30	.25
2	.80	−1.35	.10	13	1.20	.45	.20
3	1.00	−1.20	.15	14	1.20	.60	.22
4	1.00	−1.05	.15	15	1.10	.75	.22
5	1.10	−.90	.20	16	1.00	.90	.20
6	1.20	−.75	.20	17	1.00	1.05	.25
7	1.20	−.60	.22	18	.80	1.25	.25
8	1.20	−.45	.20	19	.80	1.35	.25
9	1.30	−.30	.20	20	.80	1.50	.25
10	1.40	−.15	.20	Mean	1.06	.00	.20
11	1.40	.15	.22	*SD*	.21	.95	.04

RESULTS

Table 7.4 shows the intercorrelations between the known item parameters and the estimated parameters. As past research indicates (Urry, 1976), correlations increased with increasing sample size. The correlations in Test 1 for *b* and estimates of *b* started high, at .952, and increased to an exceptionally high .992. Correlations for *a* and estimates of *a* began moderately, at .666, and climbed to .869; but the correlations of *c* and estimated *c* increased from only .031 to .115. In Test 2 much the same pattern was observed except that the correlation of *c* and estimated *c* increased from .164 to .315 as sample size increased.

Because correlations are insensitive to constant differences, as might be found if ICC parameters were either overestimated or underestimated by a constant amount, summed absolute deviations of the estimated parameters from the known parameters were computed for each parameter in each sample. Table 7.5 presents the summed absolute deviations (or summed errors) for both tests with the four sample sizes. Figure 7.2 displays this graphically.

There was a large decrease in summed error when the *a* parameter was estimated on progressively larger samples up to and including the difference between 1000 and 500 simulees. Between 1000 and 2000 simulees the difference in summed error was smaller. The relation between error and sample size for the *b* parameter was more nearly constant. That is, the line

TABLE 7.4

Intercorrelations between Known and Estimated ICC Item Parameters for Tests *T*1 and *T*2 for Both Groups with Varying Sample Sizes

Item parameter and sample size	Test 1	Test 2
a		
250	.666	.512
500	.671	.725
1000	.831	.813
2000	.869	.886
b		
250	.952	.929
500	.964	.962
1000	.980	.979
2000	.992	.987
c		
250	.031	.164
500	.035	.109
1000	−.012	.331
2000	.115	.315

TABLE 7.5

Summed Absolute Deviations ($\Sigma|\text{Error}|$) and Average Absolute Deviations ($|\overline{\text{Error}}|$) for the Three ICC Item Parameters for Tests 1 and 2, for Both Groups with Varying Item Sample Sizes

Item parameter and sample size	Test 1		Test 2	
	$\Sigma\|\text{Error}\|$	$\|\overline{\text{Error}}\|$	$\Sigma\|\text{Error}\|$	$\|\overline{\text{Error}}\|$
a				
250	30.645	.383	30.529	.382
500	22.809	.285	20.691	.259
1000	15.749	.197	16.891	.211
2000	15.598	.195	15.139	.189
b				
250	23.505	.294	20.847	.261
500	19.860	.248	16.607	.208
1000	17.689	.221	13.805	.173
2000	12.735	.159	11.513	.144
c				
250	7.736	.097	7.235	.090
500	7.360	.092	7.512	.094
1000	6.908	.086	7.318	.092
2000	6.440	.081	6.864	.086

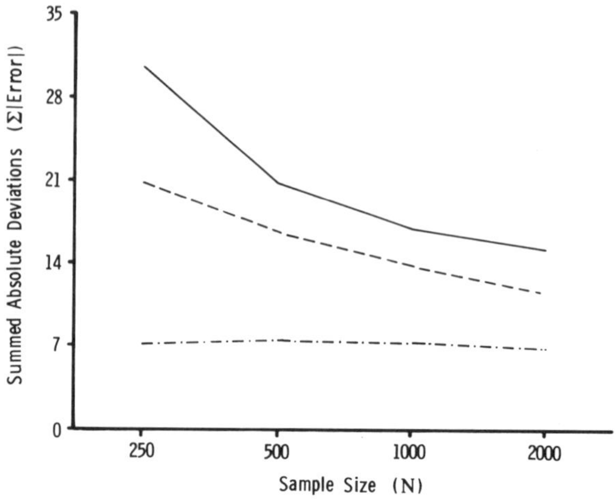

FIGURE 7.2. Errors in estimation of ICC parameters. (Solid line, *a*; dashed line, *b*; dot–dashed line, *c*.)

on the figure for estimates of *b* is generally straight, which means error tended to be reduced in direct proportion to the number of simulees. The almost-flat line for the *c* parameter indicates that virtually no reduction of error occurred with increasing sample size for that parameter. The average absolute deviation for the *c* parameter was almost one-third of the entire range of the parameter, as the *c* parameter is generally estimated between .00 and .30. Past research (Ree, 1979) indicates, however, that even for low-ability examinees, the effects of errors in the estimation of the *c* parameter are small.

Summed deviations of known ICC parameters from the equated value of the ICC parameters were computed for the *a* and *b* parameters for the 16 combinations of calibration sample size and equating sample size. Table 7.6 shows the summed absolute deviations and the per-item deviation for both parameters for the 16 combinations. The equated *a* parameter showed large summed deviations whenever the sample was limited to 250 simulees, whether in the calibration or the equating sample. The lowest error rates for the *a* parameter occurred when the anchor item values were estimated on 2000 simulees. The effects of the size of the calibration sample were not as clear. When 2000 simulees were used to estimate the anchor item ICC parameters, the magnitude of the error was approximately the same for all calibration sample sizes except 250.

With increasing calibration sample size, the error rate increased by

TABLE 7.6
Summed Absolute Deviations ($\Sigma|\text{Error}|$) and Average Absolute Deviations ($|\overline{\text{Error}}|$) for Item Parameters *a* and *b* for Various Calibration and Equating Sample Sizes

Sample size		Item parameter			
		a		*b*	
Calibration	Equating	$\Sigma\|\text{Error}\|$	$\|\overline{\text{Error}}\|$	$\Sigma\|\text{Error}\|$	$\|\overline{\text{Error}}\|$
250	2000	34.226	.428	23.368	.292
500	2000	15.128	.189	21.934	.274
1000	2000	15.987	.120	16.366	.205
2000	2000	16.596	.207	13.458	.168
250	1000	38.363	.480	25.644	.321
500	1000	17.679	.221	24.341	.304
1000	1000	19.587	.245	19.116	.239
2000	1000	21.032	.263	16.883	.211
250	500	48.611	.608	25.437	.318
500	500	24.558	.307	22.899	.286
1000	500	28.829	.360	18.187	.227
2000	500	31.209	.390	15.833	.198
250	250	44.312	.554	26.201	.328
500	250	21.577	.270	24.416	.305
1000	250	24.439	.312	19.484	.244
2000	250	27.024	.338	17.326	.217

some small amount, as indicated by the average (per-item) absolute deviation error. This is an unexpected result; an explanation may be found in the relation between the sets of estimated *a* parameters. If the estimated *a* parameters were all estimates of the same value and if the test scale were unidimensional (a basic assumption of the theory), then the estimated *a* parameters should be linear transformations of one another and should be correlated 1.0, as correlations are invariant to a linear transformation. This was not found to be the case, and Table 7.7 shows the intercorrelations of the estimated *a* parameters. Only the correlation between the estimate of *a* calculated on 1000 simulees and the estimate of *a* calculated on 2000 simulees approached this relation. This lack of linearity may be due to the assumption of normality and to the rescaling used in the calibration procedure; these may interact in such a way as to produce the anomalous results.

Table 7.7 also shows the intercorrelations of estimated *b* parameters. All exceeded .90, and the summed deviations also showed a steady decrease as sample size increased for the *b* parameter, indicating a virtually

TABLE 7.7
Intercorrelations, Means, and Standard Deviations of the Estimated *a* Parameters (Lower Triangle) and *b* Parameters (Upper Triangle) for Test 2

	Sample size				*b*	
Sample size	250	500	1000	2000	Mean	*SD*
250		.952	.940	.935	.056	.856
500	.757		.978	.969	.059	.838
1000	.690	.860		.986	.074	.870
2000	.595	.803	.926		.056	.873
a						
Mean	1.353	1.254	1.235	1.227		
SD	.484	.335	.325	.306		

linear transformation of estimated *b* parameters from sample to sample. With 500 simulees in the equating sample, however, a similar anomaly was observed, which may also be due to normal assumptions and to rescaling.

DISCUSSION

The results of the study present new evidence of the critical interrelation between item calibration and equating sample sizes and the values of ICC parameters.

Estimating and Equating *a*

For the 16 combinations of calibration sample sizes and equating sample sizes identified in Table 7.6, the least deviation of estimated *a* from its known value occurred with an equating sample size of 2000 and a calibration sample size of 500. As mentioned above, although the least error between an estimated and known *a* value was expected with a match of 2000 equating and 2000 calibrating sample sizes, error actually increased very slightly with increasing calibration sample sizes beyond 500. This discrepancy apparently resulted from a nonlinear transformation with sample sizes of 250 and 500 but tended toward linearity with sample sizes of 1000 and 2000.

During equating procedures a sample size of greater than 500 should be used to ensure an acceptable degree of confidence that the estimation of a does not significantly depart from its "true" value. In the same light, estimation of a suffers considerably using equating sample sizes of less than 500 such that equating samples of 1000 or 2000 are highly desirable to minimize error in estimating a.

Estimating and Equating *b*

Table 7.6 also shows the linear relation between error and sample size for the b parameter. The b parameter was best estimated with calibration and equating samples of 2000 each, although a calibration sample size of 1000 with an equating sample size of 500 can be tolerated without an appreciable increase in error. With all combinations of calibration and equating sample sizes, b was estimated quite well.

Estimating and Equating *c*

The flat line drawn in Figure 7.2 representing the data from Table 7.5 shows the estimation of the c parameter to be nearly insensitive to increases in sample size. As sample size increased from 250 to 2000 simulees, error decreased, but only very slightly. With c defined as the lower asymptote of the ICC and representing the probability of extremely low-ability examinees correctly answering an item, the inability to estimate c with precision could be disturbing. It has been pointed out (Lord, 1975), however, that if $a(\theta - b) < -2$, then the probability of a correct response is c. Therefore, if there are a large number of subjects with ability θ so that $\theta < -(2/a - b)$, c can be accurately estimated. If this requirement is not met, c will be poorly estimated.

CONCLUSIONS

A stable and accurate estimate of the a and b parameters requires large numbers of subjects over a broad range of ability. The estimation of c requires large numbers of subjects at very low ability levels. This holds for both equating and calibrating samples; therefore it is necessary to administer test items, whether to be calibrated or equated, to the largest samples available. See Vale, Maurelli, Gialluca, Weiss, and Ree (1981) for information on effects of additional factors on equating ICC parameters.

REFERENCES

Birnbaum, A. Some latent trait models and their use in inferring an examinee's ability. In F. M. Lord & M. R. Novick, *Statistical theories of mental test scores*. Reading, Massachusetts: Addison-Wesley, 1968.

Lord, F. M. Practical applications of item characteristic curve theory. *Journal of Educational Measurement*, 1977, *14*, 117–138.

Ree, M. J. *Adaptive testing at an AFEES*. Paper presented at the annual meeting of the Military Testing Association, San Antonio, Texas, October 1977.

Ree, M. J. *Estimating item characteristic curves* (AFHRL-TR-78-68). Brooks Air Force Base, Texas: Air Force Human Resources Laboratory, Personnel Research Division, November 1978.

Ree, M. J. *The effects of errors in the estimation of item characteristic curve parameters*. Paper presented at the annual meeting of the Military Testing Association, San Diego, California, October 1979.

Sympson, J. B., Weiss, D. J., & Ree, M. J. *Predictive validity of conventional and adaptive tests in an Air Force training environment* (AFHRL-TR-81-40). Brooks Air Force Base, Texas: Air Force Human Resources Laboratory, Manpower and Personnel Division, August 1982.

Urry, V. W. A five-year quest: Is computerized adaptive testing feasible? In C. L. Clark (Ed.), *Proceedings of the first conference on computerized adaptive testing* (U.S. Civil Service Commission, Personnel Research and Development Center, PS-75-6). Washington, D.C.: U.S. Government Printing Office, 1976. (Superintendent of Documents Stock No. 006-00940-9.)

Vale, C. D., Maurelli, V. A., Gialluca, K. A., Weiss, D. J., & Ree, M. J. *Methods for linking item parameters* (AFHRL-TR-81-10). Brooks Air Force Base, Texas: Manpower and Personnel Division, 1981.

8

A Test of the Adequacy of Curvilinear Score Equating Models

GARY L. MARCO
NANCY S. PETERSEN ELIZABETH E. STEWART

In many common testing situations it is necessary to compare the test scores of examinees who have taken different forms of a test. In practice, two forms of a test cannot be expected to be of exactly equal difficulty for examinees at all ability levels. Therefore a comparison of raw scores on two forms of a test will be unfair to the examinees who have taken the more difficult form. Statistical procedures that have been developed to deal with this problem are referred to as *equating methods*.

In an ideal psychometric world, tests on which scores need to be equated would be parallel in all important respects: An anchor test, if used, would be parallel to the total tests, and random samples on which to base the equating would always be available. In actual testing practice, however, scores must sometimes be equated under less-than-optimum conditions. This study is the first part of a larger study, the purpose of which was to examine the adequacy of score-equating models when certain sample and test characteristics are systematically varied. The emphasis in this part of the study is on curvilinear models, whereas the second part focused on linear models (Petersen, Marco, & Stewart, 1982).[1] This study is more comprehensive than previous studies of equat-

[1] Since the same methodology was used for both parts of the study, much of the material in this report appeared in the report on linear models in *Test Equating*, edited by P. Holland and D. Rubin, copyright © 1982 by Academic Press, Inc.

NEW HORIZONS IN TESTING
Latent Trait Test Theory and
Computerized Adaptive Testing

ISBN 0-12-742780-5

ing models (e.g., Levine, 1955; Rentz & Bashaw, 1975; Slinde & Linn, 1977, 1978; Tucker, 1974) in that it includes a greater variety of equipercentile, linear, and IRT models and investigates equatings based on dissimilar samples as well as on random samples.

EQUATING MODELS

An equating method is an *empirical* procedure for determining a transformation to be applied to the scores on one of two forms of a test. Its purpose is, ideally, to transform the scores in such a way that it makes no difference to the examinee which form of the test he/she takes. This ideal can be reached only if (*a*) the two forms of the test measure the same latent trait (ability or skill) and yield scores that are equally reliable and (*b*) the equating transformation is invertible.

General Description

Because an equating method is an empirical procedure, it involves a design for data collection and a rule for determining the transformation. There are three basic designs for data collection and three general rules for determining the transformation. Any of the three designs for data collection can be used with any of the three transformation rules.

The three designs for data collection are the single-group method, the equivalent-group method, and the anchor-test method (Lord, 1975). All the equating procedures used in this study assume that the data were collected using the anchor-test method. An anchor-test design requires administering one form of a test to one group of examinees, a second form to a second group of examinees, and a common anchor test to both groups. The anchor test can be either internal or external to the tests to be equated. The anchor test is used to reduce equating bias resulting from differences in ability between the two groups.

The three general rules for determining the transformation are the following:

1. *Equipercentile equating.* Choose a transformation such that scores from the two tests will be equated if they correspond to the same percentile rank in some group of examinees.
2. *Linear equating.* Choose a linear transformation such that scores from the two tests will be equated if they correspond to the same number of standard deviations from the mean in some group of examinees.

3. *Item response theory (IRT) equating*. Choose a transformation such that true scores from the two tests will be equated if they correspond to the same estimated level of the latent trait underlying both tests.

All three types of equating were represented in this study.

Equipercentile and linear methods using anchor-test data can be further classified by whether the equating is done directly or indirectly by frequency estimation. In direct equipercentile equating, scores on each test and on the anchor test are first equated separately within each group. Then scores on the two tests to be equated are said to be equivalent if they correspond to the same score on the anchor test. Frequency estimation (Angoff, 1971, pp. 581–582), on the other hand, makes use of the combined distribution of scores on the anchor test. The score distributions of the tests to be equated are estimated for the combined group of examinees, and these estimated distributions are then used as if they had been observed from a single-group design. The resulting marginal distributions on the two forms are used in the case of equipercentile equating; the resulting estimated means and standard deviations on the two forms are used in the case of linear equating.

Models Used in the Study

The various equating models used in this study are described briefly in Table 8.1. The models are categorized by whether the procedure results in a linear or curvilinear transformation between observed or true scores. The table provides references and information about the major assumption underlying each model, the kind of data required, and whether specific error variance estimates are needed. If the codes for the model differ for an external and an internal anchor test, then the formulas for computing the transformation parameters differ in the two cases. A total of 40 linear (including four based on the marginal means and standard deviations resulting from frequency estimation), two equipercentile, and two IRT equating models were used in the study.

Many of the linear methods require error variance estimates. The three methods of estimating error variances that were used in this study were the anchor-test–total-test method (Angoff, 1953), which uses anchor-test data; the part–whole method (Angoff, 1953; Feldt, 1975), which uses part-test data; and the coefficient-alpha method, which uses item response data.

One IRT procedure used the item difficulties from the one-parameter logistic test model; the other used the item parameters from the three-parameter logistic test model. The computer program LOGIST (Wood & Lord, 1976; Wood, Wingersky, & Lord, 1976) was used to estimate item

TABLE 8.1
Description of Equating Models

Major assumptions	Data	Error variances known: External anchor	Error variances known: Internal anchor	Codes: External anchor	Codes: Internal anchor	Model	Reference
I. Linear Observed-Score Models							
A. Constancy of Regression							
1. X on v and Y on v							
a. Curvilinear	Bivariate distributions	None	None	FE-O	FE-O	Frequency Estimation	Angoff, 1971
b. Linear	Cross-products matrices	None	None	T1	T1	Tucker 1	Angoff, 1971
2. X on v' and Y on v' are linear	Cross-products matrices	v_a, v_b	v_a, v_b	T20-E[a]	T20-I[b]	Tucker 2 Observed	Tucker, 1974
3. V on x' and V on y' are linear	Cross-products matrices	x_a, y_b	x_a, v_a, y_b, v_b	T30-E[a]	T30-I[b]	Tucker 3 Observed	Tucker, 1974
B. y and V and Y and V Congeneric and Constancy of Regression of X′ on v' and of Y′ on v'	Cross-products matrices	x_a, v_b, y_b, v_b	x_a, v_a, y_b, v_b	LER[a]	LER[b]	Levine Equally Reliable	Levine, 1955
C. Distributional Assumptions							
1. X|v and Y|v are normal	Cross-products matrices	None	None	PAC[h]	PAC[i]	Potthoff A	Potthoff, 1966
2. V|x and V|y are normal	Cross-products matrices	None	None	PB	PB[f,g]	Potthoff B	Potthoff, 1966
3. X and V and Y and V are bivariate normal	Cross-products matrices	None	None	PAC[h]	PAC[i]	Potthoff C	Potthoff, 1966
4. X and V and Y and V are more general than bivariate normal to allow for group differences	Cross-products matrices	None	None	PD	PD	Potthoff D	Potthoff, 1966
II. Curvilinear Observed-Score Models							
A. Constancy of Regression of X on v and Y on v are curvilinear	Bivariate distributions	None	None	FE-EQ	FE-EQ	Frequency Estimation	Angoff, 1971
B. No explicit assumptions	Marginal distributions	None	None	EQ	EQ	Equipercentile	
III. Linear True Score Models							
A. Constancy of Regression							
1. X′ on v' and Y′ on v'							
a. Curvilinear	Bivariate distributions	x_a, y_b	x_a, y_b	FE-T	FE-T	Frequency Estimation	Angoff, 1971

b. Linear	Cross-products matrices	x_a, v_a, y_b, v_b	x_a, v_a, y_b, v_b	T2T-E[c]	T2T-I[d]	Tucker 2 True	Tucker, 1974
2. V′ on x' and V′ on y' are linear	Cross-products matrices	x_a, y_b	x_a, v_a, y_b, v_b	T3T-E[c]	T3T-I[d]	Tucker 3 True	Tucker, 1974
B. X and V and Y and V Congeneric							
1. Constancy of regression of X′ on v' and of Y′ on v'	Cross-products matrices	x_a, v_a, y_b, v_b	x_a, v_a, y_b, v_b	LUR[e]	LUR[f]	Levine Unequally Reliable	Levine, 1955
2. Observed scores on test form are linear function of observed scores on ideal test. True scores on ideal test are multiplicative function of true scores on anchor test	Cross-products matrices	x_a, v_a, y_b, v_b	None	TML-E[e]	TML-E[f,g]	Tucker Modified Levine	Tucker, 1974
3. True scores on test form are linear function of true scores on anchor test							
a. As stated above	Cross-products matrices	x_a, y_b	$x_a, y_b,$	LXY-E[e]	LXY-I[f]	Lord X Y	Lord, 1976
b. As stated above	Cross-products matrices	v_a, v_b	v_a, v_b	LV-E[e]	LV-I[f]	Lord V	Lord, 1976
c. As stated above	Cross-products matrices	x_a, v_a, y_b, v_b	x_a, v_a, y_b, v_b	LML-E[e]	LML-I[f]	Lord Maximum Likelihood	Lord, 1976 Birch, 1964
4. True scores on part test are linear function of true scores on another test	Cross-products matrices	None		LCS		Lord Congeneric Subtests	Lord, 1976
IV. Curvilinear True Score Models							
A. X and V and Y and V Congeneric							
1. Item response function represented by a 1-parameter logistic function	Item responses	None	None	ICC-1	ICC-1	1-Parameter Logistic ICC	Lord, 1975
2. Item response function represented by a 3-parameter logistic function	Item responses	None	None	ICC-3	ICC-3	3-Parameter Logistic ICC	Lord, 1975

[a–f] These models yield identical conversion parameters when the anchor-test–total-test method of estimating error variances is used.
[g–i] These models always yield identical conversion parameters.
Notes. x, y, and v are the observed score on the new form, the old form, and the anchor test, respectively.
x', y', and v' are, respectively, the true scores.
a, b, and c are the group taking tests X and V, the group taking tests Y and V, and the combined group, respectively.

parameters and examinee abilities. For both models, true formula scores $[R - (W/4)]$ were equated by calculating the true formula scores on each test form corresponding to selected ability levels (Lord, 1975) and interpolating as necessary. Since for either model there is a functional relation between ability and true score, the true formula score is readily computed by the equation

$$R_g - (N - R_g)/(A - 1), \tag{8.1}$$

where R_g is the true number-correct score at ability g computed by summing the item proportions correct under the model, N the number of test items, and A the number of response options for the items in the test (five choices in all cases).

STUDY DESIGN

Computer files for two national administrations (April 1975 and November 1975) of the verbal portion of the College Board Scholastic Aptitude Test were obtained from Educational Testing Service (ETS). Fourteen pairs of total test scores were equated on the basis of the data from each of the two administrations. Records of test scores, responses to test items, and responses to a Student Descriptive Questionnaire (SDQ) were used to construct samples having specified characteristics, to extract or to calculate selected scores for a number of special purpose total tests and anchor tests, and to compute items statistics and other data needed for the various equating models.

Equating Design

The combinations of total test and anchor test used in the various equatings for each of the two administrations can be classified into five categories, referred to here as test variations. (There were 11 test variations used in the full study of which this study is a part.)

EQUATING A TEST TO ITSELF

In this part of the study (see Table 8.2), a single form of the operational verbal portion of the SAT (SAT-V) was treated as if it represented two different forms; that is, it was to be equated to itself. (This type of design was first used by Levine, 1955.) This part of the study was designed to investigate the effects of varying (a) the relative difficulty levels of the total test and the anchor test and (b) the degree of similarity between the two samples on which the equating operations were based.

TABLE 8.2
Design for Equating a Medium-Difficulty Test to Itself through an Anchor Test of Similar Content

Test variation	Anchor test: Location	Anchor test: Difficulty	Relation between samples	SAT-Verbal score level: New form sample	SAT-Verbal score level: Old form sample
1	External	Medium	Random	Middle	Middle
			Dissimilar	Middle	High
4	Internal	Medium	Random	Middle	Middle
			Dissimilar	Middle	High
5	Internal	Easy	Random	Middle	Middle
			Dissimilar	Middle	High
	Internal	Difficult	Random	Middle	Middle
			Dissimilar	Middle	High

SAT-V was equated to itself through two anchor tests that, like the total test, were of medium difficulty. One was external (Test Variation 1), and one was internal (Test Variation 4). SAT-V was also equated to itself through two internal anchor tests that differed from it in difficulty (Test Variation 5). One of the internal anchor tests was easier than SAT-V; the other was more difficult.

For each administration, a single pair of random samples and a single pair of dissimilar samples were used for all equatings of SAT-V to itself. The dissimilar samples, by virtue of the sample selection procedure, were expected to be of middle and high verbal ability, respectively.

EQUATING A TEST TO A DIFFERENT TEST

In this part of the study (see Table 8.3), three total tests were constructed for each administration from a pool comprising the operational SAT-V items and the items in a nonoperational section of verbal material. The purpose of this part of the study was to examine the effects of varying (*a*) the relative difficulty levels of the two total tests on which scores were to be equated and (*b*) the degree of similarity between the two equating samples.

Pairs of total tests constructed to differ in difficulty were equated through an internal anchor test of medium difficulty. The random and dissimilar samples used in equating SAT-V to itself were used in these equatings also, along with an additional sample expected to be of low verbal ability.

In Test Variation 8 an easy test was equated to a medium-difficulty test and the medium-difficulty test was equated to a difficult test. In Test

TABLE 8.3
Design for Equating Tests That Differed Only in Difficulty through an Internal Anchor Test of Similar Content and Medium Difficulty

Test variation	Total test difficulty		Relation between samples	SAT-Verbal score level	
	New form	Old form		New form sample	Old form sample
8	Easy	Medium	Random	Middle	Middle
			Dissimilar	Low	Middle
	Medium	Difficult	Random	Middle	Middle
			Dissimilar	Middle	High
9	Easy	Difficult	Random	Middle	Middle
			Dissimilar	Low	High

Variation 9 the easy test was equated to the difficult test. For equatings based on dissimilar samples, data from the low-ability sample were used for the easy test, from the middle-ability sample for the medium-difficulty test, and from the high-ability sample for the difficult test.

Tests

Several scores calculated as part of the normal processing for SAT administrations were included in the study. Included also were scores computed on a number of tests constructed retrospectively solely for use in the study. The general approach followed in constructing the special-purpose tests entailed (*a*) developing content and statistical specifications for all special purpose tests; (*b*) identifying sets of items in accordance with those specifications; and (*c*) identifying subsets of items on which separate scores were to be obtained for use in calculating one of the three sets of reliability estimates required for the equating analyses.

Test content was specified only in terms of distributions of item types, although more detailed specifications are followed in developing operational forms of SAT-V. SAT-V is composed of three types of discrete five-choice items—antonyms (25 items), analogies (20 items), and sentence completions (15 items)—and of five reading comprehension passages, each of which is followed by 5 five-choice items based on the passage.

Statistical specifications were stated in terms of the item statistics conventionally used in the development of ETS tests (described by Angoff & Dyer, 1971, pp. 9–10). The equated delta Δ_e served as the index of item difficulty, and the biserial correlation r_b of the item score with the total score on the operational test in which the item appeared as the index of item discrimination. The statistic Δ_e is an estimate of the difficulty of the item for a standard reference group. It ranges in value from about 6 (very easy) to 18 (very difficult). If a test composed of five-choice items were of middle difficulty for the reference group, its mean Δ_e would be 12.0.

The item statistics used to construct the tests were taken from the results of item analyses routinely conducted for each new form of the SAT-V. The analyses were based on systematic samples of 1700–2000 examinees each. After all full-length total tests and anchor tests had been identified, part tests for use in reliability estimation were created. Each part test was parallel, except for length, to the full-length test from which it was derived.

TESTS USED IN EQUATING A TEST TO ITSELF

For each administration, scores were available on an external anchor test (a nonoperational section of verbal material) similar in content to the SAT-V. The external anchor tests each contained equal numbers of items of the four types included in the SAT-V. The difficulties of the external anchor tests were not subject to experimentation. The mean difficulties of the external anchor tests were within about one-half a Δ_e point of the mean for SAT-V, but both the standard deviations of the Δ_e (σ_{Δ_e}) and the mean of the r_b tended to be somewhat lower for the anchor tests.

For each administration, three internal anchor tests for equating SAT-V to itself were specially constructed for the study from the pool of 85 operational items in each form. The internal anchor tests were constructed to be similar in content to SAT-V but to vary with regard to each other in mean difficulty. Internal anchor tests constructed from the April 1975 item pool each contained 10 antonym, 6 analogy, 8 sentence completion, and 10 reading comprehension items. The number of items in these respective categories from the November 1975 item pool were 10, 8, 8, and 10. For the medium-difficulty anchor tests, $\overline{\Delta}_e$, σ_{Δ_e}, and $\bar{r}_b$ were made to match the corresponding values for SAT-V as closely as possible, given the prescribed content distribution. The σ_{Δ_e} for the easy and difficult anchor tests were, of necessity, smaller than those for SAT-V and the medium-difficulty anchor test. The item summary statistics and identification codes for the total tests and the anchor tests used in equating a test to itself are given in Table 8.4.

TESTS USED IN EQUATING A TEST TO A DIFFERENT TEST

For each administration, an expanded item pool consisting of the 85 SAT-V items and the 40 items in the verbal external anchor test was used

TABLE 8.4
Item Summary Statistics and Identification Codes for Total Tests and Anchor Tests Used in Equating at Two Administrations

Administration	Test description	ID	n	$\bar{\Delta}_e$	σ_{Δ_e}	$\bar{r}_b$
Equating a test to itself						
April 1975	Total Test					
	SAT-V	FT2XX	85	11.40	3.38	.47
	Anchor Test					
	External	FE2FM	40	11.14	3.06	.43
	Internal					
	Easy	FE2DE	34	9.39	2.90	.49
	Medium	FE2DM	34	11.40	3.26	.49
	Difficult	FE2DH	34	12.96	2.49	.47
November 1975	Total Test					
	SAT-V	FT4XX	85	11.36	3.40	.48[a]
	Anchor Test					
	External	FE4FG	40	12.01	2.95	.43
	Internal					
	Easy	FE4DE	36	9.40	2.68	.51
	Medium	FE4DM	36	11.44	3.43	.49
	Difficult	FE4DH	36	13.29	2.18	.48
Equating different tests						
April 1975	Total Tests					
	Easy	FT2DE	54	9.34	2.89	.46
	Medium	FT2DM	54	11.34	2.80	.46
	Difficult	FT2DH	54	13.26	2.95	.44
	Anchor Test					
	Internal	FE2PA	20	11.31	3.26	.46
November 1975	Total Tests					
	Easy	FT4DE	54	9.59	2.94	.50
	Medium	FT4DM	54	11.69	2.49	.46
	Difficult	FT4DH	54	13.51	2.77	.45
	Anchor Test					
	Internal	FE4PA	20	11.58	3.05	.47

[a] Based on 84 of the 85 items.

for creating three total tests. The tests within each set were systematically different in average difficulty but were similar in content and equal in length. The total tests constructed from the April 1975 and the November 1975 item pools each contained 15 antonym, 13 analogy, 11 sentence completion, and 15 reading comprehension items.

For each administration, a 20-item internal anchor test of medium difficulty, similar in content to SAT-V, was constructed for use in equating the special-purpose total tests. The internal anchor tests constructed from the

April 1975 and the November 1975 item pools each contained 6 antonym, 5 analogy, 4 sentence completion, and 5 reading comprehension items. The item summary statistics and identification codes for the total tests and the anchor tests used in equating different tests are given in Table 8.4.

Samples

Two base samples were used for the study, one each from the April 1975 and the November 1975 Saturday administrations of the SAT-V. The April base sample (No. 32) was selected from those candidates who took Verbal Equating Test FM; the November base sample (No. 44), from those who took Verbal Equating Test FG. (Six base samples were used in the full study.)

Each base sample consisted of 4731 cases, from which five subsamples of 1577 cases each were created in two different ways. Two nonoverlapping subsamples ("random" samples) were selected by use of an IBM recursive random number generator. Three nonoverlapping subsamples ("dissimilar" samples) were selected by an algorithm designed to yield samples dissimilar in mean verbal ability. Two variables from the SDQ—level of educational aspiration and amount of high-school foreign language training—were used to select the dissimilar samples. These variables were known from prior information to be related to SAT-V scores.

Thus a total of 10 subsamples of 1577 cases each, 5 for each of the 2 base samples, were used in the study. Tables 8.5 and 8.6 give the summary statistics for the total tests and the anchor tests used in the various equating analyses.

Evaluative Procedures

DISCREPANCY INDICES

For each raw score x, there is a corresponding criterion score t and an estimated criterion score t' derived from a specific equating model. The smaller the difference d between t' and t, the smaller the equating error and the more appropriate the equating model.

The standardized weighted mean square difference and squared bias were selected as the most useful summary indices for evaluating the effectiveness of the various models. The weighted mean square difference gives the greatest weight to those values of x that are most likely to occur and is consistent with what is used to represent total error in the statistical literature. The indices were standardized (expressed as a proportion of the criterion standard deviation) so that results could be compared across equating situations as well as across equating models.

TABLE 8.5
Formula Score Means, Standard Deviations, and Correlations[a] between Anchor Test and Total Test Scores for Base Sample No. 32 (April 1975)

Test	Number of items	Samples[b] Random 321XX	322XX	Dissimilar 327XX	328XX	329XX
Equating a test to itself						
Total test						
FT2XX	85					
Mean		34.47	34.67		32.95	40.28
SD		15.48	15.57		14.47	14.82
Anchor test						
FE2DE	34					
Mean		19.91	19.96		19.43	22.27
SD		6.78	6.72		6.46	6.00
r		.93	.93		.92	.92
FE2DM	34					
Mean		14.00	14.12		13.39	16.56
SD		6.98	6.98		6.58	6.66
r		.94	.94		.93	.94
FE2DH	34					
Mean		9.15	9.27		8.45	11.83
SD		7.35	7.36		6.90	7.37
r		.94	.94		.93	.95
FE2FM	40					
Mean		16.99	16.63		16.01	19.52
SD		8.06	8.06		7.78	7.86
r		.87	.87		.85	.86
Equating a test to a different test						
Total test						
FT4DE	54					
Mean		31.61		27.94		
SD		10.17		10.20		
r		.88		.86		
FT4DM	54					
Mean		21.26	21.14	21.47		
SD		11.67	11.08	10.99		
r		.90	.89	.89		
FT4DH	54					
Mean		11.89				16.17
SD		9.53				10.89
r		.89				.89
Anchor test						
FE2PA	20					
Mean		8.46	8.42	7.25	8.02	9.72
SD		4.15	4.17	3.97	3.99	4.07

[a] Correlations are between the indicated anchor test score and total test score FT2XX in the case of a test being equated to itself, and between anchor test score FE2PA and the indicated total test score in the case of a test being equated to a different test. All anchor test scores were included in the total test score except FE2FM.

[b] The last two digits of the sample number refer to the total test and anchor test score combination used for a particular equating.

TABLE 8.6
Formula Score Means, Standard Deviations, and Correlations[a] between Anchor Test and Total Test Scores for Base Sample No. 44 (November 1975)

Test	Number of items	Samples[b]				
		Random		Dissimilar		
		441XX	442XX	447XX	448XX	449XX
Equating a test to itself						
Total test						
FT4XX	85					
Mean		36.34	35.97		36.70	42.18
SD		15.77	14.92		14.74	15.10
Anchor test						
FE4DE	36					
Mean		21.82	21.74		22.15	24.19
SD		7.13	7.11		6.76	6.43
r		.93	.93		.93	.92
FE4DM	36					
Mean		15.31	15.23		15.47	17.81
SD		7.13	6.85		6.75	6.89
r		.95	.95		.94	.95
FE4DH	36					
Mean		9.19	8.98		9.12	12.02
SD		7.85	7.29		7.56	8.06
r		.94	.93		.93	.95
FE4FG	40					
Mean		14.38	14.15		14.57	17.22
SD		8.32	8.02		7.80	8.24
r		.87	.86		.85	.87
Equating a test to a different test						
Total Test						
FT2DE	54					
Mean		31.88		28.61		
SD		10.05		10.25		
r		.89		.87		
FT2DM	54					
Mean		21.99	22.07	20.85		
SD		10.95	10.92	10.30		
r		.91	.90	.89		
FT2DH	54					
Mean		12.94				16.47
SD		10.21				10.45
r		.89				.89
Anchor test						
FE4PA	20					
Mean		8.12	7.93	6.66	8.08	9.48
SD		4.36	4.13	3.89	4.07	4.39

[a] Correlations are between the indicated anchor test score and total test score FT4XX in the case of a test being equated to itself, and between anchor test score FE4PA and the indicated total test score in the case of a test being equated to a different test. All anchor test scores were included in the total test score except FE4FG.

[b] The last two digits of the sample number refer to the total test and anchor test score combination used for a particular equating.

The standardized weighted mean square difference or total error is equal to the variance of the difference plus the squared bias, that is,

$$\frac{\Sigma_i f_i d_i^2}{n s_t^2} = \frac{\Sigma_i f_i (d_i - \bar{d})^2}{n s_t^2} + \frac{\bar{d}^2}{s_t^2}, \tag{8.2}$$

where $d_i = (t_i' - t_i)$; t_i' is the estimated criterion score for raw score x_i; t_i is the criterion score for x_i; $\bar{d} = \Sigma_i f_i d_i / n$; s_t is the standard deviation of the criterion scores t_i; f_i is the frequency of x_i; $n = \Sigma_i f_i$; and the summation was over that range of x for which extrapolation was unnecessary for any of the models studied. If the ratio of the squared bias to the total error is 1, then the criterion line and the conversion line are parallel. If the difference is less than 1, then there is an interaction between the model and the criterion.

CRITERION EQUATINGS

In the case in which a test was equated to itself, the criterion for the various equatings was the test score itself. The new and old forms were treated as different tests, when in reality they were the same test. The ideal equating would reproduce the score on the old form exactly; that is, the conversions from raw to scaled scores would be the same for the new form as for the old form. The criterion was not so simply established in the case in which a test was equated to a different test. In these instances, it was necessary to calculate "true" conversions by equating the tests in as ideal a manner as possible.

The criterion equatings were accomplished using data from all of the cases in the two base samples (Nos. 32 and 44) from which subsamples were selected. Since all 4731 cases in each base sample had scores on the tests being equated, it was possible to equate the scores using a single sample, an ideal equating situation. The scores could be linked directly without involving an anchor test.

Two equating methods were used to establish the two criteria against which to compare the results of the experimental equatings: equipercentile equating of estimated true scores derived from the three-parameter logistic test model (the IRT equipercentile criterion) and equipercentile equating of observed scores (the direct equipercentile criterion). Although the criterion equatings using estimated true scores were based on IRT methodology, the method was different from that used in the experimental equatings. Nevertheless, the IRT equipercentile criterion could be biased in favor of IRT equating methods. Thus, the direct equipercentile criterion was also used. This criterion might be biased in favor of equipercentile, but not IRT, methods. For the study of the curvilinear methods reported here, both methods were appropriate, since the total tests were equal in length and yielded scores with nearly equal reliabilities. (True score equating will generally yield better conversions than observed score equating when the new and old form scores have unequal reliabilities.)

To determine the IRT equipercentile criterion, the three-parameter logistic test model was applied separately to Reading (reading comprehension and sentence completion) and Vocabulary (antonym and analogy) items, so that unidimensionality did not have to be assumed across all item types. The following scores were equated:

Base Sample No. 32		Base Sample No. 44	
New Form	Old Form	New Form	Old Form
FT2DE	FT2DM	FT4DE	FT4DM
FT2DE	FT2DH	FT4DE	FT4DH
FT2DM	FT2DH	FT4DM	FT4DH

The following steps were required to accomplish the three-parameter logistic IRT criterion equatings:

1. LOGIST was used to calculate item parameter estimates and examinee ability estimates separately for each set of relatively homogeneous items (Reading and Vocabulary).
2. For each of the three scores, for each examinee, an estimated true raw $(R - W/4)$ score was calculated across the item types represented in each score. The true raw score (R_a) was estimated as

$$R_a = \sum_i P_i(\hat{\theta}_{ia}) - \left[\sum Q_i(\hat{\theta}_{ia})\right] \Big/ 4 \tag{8.3}$$

   where $\hat{\theta}_{ia}$ is the estimated ability of examinee a on the item type represented by item i (if item i is a Reading item, then $\hat{\theta}_{ia}$ is the estimate of the examinee's reading ability, and so forth); P_i is the probability of examinee a answering item i correctly (as calculated from the three-parameter logistic test model); Q_i is $1 - P_i$; and one fourth is the correction factor for guessing for five-choice multiple-choice items. (Each summation was over only those items to which examinee a actually responded.)
3. For each of the three pairs of scores, the scores were directly equated by the equipercentile method. Raw and scaled score equivalents were generated for each integral score on the new form for which it was possible to establish a conversion. (The true raw scores did not usually extend over the possible score range, thus making it impossible to establish a conversion for some scores.)

RESULTS AND DISCUSSION

Comparisons of Equating Models

Tables 8.7–8.12 and Figures 8.1–8.6 summarize discrepancies between the results of the experimental equatings and the criterion equatings. The discrepancies are stated as mean square error and squared bias. To make the results comparable for different tests, the discrepancies are expressed on a scale on which the standard deviation of the criterion scores would be 100 for a standard reference group. The discrepancy indices for Tables 8.9 and 8.11 and Figures 8.3 and 8.5 were calculated in relation to criterion equatings in which the three-parameter logistic test model was used (the IRT equipercentile criterion). The discrepancy indices for Tables 8.10 and 8.12 and Figures 8.4 and 8.6 were calculated in relation to criterion equatings in which the equipercentile equating method was used (the direct equipercentile criterion).

The test variations represented in Tables 8.7–8.12 are characterized in Tables 8.2 and 8.3. The statistical characteristics of the tests on which the total test scores and the anchor test scores were based are described in Table 8.4. The first three digits of the numeric codes in the column labeled "Sample Number" identify the subsamples on which the equatings were based (see Tables 8.5 and 8.6 for sample statistics).

The column labeled "Best Lin" identifies the linear model that had the smallest mean square error under the condition specified. The linear models considered are enumerated in Sections I (observed score models) and III (true score models) of Table 8.1, which also gives the identification codes for all models. The models identified in Tables 8.7–8.12 as Equi% (Dir), Equi% (FE), 3-Par IRT, and 1-Par IRT correspond, respectively, to Sections IIB, IIA, IVA2, and IVA8.1 of Table 8.1.

Figures 8.1–8.6 correspond sequentially to Tables 8.7–8.12. It is essential to note that *the size of the scale units represented on the y axis is different in different figures*. The scale on the left-hand side of each figure is in terms of standard deviation units and is the square root of the linear scale of mean square error shown on the right-hand side of the figure. Contiguous bars represent the results for one model or, in the case of "Best Lin," for the best of a set of models. The results for random samples are shown in the top half of the figure, and the results for dissimilar samples in the bottom half. The outer bars represent total error; and the inner shaded bars, bias.

The bars in each set, from left to right, appear in the successive rows in the comparable area of the corresponding table. For example, the first bar in each set in Figure 8.1 represents the results obtained with an internal anchor test for a pair of April 1975 samples, which were either random or of middle and high ability, respectively; the second bar, an external anchor test for the same April samples; the third bar, an internal anchor test for a pair of November 1975 samples; and the fourth bar, an external anchor test for the same November samples.

EQUATING A TEST TO ITSELF

Test Variations 1 and 4. The data in Table 8.7 (and Figure 8.1) show that for both similar and dissimilar samples the best linear model had the smallest amount of total error across replications, followed by the one-parameter IRT and three-parameter IRT models. The direct equipercentile and frequency estimation equipercentile methods had relatively more error. Surprisingly, the total error for dissimilar samples was very similar to the total error for random samples, implying, perhaps, that if the anchor test is nearly parallel to the total tests, the differences between samples are

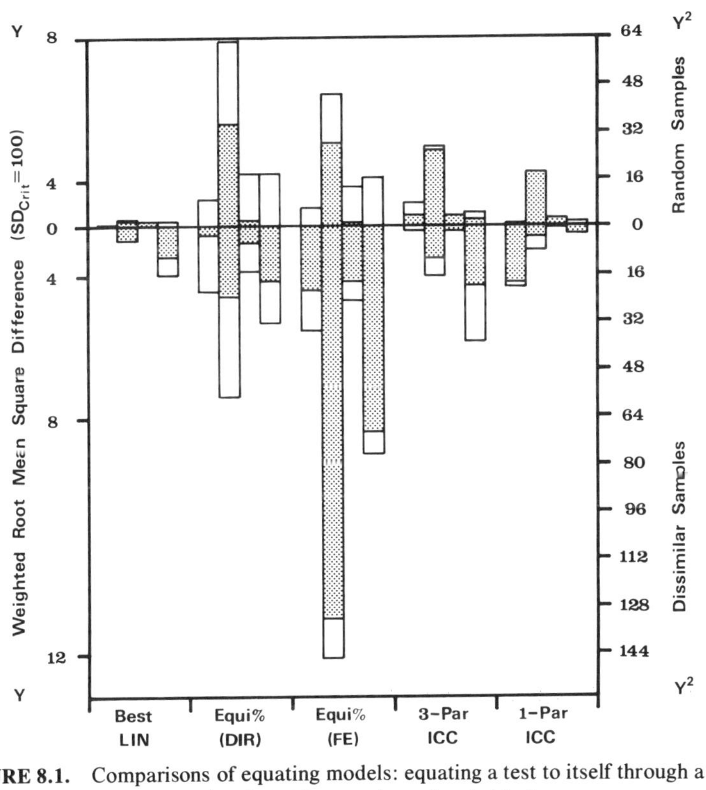

FIGURE 8.1. Comparisons of equating models: equating a test to itself through a medium-difficulty anchor test (outer bar is total error; inner bar is bias).

TABLE 8.7
Mean Square Errors and Squared Biases for Comparing Equating Models Where a Test Is Equated to Itself through a Medium-Difficulty Equating Test (Test Variations 1 and 4)

	New form		Old form				Equating model				
Test var.	Total score[a]	Sample number	Total score[a]	Sample number	Equating score[a]	Best linear model	Best lin	Equi% (Dir)	Equi% (FE)	3-Par ICC	1-Par ICC
Random Samples—Mean Square Error											
4	FT2XX	32102	FT2XX	32202	FE2DM	FE-T-A	.29	9.30	6.25	7.67	.56
1	FT2XX	32110	FT2XX	32210	FE2FM	PAC	2.34	63.20	45.02	26.83	18.40
4	FT4XX	44102	FT4XX	44202	FE4DM	LXY-IF	1.59	17.98	13.40	3.72	2.56
1	FT4XX	44110	FT4XX	44210	FE4FG	PB	1.74	17.89	16.56	4.54	1.06
Dissimilar Samples—Mean Square Error											
4	FT2XX	32802	FT2XX	32902	FE2DM	LML-IF	.13	22.28	35.76	1.72	20.98
1	FT2XX	32810	FT2XX	32910	FE2FM	PB	4.88	58.06	146.65	17.14	8.53
4	FT4XX	44802	FT4XX	44902	FE4DM	T20-IF	.18	15.44	25.20	2.19	.85
1	FT4XX	44810	FT4XX	44910	FE4FG	LV-IR1	16.65	33.18	77.44	39.31	2.66
Random Samples—Squared Bias[b]											
1	FT4XX	44110	FT4XX	44210	FE4FG	PB	(+) .65	(+) .24	(+) .32	(−) 2.38	(−) .39
Dissimilar Samples—Squared Bias[b]											
4	FT2XX	32802	FT2XX	32902	FE2DM	LML-IF	(−) .01	(+) 3.02	(+) 22.00	(−) .06	(−)19.51
1	FT2XX	32810	FT2XX	32910	FE2FM	PB	(−) 4.83	(+)24.08	(+)133.29	(+)11.14	(+) 4.03
4	FT4XX	44802	FT4XX	44902	FE4DM	T20-IF	(+) .18	(+) 5.82	(+) 18.83	(+) 1.94	(−) .15
1	FT4XX	44810	FT4XX	44910	FE4FG	LV-IR1	(+)10.48	(+)18.65	(+) 69.88	(+)20.36	(+) 2.53

[a] Scores were based on the following numbers of items: FT2XX and FT4XX, 85; FE2FM and FE4FG, 40; FE2DM, 34; FE4DM, 36.

[b] The sign in parentheses indicates the direction of the bias; a minus sign indicates that the equating model mean was lower than the criterion mean; a plus sign, that the mean was higher.

not so important. The only method showing noticeably more error for dissimilar samples was the frequency estimation equipercentile method. In this instance, bias accounted for a large proportion of the error. For dissimilar samples, the equatings through an external anchor had noticeably more error for all but the one-parameter IRT model.

Test Variation 5. For random samples the total errors for Test Variation 5 (Table 8.8 and Figure 8.2) were similar to the total errors for Test Variations 1 and 4. That the total error for the best linear model was small for random samples suggests that the curvilinear relation of the anchor test and the total test had little effect on the equating.

The total error for the best linear model was substantially larger when dissimilar samples were used. The IRT models were noticeably superior to the other models in this case, suggesting that these models are relatively robust when the anchor test is different in difficulty from the total test and the samples differ in ability.

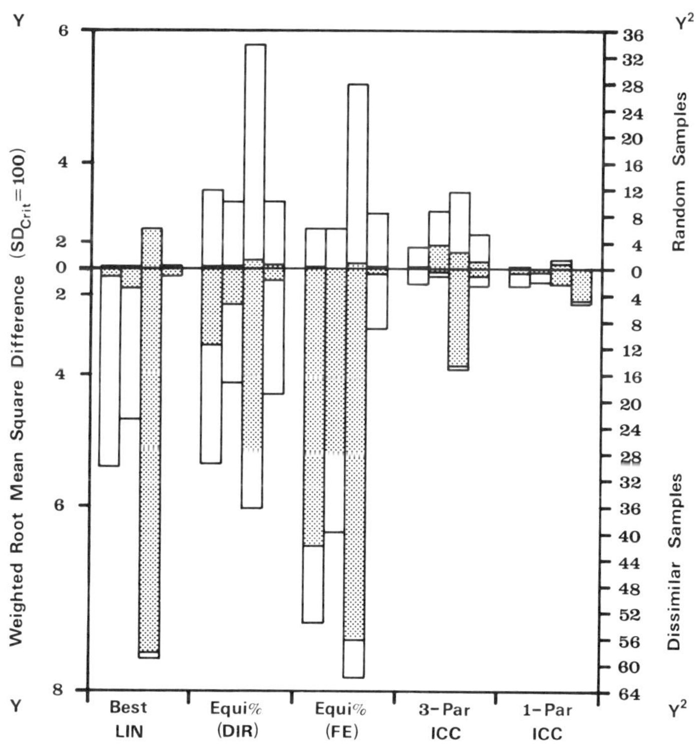

FIGURE 8.2. Comparisons of equating models: equating a test to itself through an easy or difficult anchor test (outer bar is total error; inner bar is bias).

TABLE 8.8
Mean Square Errors and Squared Biases for Comparing Equating Models Where a Test Is Equated to Itself through an Easy or a Difficult Test (Test Variation 5)

New form		Old form				Equating model				
Total score[a]	Sample number	Total score[a]	Sample number	Equating score[a]	Best linear model	Best lin	Equi% (Dir)	Equi% (FE)	3-Par ICC	1-Par ICC
				Random Samples—Mean Square Error						
FT2XX	32101	FT2XX	32201	FE2DE	LXY-IR1	.22	11.90	6.05	3.28	.34
FT2XX	32103	FT2XX	32203	FE2DH	T1	.31	10.11	6.15	8.76	.20
FT4XX	44101	FT4XX	44201	FE4DE	PAC	6.10	33.87	27.98	11.63	1.37
FT4XX	44103	FT4XX	44203	FE4DH	FE-T-F	.45	10.18	8.35	5.20	.11
				Dissimilar Samples—Mean Square Error						
FT2XX	32801	FT2XX	32901	FE2DE	LXY-IF	30.14	29.70	53.73	2.34	2.69
FT2XX	32803	FT2XX	32903	FE2DH	LV-IR1	22.85	17.39	39.94	1.28	2.07
FT4XX	44801	FT4XX	44901	FE4DE	FE-T-A	59.14	36.36	61.94	15.37	2.40
FT4XX	44803	FT4XX	44903	FE4DH	FE-O	1.19	19.10	9.18	2.72	5.34
				Random Samples—Squared Bias[b]						
FT2XX	32101	FT2XX	32201	FE2DE	LXY-IR1	(+) .21	(+) .31	(+) .40	(+) .26	(+) .27
FT2XX	32103	FT2XX	32203	FE2DH	T1	(−) .08	(−) .41	(−) .21	(−) 3.63	(−) .16
FT4XX	44101	FT4XX	44201	FE4DE	PAC	(−) 6.07	(−) 1.36	(−) .88	(−) 2.53	(−) .82
FT4XX	44103	FT4XX	44203	FE4DH	FE-T-F	(−) .45	(−) .64	(−) .37	(−) 1.07	(−) .06
				Dissimilar Samples—Squared Bias[b]						
FT2XX	32801	FT2XX	32901	FE2DE	LXY-IF	(−) 1.27	(+)11.66	(+)41.91	(+) .17	(−) .65
FT2XX	32803	FT2XX	32903	FE2DH	LV-IR1	(+) 3.04	(+) 5.44	(+)28.10	(+) .60	(−) .59
FT4XX	44801	FT4XX	44901	FE4DE	FE-T-A	(+)58.19	(+)27.93	(+)56.22	(+)14.70	(+)2.33
FT4XX	44803	FT4XX	44903	FE4DH	FE-O	(+) .90	(−) 1.78	(+) .87	(−) 1.16	(−)4.88

[a] Scores were based on the following numbers of items: FT2XX and FT4XX, 85; FE2FM and FE4FG, 40; FE2DM, 34; FE4DM, 36.

[b] The sign in parentheses indicates the direction of the bias; a minus sign indicates that the equating model mean was lower than the criterion mean; a plus sign, that the mean was higher.

EQUATING A TEST TO A DIFFERENT TEST

Test Variations 8 and 9. The introduction of some curvilinearity in the relation between the scores of the tests being equated (due to differences in difficulty) resulted, as expected, in a very large total error for the best linear model (see Tables 8.9 through 8.12 and Figures 8.3 to 8.6). For the most part, sample variation seemed to have little effect on total error for the curvilinear models. All the curvilinear models had noticeably less error than the best linear model, but the one-parameter IRT model had substantially more error than the other curvilinear models. For most equatings, the model with the smallest total error was the three-parameter IRT model. For the most part, when greater differences in total test difficulty were introduced (greater curvilinearity), the total error tended to increase substantially for all models, becoming exceedingly large for the best linear model.

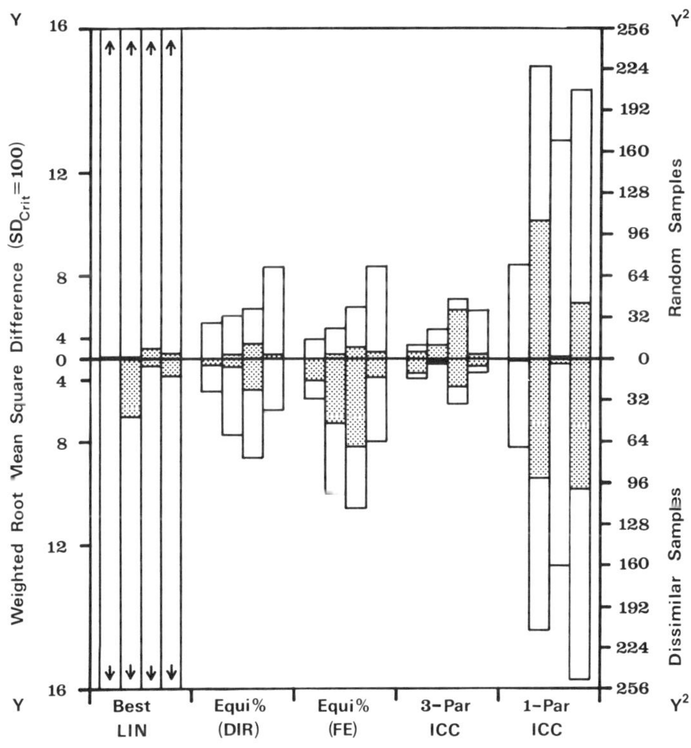

FIGURE 8.3. Comparisons of equating models using IRT equipercentile criterion: equating a test to a different test using easy and medium or medium and difficult tests (outer bar is total error; inner bar is bias).

TABLE 8.9
Mean Square Errors and Squared Biases (IRT Equipercentile Criterion) for Comparing Equating Models Where a Test is Equated to a Different, More Difficult Test through a Medium-Difficulty Equating Test (Test Variation 8)

New form		Old form				Equating model				
Total score[a]	Sample number	Total score[a]	Sample number	Equating score[a]	Best linear model	Best lin	Equi% (Dir)	Equi% (FE)	3-Par IRT	1-Par IRT
				Random Samples—Mean Square Error						
FT2DE	32115	FT2DM	32217	FE2PA	TML-IA[b]	411.28	27.77	15.37	10.82	72.93
FT2DM	32117	FT2DH	32216	FE2PA	LV-IF	547.56	33.18	23.91	22.94	225.90
FT4DE	44115	FT4DM	44217	FE4PA	FE-O	656.90	38.94	40.32	45.97	169.00
FT4DM	44117	FT4DH	44216	FE4PA	LV-IF	469.59	71.23	71.57	37.09	207.94
				Dissimilar Samples—Mean Square Error						
FT2DE	32715	FT2DM	32817	FE2PA	LV-IR1	417.79	25.50	31.36	15.84	68.89
FT2DM	32817	FT2DH	32916	FE2PA	FE-T-A	608 61	59.14	105.06	4.49	211.12
FT4DE	44715	FT4DM	44817	FE4PA	LV-IF	679.12	77.44	116.21	35.40	160.53
FT4DM	44817	FT4DH	44916	FE4PA	LV-IF	482.68	40.07	64.80	11.36	249.32
				Random Samples—Squared Bias[c]						
FT2DE	32115	FT2DM	32217	FE2PA	TML-IA	(+) 1.47	(+) .77	(+) .37	(−) 5.79	(+) .02
FT2DM	32117	FT2DH	32216	FE2PA	LV-IF	(−) 1.85	(−) 3.78	(−) 3.86	(−)10.95	(−)107.26
FT4DE	44115	FT4DM	44217	FE4PA	FE-O	(+) 8.43	(+)12.08	(+) 9.49	(+)38.10	(+) 2.21
FT4DM	44117	FT4DH	44216	FE4PA	LV-IF	(−) 4.38	(−) 3.24	(−) 5.34	(−) 3.85	(−) 43.03
				Dissimilar Samples—Squared Bias[c]						
FT2DE	32715	FT2DM	32817	FE2PA	LV-IR1	(−) .36	(+) 5.36	(+) 17.27	(−)11.69	(+) 2.14
FT2DM	32817	FT2DH	32916	FE2PA	FE-T-A	(+) 45.20	(+) 6.74	(−) 50.63	(−) 2.29	(−) 93.00
FT4DE	44715	FT4DM	44817	FE4PA	LV-IF	(−) 6.11	(+)24.56	(+) 68.33	(+)22.42	(+) 4.41
FT4DM	44817	FT4DH	44916	FE4PA	LV-IF	(−) 13.45	(+) 1.06	(+) 15.01	(−) 6.29	(−)101.20

[a] Scores were based on the following numbers of items: FT2DE, FT2DM, FT2DH, FT4DE, FT4DM, and FT4DH, 54; FE2PA and FE4PA, 20.
[b] Also TML-IF, TML-IR1, LUR-IA, LXY-1A, LV-IA, LML-IA, and PB.
[c] The sign in parentheses indicates the direction of the bias; a minus sign indicates that the equating model mean was lower than the criterion mean; a plus sign, that the mean was higher.

TABLE 8.10

Mean Square Errors and Squared Biases (Direct Equipercentile Criterion) for Comparing Equating Models Where a Test Is Equated to a Different, More Difficult Test through a Medium-Difficulty Equating Test (Test Variation 8)

New form		Old form				Equating model				
Total score[a]	Sample number	Total score[a]	Sample number	Equating score[a]	Best linear model	Best lin	Equi% (Dir)	Equi% (FE)	3-Par IRT	1-Par IRT
				Random Samples—Mean Square Error						
FT2DE	32115	FT2DM	32217	FE2PA	LML-IF	337.82	13.54	7.45	23.04	64.48
FT2DM	32117	FT2DH	32216	FE2PA	LV-IR1	361.00	8.01	9.92	34.57	102.62
FT4DE	44115	FT4DM	44217	FE4PA	FE-O	528.08	21.72	15.76	48.72	129.73
FT4DM	44117	FT4DH	44216	FE4PA	PD	273.90	13.54	11.90	36.36	100.00
				Dissimilar Samples—Mean Square Error						
FT2DE	32715	FT2DM	32817	FE2PA	LV-IR1	344.10	15.92	20.52	29.81	58.06
FT2DM	32817	FT2DH	32915	FE2PA	LV-IR1	431.39	43.30	114.28	19.62	89.11
FT4DE	44715	FT4DM	44817	FE4PA	LV-IF	543.82	45.83	85.01	31.47	118.81
FT4DM	44817	FT4DH	44916	FE4PA	LV-IF	275.23	25.10	44.22	34.57	129.73
				Random Samples—Squared Bias[b]						
FT2DE	32115	FT2DM	32217	FE2PA	LML-IF	(+) .80	(+) .28	(+) .07	(−) 7.68	(−) .05
FT2DM	32117	FT2DH	32216	FE2PA	LV-IR1	(+) 2.53	(+) 1.06	(+) 1.02	(−) .11	(−) 54.97
FT4DE	44115	FT4DM	44217	FE4PA	FE-O	(+) 8.99	(+)12.84	(+) 10.12	(+)39.86	(+) 2.47
FT4DM	44117	FT4DH	44216	FE4PA	PD	(−) .01	(+) .36	(+) .01	(+) .20	(−) 16.88
				Dissimilar Samples—Squared Bias[b]						
FT2DE	32715	FT2DM	32817	FE2PA	LV-IR1	(−) .91	(+) 3.92	(+) 14.72	(−)14.36	(+) 1.26
FT2DM	32817	FT2DH	32916	FE2PA	LV-IR1	(−) 2.59	(+)31.30	(+)102.63	(+) 2.15	(−) 44.89
FT4DE	44715	FT4DM	44817	FE4PA	LV-IF	(−) 5.97	(+)25.78	(+) 71.08	(+)23.59	(+) 4.77
FT4DM	44817	FT4DH	44916	FE4PA	LV-IF	(−) 1.56	(+)11.58	(+) 38.65	(−) .01	(−) 57.31

[a] Scores were based on the following numbers of items: FT2DE, FT2DM, FT2DH, FT4DE, FT4DM, and FT4DH, 54; FE2PA and FE4PA, 20.

[b] The sign in parentheses indicates the direction of the bias; a minus sign indicates that the equating model mean was lower than the criterion mean; a plus sign, that the mean was higher.

TABLE 8.11

Mean Square Errors and Squared Biases (IRT Equipercentile Criterion) for Comparing Equating Models Where a Test Is Equated to a Different, Much More Difficult Test through a Medium-Difficulty Equating Test (Test Variation 9)

New form		Old form				Equating model				
Total score[a]	Sample number	Total score[a]	Sample number	Equating score[a]	Best linear model	Best lin	Equi% (Dir)	Equi% (FE)	3-Par IRT	1-Par IRT
				Random Samples—Mean Square Error						
FT2DE	32115	FT2DH	32216	FE2PA	PAC	1,710.65	84.46	70.22	61.47	412.09
FT4DE	44115	FT4DH	44216	FE4PA	LXY-IR1	1,921.95	131.56	137.36	23.23	547.09
				Dissimilar Samples—Mean Square Error						
FT2DE	32715	FT2DH	32916	FE2PA	PD	1,704.04	88.55	178.49	30.80	327.61
FT4DE	44715	FT4DH	44916	FE4PA	PD	1,927.21	166.15	241.80	30.58	495.95
				Random Samples—Squared Bias[b]						
FT2DE	32115	FT2DH	32216	FE2PA	PAC	(−) 12.94	(−) 12.39	(−) 13.41	(−)43.24	(−)188.18
FT4DE	44115	FT4DH	44216	FE4PA	LXY-IR1	(−) 3.17	(−) 5.28	(−) 7.55	(−) .43	(−)137.14
				Dissimilar Samples—Squared Bias[b]						
FT2DE	32715	FT2DH	32916	FE2PA	PD	(+) .37	(+) 17.46	(+)101.59	(−)24.54	(−) 97.89
FT4DE	44715	FT4DH	44916	FE4PA	PD	(+) .38	(+) 19.12	(+)117.57	(+) .85	(−) 94.08

[a] Scores were based on the following numbers of items: FT2DE, FT2DM, FT4DE, FT4DE, FT4DH, 54; FE2PA and FE4PM, 20.

[b] The sign in parentheses indicates the direction of the bias; a minus sign indicates that the equating model mean was lower than the criterion mean; a plus sign, that the mean was higher.

TABLE 8.12

Mean Square Errors and Squared Biases (Direct Equipercentile Criterion) for Comparing Equating Models Where a Test Is Equated to a Different, Much More Difficult Test through a Medium-Difficulty Equating Test (Test Variation 9)

New form		Old form				Equating model				
Total score[a]	Sample number	Total score[a]	Sample number	Equating score[a]	Best linear model	Best lin	Equi% (Dir)	Equi% (FE)	3-Par IRT	1-Par IRT
Random Samples—Mean Square Error										
FT2DE	32115	FT2DH	32216	FE2PA	PAC	1,253.16	10.76	12.89	101.40	247.43
FT4DE	44115	FT4DH	44216	FE4PA	LXY-IF	1,326.42	16.89	16.08	87.24	302.76
Dissimilar Samples—Mean Square Error										
FT2DE	32715	FT2DH	32916	FE2PA	TML-IA[b]	1,254.58	69.39	191.27	49.98	171.61
FT4DE	44715	FT4DH	44916	FE4PA	T20-IR1	1,334.44	85.75	215.50	53.29	254.40
Random Samples—Squared Bias[c]										
FT2DE	32115	FT2DH	32216	FE2PA	PAC	(−) .15	(−) .09	(−) .19	(−) 11.71	(−)114.61
FT4DE	44115	FT4DH	44216	FE4PA	LXY-IF	(+) 1.42	(+) .43	(+) .04	(+) 13.37	(−) 79.15
Dissimilar Samples—Squared Bias[c]										
FT2DE	32715	FT2DH	32916	FE2PA	TML-IA	(−) 24.93	(+)57.04	(+)184.43	(−) 3.11	(−) 46.30
FT4DE	44715	FT4DH	44916	FE4PA	T20-IR1	(−) 7.53	(+)55.03	(+)195.79	(+) 15.41	(−) 47.08

[a] Scores were based on the following numbers of items: FT2DE, FT2DH, FT4DE, and FT4DM, 54; FE2PA and FE4PM, 20.

[b] Also TML-IF, TML-IR1, LUR-IA, LXY-IA, LV-IA, LML-IA, and PB.

[c] The sign in parentheses indicates the direction of the bias; a minus sign indicates that the equating model mean was lower than the criterion mean; a plus sign, that the mean was higher.

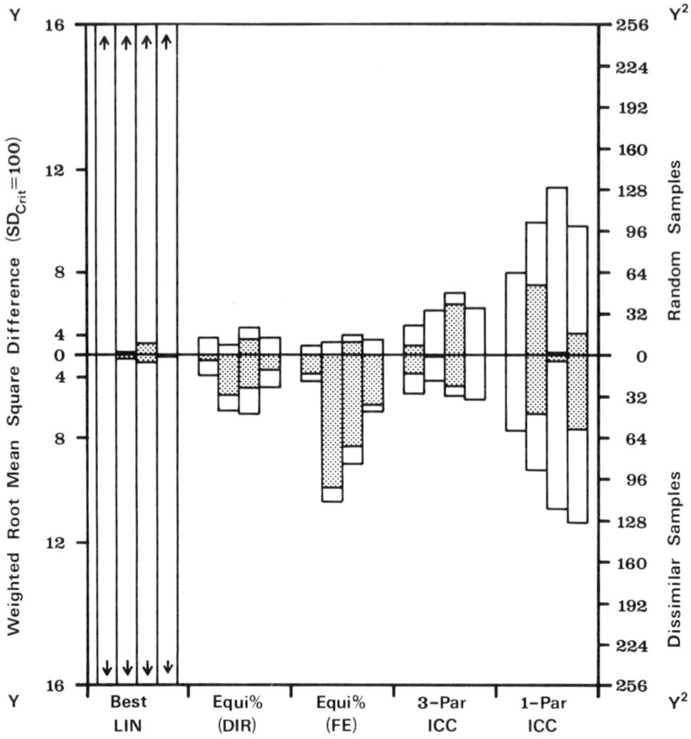

FIGURE 8.4. Comparisons of equating models using direct equipercentile criterion: equating a test to a different test using easy and medium or medium and difficult total tests (outer bar is total error; inner bar is bias).

Possible Criterion Bias

EQUATING A TEST TO A DIFFERENT TEST

The total error and bias for the two criteria for Test Variations 8 and 9 can be compared in Figures 8.3 and 8.4 (Tables 8.9 and 8.10) and in Figures 8.5 and 8.6 (Tables 8.11 and 8.12). It may be noted that for random samples, the experimental equipercentile equatings had less error than the three-parameter IRT model when the equipercentile observed-score criterion was used, suggesting that the equipercentile criterion equatings based on true scores may be biased in favor of the three-parameter IRT model, just as the equipercentile criterion equatings based on observed scores seem to be biased in favor of the equipercentile models. Interestingly enough, the three-parameter IRT model had less total error than the one-parameter IRT model regardless of which criterion was used, although the

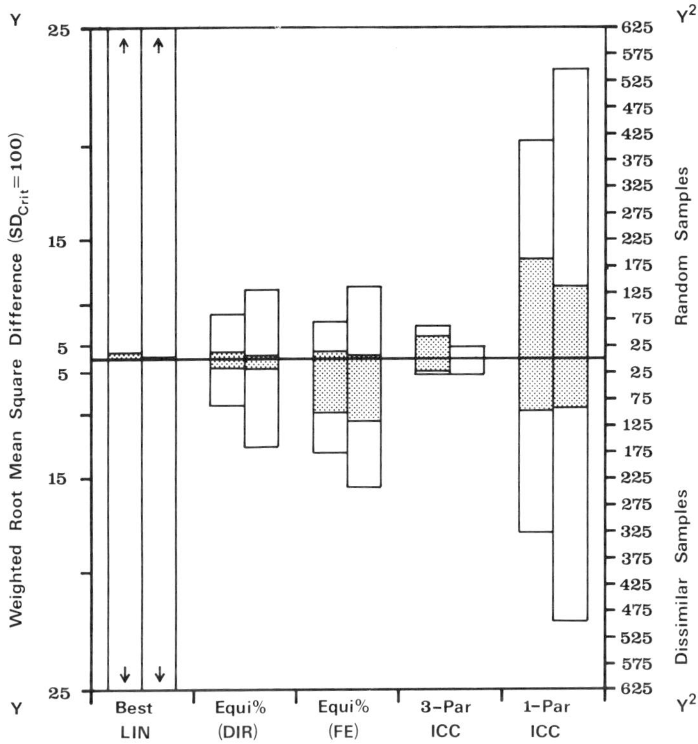

FIGURE 8.5. Comparisons of equating models using IRT equipercentile criterion: equating a test to a different test using easy and difficult total tests (outer bar is total error; inner bar is bias).

difference between the models was less for the direct equipercentile criterion.

The criterion bias was less obvious in the case of dissimilar samples; in fact, the rank ordering of the models in terms of total error was not affected by the choice of criterion. The three-parameter IRT model had the smallest error under both criteria, but the size of the error decreased for the equipercentile and one-parameter IRT models and increased for the three-parameter IRT model when the direct equipercentile criterion was used.

EQUATING A TEST TO ITSELF

The criterion would seem to be well established when a test is equated to itself. It is possible, however, that the criterion in this case is biased in favor of a model that comes closest to fixing all the item parameters at the same values, in this case, the one-parameter IRT model.

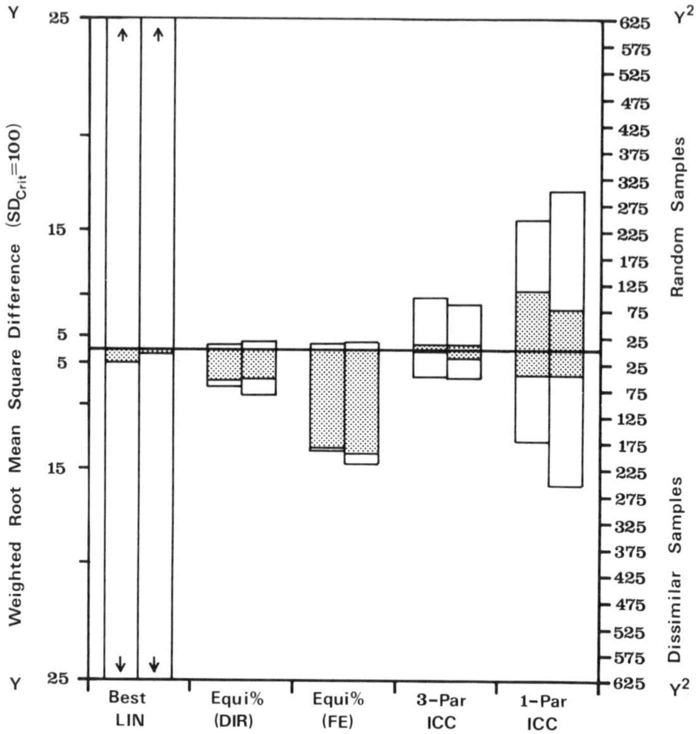

FIGURE 8.6. Comparisons of equating models using direct equipercentile criterion: equating a test to a different test using easy and difficult total tests (outer bar is total error; inner bar is bias).

It is easily seen that if the *a*s, *b*s, and *c*s in the three-parameter logistic test model are fixed at some constant values, the true raw scores for the old and new forms corresponding to a given ability level have to be the same, since the probability of getting a particular item correct is then a function of only the item parameters and ability.

In the one-parameter logistic test model, *c* is fixed at 0 and *a* at 1 for all items. Since only the *b*s are estimated, it might be expected that the one-parameter IRT model is more likely than the three-parameter IRT model to yield the appropriate conversions. Is it of any consequence, then, if it is found that the one-parameter IRT model seems to be superior to the three-parameter IRT model in equating a test to itself? What is really desired is to know which of the models works best when a test is equated to a different test, particularly when a test is equated to a parallel test. It would seem, then, that there may be a natural bias in favor of the one-parameter IRT model when a test is equated to itself, but this probably does not affect the other models.

CONCLUSIONS

Several conclusions can be drawn from the study. They should be considered tentative because of possible criterion bias, which has been only partially controlled; because it was not possible to study all test variations in the full design, in particular, the variations in which parallel, but not identical, tests were equated; and because the results showed occasional inconsistencies that have not yet been explained:

1. When a test is equated to itself (or, to generalize, to a test like itself) through a parallel anchor test, a linear model yields very good results regardless of the type of samples used. Whether any particular linear model consistently gives satisfactory results, however, requires further study. The best linear model for a particular test variation and type of sample was used here.
2. Curvilinear models give results nearly comparable to those of the best linear model when a test is equated to a test like itself through an internal anchor. When an external anchor is used, of the curvilinear models, the IRT models (particularly the one-parameter model) give relatively better results. The criterion may be biased, however, in favor of the one-parameter model.
3. The kinds of samples have a relatively small and unsystematic effect on the quality of the equating results if the anchor test is similar in content and in difficulty to the total tests. The one exception is the frequency estimation procedure, which seems not to perform well when an external anchor and dissimilar samples are used.
4. The equatings involving an internal anchor have less total error than comparable equatings with an external anchor. Whether or not this inconsistency would obtain when a test is equated to a test of different difficulty needs to be studied. The possibility that the criterion is biased in favor of an internal anchor also needs investigation. It may, however, simply be because the external anchor tests were not quite so similar to the total tests in content and statistical characteristics as were the internal anchor tests.
5. When a test is equated to a test like itself through an easy or difficult anchor test with random samples, all the models have a small mean square error. When dissimilar samples are used, however, the IRT models give clearly superior results.
6. When total tests differ considerably in difficulty, linear models yield unsatisfactory results in that the mean square error becomes very large; but they tend to yield better estimates of the mean than the curvilinear models. The one-parameter IRT model and the frequency estimation method also give unacceptable results in many

instances. This result is consistent with the Slinde and Linn (1978) findings that the Rasch model yielded poor results for vertical equating.

7. The three-parameter IRT model is the best equating model when total tests of unequal difficulty are equated through a medium-difficulty anchor test with dissimilar samples. It would appear that the three-parameter model is the equating model most likely to yield acceptable results under unusual or extreme conditions.

This study, though comprehensive in its coverage, is limited in that SAT-V items made up the item pool. These items are known to be relatively homogeneous and somewhat difficult for the current test-taking population. Similar studies are needed in situations where the content of the anchor test is allowed to depart by varying degrees from the content of the total tests and where the test-taking population is from a different age group.

ACKNOWLEDGMENTS

The authors are listed alphabetically. This study was funded jointly by the College Entrance Examination Board and Educational Testing Service. The authors gratefully acknowledge the advice and assistance provided by William H. Angoff, Rebecca Bullock, Samuel A. Livingston, Frederic M. Lord, Ruth C. Mroczka, Joanne S. Peng, William B. Schrader, Ledyard, R Tucker, Marilyn S. Wingersky, and R L Wood.

REFERENCES

Angoff, W. H. Test reliability and effective test length. *Psychometrika,* 1953, *18,* 1–14.

Angoff, W. H. Scales, norms, and equivalent scores. In R. L. Thorndike (Ed.), *Educational measurement* (2nd ed.). Washington, D.C.: American Council on Education, 1971.

Angoff, W. H., & Dyer, H. S. The admissions testing program. In W. H. Angoff (Ed.), *The College Board admissions testing program: A technical report on research and development activities relating to the Scholastic Aptitude Test and Achievement Tests.* New York: College Entrance Examination Board, 1971.

Birch, M. W. A note on the maximum likelihood estimation of a linear structural relationship. *Journal of the American Statistical Association,* 1964, *54,* 1175–1178.

Feldt, L. S. Estimation of the reliability of a test divided into two parts of unequal length. *Psychometrika,* 1975, *40,* 557–561.

Levine, R. S. Equating the score scales of alternate forms administered to samples of different ability (ETS RB-55-23). Princeton, New Jersey: Educational Testing Service, 1955.

Lord, F. M. A survey of equating methods based on item characteristic curve theory (ETS RB-75-13). Princeton, New Jersey: Educational Testing Service, 1975.

Lord, F. M. Personal communication, 24 February, 1976.

Petersen, N. S., Marco, G. L., & Stewart, E. E. *A test of the adequacy of linear score equating models*. In P. W. Holland & D. B. Rubin (Eds.), *Test equating*. New York: Academic Press, 1982.

Potthoff, R. F. Equating of grades or scores on the basis of a common battery of measurements. In P. R. Krishnaiah (Ed.)., *Multivariate analysis*. New York: Academic Press, 1966.

Rentz, R. R., & Bashaw, W. L. *Equating reading tests with the Rasch model* (Vols. 1 & 2). Athens: University of Georgia, College of Education, Educational Research Laboratory, September 1975.

Slinde, J. A., & Linn, R. L. Vertically equated tests: Fact or phantom? *Journal of Educational Measurement*, 1977, *14*, 23–32.

Slinde, J. A., & Linn, R. L. An exploration of the adequacy of the Rasch model for the problem of vertical equating. *Journal of Educational Measurement*, 1978, *15*, 23–35.

Tucker, L. R. Personal communication, 7 August 1974.

Wood, R L, & Lord, F. M. A user's guide to LOGIST (ETS RM-76-4). Princeton, New Jersey: Educational Testing Service, 1976.

Wood, R L, Wingersky, M. S., & Lord, F. M. *LOGIST: A computer program for estimating examinee ability and item characteristic curve parameters* (ETS RM-76-6). Princeton, New Jersey: Educational Testing Service, 1976.

9

Timed Testing: An Approach Using Item Response Theory

DAVID THISSEN

Response latency, which currently provides the basic data for research in many areas of cognitive psychology, is presumed to be related to important aspects of cognitive processing. Nevertheless, latency has seen little use as a response variable in the context of the routine measurement of individual cognitive abilities with psychological tests. Practical obstacles to both the administration and scoring of timed tests have contributed to this delay in the use of latency data in testing, but those obstacles are being removed. With the increasing use of electronic means of test-item presentation and recording of responses, latency data have become as easy to obtain and to store as the item responses themselves. This chapter describes the use of an item response model for item analysis and test scoring in the context of timed testing.

The model described here is a revision of a model proposed by Furneaux (1961). Furneaux described the outline of the model and provided some illustration of the performance of parts of the system with a small set of test data, using parameters estimated by various heuristic procedures. Here, using a slightly different model and an integrated parameter estimation scheme, the outcome of timed testing with several different kinds of test items is described.

Although alternative models for timed testing have been proposed,

NEW HORIZONS IN TESTING
Latent Trait Test Theory and
Computerized Adaptive Testing

ISBN 0-12-742780-5

none is completely satisfactory as a general model in the tradition of item response theory. White (1973, 1979) proposed a complete system for the use of timed test data to estimate three individual parameters: ability, speed, and persistence. In that system, however, latency is not treated as a response variable; rather, it is used as a fixed quantity in the model for the prediction of the item response. Thus White's system is not an item response model for the data that make up the item response in timed testing: the response and its latency.

Tatsuoka and Tatsuoka (1979) have provided a model for incorporating response-time data in scoring achievement tests; that model treats latency as a random variable depending on person and item parameters. However, it does not consider item responses simultaneously, nor does the fairly similar latency model proposed by Scheiblechner (1979).

A complete system for scoring timed tests requires an item response model for the data of timed testing: the responses *and* the associated latencies. The purpose of this chapter is to describe such a model.

THE MODEL

The item response model presented here was developed primarily on the basis of exploratory data analysis of timed test data by Furneaux (1961) and Thissen (1977). It is more straightforward, however, to describe the model first without reference to data and then proceed to examine the goodness of fit; that approach will be taken here.

Item response theory for categorical data is well established. When the items are scored dichotomously, the logistic model described by Birnbaum (1968) has been useful in many situations, so that model will be used for the item response data obtained in the timed testing situation. If the item responses r_{ij} for person i responding to item j are

$$r_{ij} = \begin{cases} 1 & \text{if the response of person } i \text{ to item } j \text{ is correct} \\ 0 & \text{otherwise,} \end{cases}$$

then the logistic model is

$$P(r_{ij} = 1) = 1/[1 + \exp(-z_{ij})], \qquad (9.1)$$

where

$$z_{ij} = a_j\theta_i + c_j \qquad (9.2)$$

and

$$P(r_{ij} = 0) = 1 - P(r_{ij} = 1). \qquad (9.3)$$

θ_i could be called the "effective ability" of person i (after White, 1973) to distinguish this ability estimate, obtained under circumstances allowing the examinee any amount of time, from more conventional ability estimates obtained with speeded tests. The parameter a_j is the discrimination parameter, or "slope," of item j; it reflects the degree of relationship between the item and the trait being measured. The parameter c_j is the easiness of item j.

In item response theory, the response time of person i on item j, t_{ij}, must also be a function of parameters describing characteristics of person i and item j. Those parameters could be (in part) the same as the parameters that are used to describe the item responses; that would mean (a) that the variations in item responses and response times were attributable to the same sources, and (b) that the responses and latencies could both be used in the measurement of the ability of each examinee and the easiness of each item. Latency should clearly be a decreasing function of z_{ij}; that is, increases in z_{ij} (through increased person ability or item easiness) should be related to decreases in response latency. The form of the random error involved in the measurement of the response latency must also be specified; analyses by Furneaux (1961) and Thissen (1977) suggest that response times are frequently approximately log-normally distributed, so a linear model for the logarithm of latency could be assumed to include Gaussian error. Thus, the following model results:

$$\log(t_{ij}) = v - bz_{ij} + \varepsilon_{ij}, \qquad \varepsilon_{ij} \sim N(0, \sigma^2), \tag{9.4}$$

in which v is the overall mean log response time and b is a regression parameter reflecting the relationship of effective ability and easiness (in z_{ij}) with latency. Of course, the parameters of z_{ij} in Eq. (9.4) are as in Eq. (9.2).

The model described by Eq. (9.4) is unlikely to provide a very good fit to observed data. There are likely to be attributes of both the examinees and the items that contribute consistently to latency but that are unrelated to the trait the items are intended to measure. For verbal items, the length of the item (and associated reading time) or the number of attempts required to obtain the needed semantic interpretations may contribute to response latency; but those factors could be unrelated to the easiness of that item (and the probability of a correct response). Some examinees may press the response keys more slowly than others in a pattern that is consistent but unrelated to their effective ability. The addition of person and item slowness parameters, s_i and u_j, to Eq. (9.4) results in the model

$$\log(t_{ij}) - v + s_i + u_j - bz_{ij} + \varepsilon_{ij}, \qquad \varepsilon_{ij} \sim N(0, \sigma^2). \tag{9.5}$$

The model formed by the combination of Eq. (9.1) for the item response data and Eq. (9.5) for the response times will be used below in the analysis of timed test data. Two comments are in order about the functions included in the model. First, after the parameters have been estimated, the appropriateness of the model may be evaluated to some extent by an examination of the goodness of fit of its various features. For instance, the empirical proportions of correct responses at various levels of ability may be compared to the logistic predictions, and the normality of the residuals of $\log(t)$ from the linear model may be examined. To the extent that the model follows the observed data, the parameters should provide a useful summary.

Second, this model is not being proposed as a process model for the psychological description of item response behavior. Although such process models have been developed for choice behavior (see Audley, 1960, 1973; Laming, 1968; Luce, 1960) and are currently being developed for specific types of complex cognitive abilities (see Sternberg, 1977; Whitely, 1981), those models tend to be too complex for current application in practical testing and measurement. The model proposed here follows an earlier tradition of test theory in an attempt to describe responses to a variety of possible kinds of test items. The parameters of the timed testing model may be useful in some psychological process research, as summaries of attributes of test items and examinees that may be compared to theoretical predictions. Beyond that, the current model is meant to be a general model for the measurement of examinees and the calibration of test items; such a model may be superseded by more complete psychological models for cognitive processes when those become available for cognitive tasks in the domain of ability measurement.

ESTIMATION

The parameters of the timed testing model form the set $\xi = \{\theta, s, a, c, u, v, b, \sigma^2\}$; there are $2N + 3n$ independent parameters (for N persons and n items), since the location and scale of θ must be set arbitrarily, as well the location of the slowness parameters. With the usual assumption of local independence extended to include the assumption that the error component of latency is independent of the response to the same item, conditional on the parameters, the likelihood of the entire set of data, given the parameters, may be written

$$L(R,T|\xi) = \prod_i \prod_j P(r_{ij} = 1)^{r_{ij}} P(r_{ij} = 0)^{(1-r_{ij})} H_{ij}, \tag{9.6}$$

in which $R = [r_{ij}]$, $T = [t_{ij}]$, $P(r_{ij} = 1)$ is given in Eq. (9.1), and

$$H_{ij} = \Phi(d_{ij}), \tag{9.7}$$

where $\Phi(d)$ is the standard normal density and

$$d_{ij} = [\log(t_{ij}) - (v + s_i + u_j - bz_{ij})]/\sigma. \tag{9.8}$$

The log likelihood is

$$l = \sum_i \sum_j [r_{ij} \log P(r_{ij} = 1) + (1 - r_{ij}) \log P(r_{ij} = 0) + \log H_{ij}]. \tag{9.9}$$

The stationary equations to be solved for l to be maximized are ($P_{ij} = P(r_{ij} = 1)$ and $Q_{ij} = 1 - P_{ij}$)

$$\frac{\partial l}{\partial \theta_i} = \sum_j \left[\left(\frac{r_{ij}}{P_{ij}}\right)\left(\frac{\partial P_{ij}}{\partial \theta_i}\right) - \left(\frac{1 - r_{ij}}{Q_{ij}}\right)\left(\frac{\partial P_{ij}}{\partial \theta_i}\right) + \left(\frac{1}{H_{ij}}\right)\left(\frac{\partial H_{ij}}{\partial \theta_i}\right)\right], \tag{9.10}$$

$$\frac{\partial l}{\partial s_i} = \sum_j \left(\frac{\partial H_{ij}}{\partial s_i}\right)\left(\frac{1}{H_{ij}}\right), \tag{9.11}$$

$$\frac{\partial l}{\partial a_j} = \sum_i \left[\left(\frac{r_{ij}}{P_{ij}}\right)\left(\frac{\partial P_{ij}}{\partial a_j}\right) - \left(\frac{1 - r_{ij}}{Q_{ij}}\right)\left(\frac{\partial P_{ij}}{\partial a_j}\right) + \left(\frac{1}{H_{ij}}\right)\left(\frac{\partial H_{ij}}{\partial a_j}\right)\right], \tag{9.12}$$

$$\frac{\partial l}{\partial c_j} = \sum_i \left[\left(\frac{r_{ij}}{P_{ij}}\right)\left(\frac{\partial P_{ij}}{\partial c_j}\right) - \left(\frac{1 - r_{ij}}{Q_{ij}}\right)\left(\frac{\partial P_{ij}}{\partial c_j}\right) + \left(\frac{1}{H_{ij}}\right)\left(\frac{\partial H_{ij}}{\partial c_j}\right)\right], \tag{9.13}$$

$$\frac{\partial l}{\partial u_j} = \sum_i \left(\frac{\partial H_{ij}}{\partial u_j}\right)\left(\frac{1}{H_{ij}}\right), \tag{9.14}$$

$$\frac{\partial l}{\partial b} = \sum_i \sum_j \left(\frac{\partial H_{ij}}{\partial b}\right)\left(\frac{1}{H_{ij}}\right), \tag{9.15}$$

$$\frac{\partial l}{\partial \sigma} = \sum_i \sum_j \left(\frac{\partial H_{ij}}{\partial \sigma}\right)\left(\frac{1}{H_{ij}}\right). \tag{9.16}$$

The required partial derivatives of the terms in the model are

$$\frac{\partial P_{ij}}{\partial \theta_i} = a_j[P_{ij}Q_{ij}], \tag{9.17}$$

$$\frac{\partial H_{ij}}{\partial \theta_i} = \frac{-ba_jd_{ij}H_{ij}}{\sigma}, \tag{9.18}$$

$$\frac{\partial H_{ij}}{\partial s_i} = \frac{d_{ij}H_{ij}}{\sigma}, \tag{9.19}$$

$$\frac{\partial P_{ij}}{\partial a_j} = \theta_i[P_{ij}Q_{ij}], \tag{9.20}$$

$$\frac{\partial P_{ij}}{\partial c_j} = P_{ij}Q_{ij}, \tag{9.21}$$

$$\frac{\partial H_{ij}}{\partial a_j} = \frac{-b\theta_id_{ij}H_{ij}}{\sigma}, \tag{9.22}$$

$$\frac{\partial H_{ij}}{\partial c_j} = \frac{-bd_{ij}H_{ij}}{\sigma}, \tag{9.23}$$

$$\frac{\partial H_{ij}}{\partial u_j} = \frac{d_{ij}H_{ij}}{\sigma}, \tag{9.24}$$

$$\frac{\partial H_{ij}}{\partial b} = \frac{-z_{ij}d_{ij}H_{ij}}{\sigma}, \tag{9.25}$$

$$\frac{\partial H_{ij}}{\partial \sigma} = \frac{(d_{ij}^2 - 1)H_{ij}}{\sigma}. \tag{9.26}$$

A two-stage procedure for locating the maximum of $L(R,T|\xi)$ is described in detail by Thissen (1977). The first stage of the procedure makes use of the conjugate gradient method (Fletcher & Reeves, 1964) to approach the joint maximum of the log likelihood in the $(2N + 3n)$-dimensional parameter space. This procedure requires about twice as many iterations as there are parameters to be estimated, but it proceeds quickly. As the maximum is approached, the conjugate gradient algorithm may not locate an appropriate direction in the parameter space. Then, a second stage of the estimation procedure is entered. It is a cyclic procedure in which each person's parameters are estimated individually using the current values of the item and test parameters; next, each item's parameters are similarly estimated; the test parameters are then revised; and, finally, the procedure is repeated. In the second stage, each person's and item's parameters are estimated using a conditioned Newton–Raphson algorithm programmed by Haberman (1974). By the time the

Newton–Raphson corrections are on the order of 10^{-3} for each parameter, there is generally no appreciable change between cycles, and the procedure is stopped.

A useful feature of the second-stage estimation procedure is that since the Newton iterations require the matrices of second partial derivatives for the person parameters (within each person) and the item parameters (within each item), those matrices may be treated as information matrices and inverted to give estimated standard errors for the parameters. The resulting estimates for the standard errors unfortunately underestimate the true values, as their computation ignores the estimation of all of the other person (or item) parameters. Nevertheless, they give some idea of the error of estimation of the parameters.

APPLICATIONS: I. SPATIAL TEST DATA

Method

This model has been applied to a set of data gathered using the Bock–Fitzgerald test of spatial visualizing ability.[1] The responses were those of 60 suburban London high school students. The test consisted of 33 two-dimensional rotation problems, using abstract geometrical figures as stimuli. The students were tested individually. The test items were presented one at a time using a slide projector; response times were obtained with a multichannel event recorder connected to the response keys.

Starting values were used both as a first trial for the iterative estimation procedure and to trim outliers from the data. The starting values for the slowness parameters were the (centered) logs of the median response times for each person and item. Standardized raw scores were used as initial values for θ. The transformed item point-biserial correlation was used to obtain an initial value for a_j, and c_j was estimated from the logit transformed proportion responding correctly to each item. The parameter b was estimated using the regression of the log response times on the starting value logits.

Before iterative estimation of the parameters, extreme outliers were trimmed from the data. This was done using the starting values for the parameters of the timed testing model; observations for which observed latency differed from its estimated value by more than three standard deviations were trimmed. These responses were treated as missing data in subsequent computations. Most responses trimmed were very long

[1] R. Darrell Bock collected these spatial test data, with generous help from R. D. Audley.

latencies—more than 1 min in many cases; .33% of the data were set aside by this trimming procedure.

Results

The parameters of the model were estimated by joint maximum likelihood, as described above. The goodness of fit of the model was examined in a variety of ways. Figure 9.1 shows the fit of the standardized deviations of log latency from the model to the normal curve; the deviations have been divided into fractiles and the histobars hung from the theoretical curve so that discrepancies between the model and data are shown at the bottom, as deviations around a straight line. A χ^2 statistic for the goodness of fit of the proportions in these 10 fractiles to the assumed normal distribution was 5.8 (with seven degrees of freedom), which is nonsignificant.

Similar goodness-of-fit χ^2s were available for the proportion of correct responses as a function of ability. The logistic item response model gives a probability of a correct response at any value of θ, so the examinees may be divided into relatively homogeneous ability groups and the proportion of correct responses in each of the ability groups may be compared to the proportion expected. This procedure yields a contingency table for each item and an approximate χ^2 test of the goodness of fit of the model. In this analysis, the examinees were divided into six equal groups and the resulting (approximately) four degree-of-freedom χ^2 tests were computed. For the entire test, the total of these χ^2s was 157.8 (with 128 degrees of freedom), which is nonsignificant. Although these tests cannot establish that

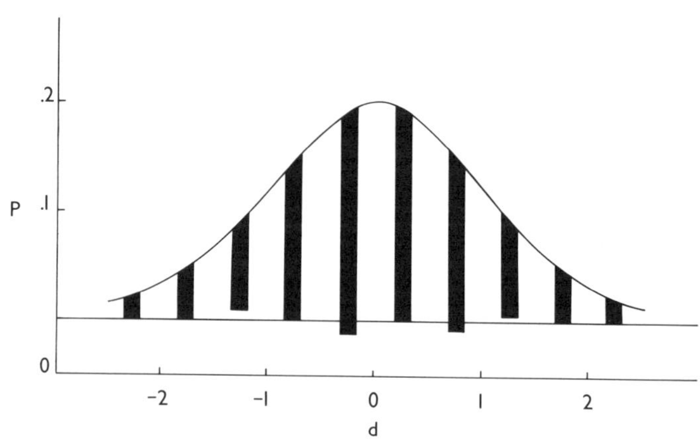

FIGURE 9.1. Hanging histogram of the standardized deviations of log latency from the model.

the model is "correct," they do establish that the error variation is approximately as predicted by the model.

Some of the results of this application are illustrated in Figure 9.2, which displays slowness plotted against effective ability. The first outstanding feature of the plot is that the variables are highly correlated, $r = .82$. As slowness increases, effective ability increases. An interesting point comes in the analysis for gender differences. A gender difference is usually found on speeded spatial tests: males have higher mean ability. In this case, neither slowness nor effective ability exhibited a significant gender difference. However, the bivariate test of the gender difference was significant (multivariate $F = 3.58$, $p < .05$). Discriminant analysis revealed that the dimension of the difference was a contrast between effective ability and slowness; in other words, correct responses per unit time. This analysis suggests that what have been called individual differences in ability (using speeded tests) are actually differences in rate of processing. This is presumably why all "good" spatial tests have been speeded.

APPLICATIONS: II. THREE TESTS

Subjects and Tests

These data were the responses of 78 University of Kansas undergraduate students to three tests of cognitive ability. The students were se-

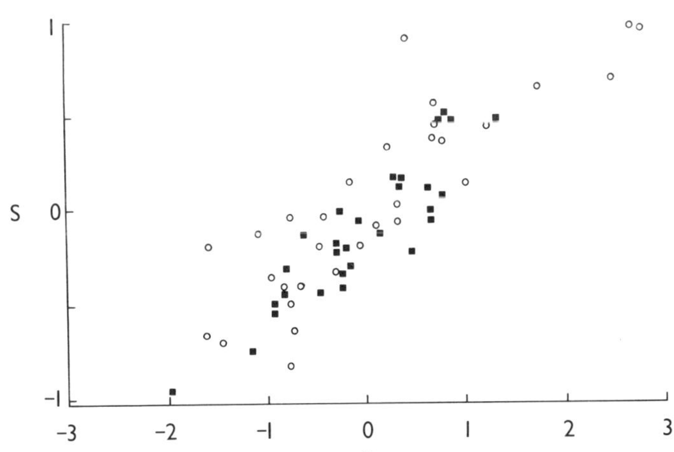

FIGURE 9.2. Effective ability and slowness plotted for the males and females in the London sample (open circles for females, closed squares for males).

lected to participate in a study of individual differences in laterality; therefore left-handed individuals were substantially overrepresented in the sample. The students partially satisfied a research participation requirement in an Introductory Psychology course by their participation in the study. The students were largely college freshman: 36 were male and 42 were female.

The Verbal Analogies test consisted of a 27-item subset of the verbal analogies described by Whitely (1977) and Tinsley (1971). These analogies are quite difficult for college students; the raw scores ranged from 4 to 24, and the average proportion correct was .57. No item was answered correctly by all students.

The Progressive Matrices test consisted of Sets B, C, and D of the Raven (1956) Colored (Set B) and Standard (Sets C and D) Progressive Matrices. There were a total of 36 items, of which two (B-1 and B-3) were answered correctly by all but one of the examinees; those two items were omitted from further analyses. The Progressive Matrices were easier for this group than the Verbal Analogies; the average proportion correct was .72. The raw scores ranged from 12 to 34.

The Clocks test consisted of 19 items drawn from a set of color slides made to resemble items on the "clocks" test of spatial visualizing ability from the Guilford–Zimmerman (1953) Aptitude Survey, Part VI. The items give pictorial instructions about the rotation in three dimensions of a large, old-fashioned alarm clock, and the examinee is required to select from a set of alternatives a view of the clock as it would appear after the rotations. Two of the items were too easy for this group and were omitted from the analyses; therefore, Clocks was a 17-item test on which the raw scores ranged from 5 to 16.

The test items were presented using a slide projector onto a rear-projection screen immediately in front of the examinee. All test items were in multiple-choice format, with from four to eight alternatives; the examinees responded by pressing a numbered response key on a calculator-like keyboard. The presentation of the slide triggered a light sensor, which started the timer; the response of the examinee stopped the timer and initiated a display of the response and latency for the examiner.

Results

Before the iterative estimation of the parameters of the latent trait model was begun, the data were trimmed of extreme outliers, following the procedure described earlier. Less than .5% of the observations were trimmed in this way.

VERBAL ANALOGIES

Estimation proceeded just as for the spatial test data. The goodness of fit of the data to the timed testing model was measured in two ways. The probability of a correct response was predicted well from the model; only one of the 27 item χ^2s (described above) was significant at the $p = .05$ level. The total of the likelihood ratio χ^2 values was 131.2 with 108 degrees of freedom ($.10 > p > .05$), which was nonsignificant and indicates that the probabilities of correct response were predicted well by the model.

The assumption of normality of the error variation of $\log(t)$ was examined by dividing the residuals into fractiles and comparing their distribution to Gaussian expectation. In this case, $\chi^2 = 3.7$ with seven degrees of freedom, which was nonsignificant. This result masks a small violation of the assumptions of the model: correct responses were slightly but significantly faster than incorrect responses; the difference was .22 standard deviation units in log time.

The goodness of fit of the model for the latencies may also be examined using data from individual examinees and items. Figure 9.3 shows a scatterplot of the observed log response times t^* and the log response time estimated from the model $\hat{t}^*$ for an illustrative examinee across the 27 items. The correlation was .59. This examinee was chosen for graphic display because he/she obtained near average estimates on θ and s on all three tests; the examinee is "typical" in the sense of "average."

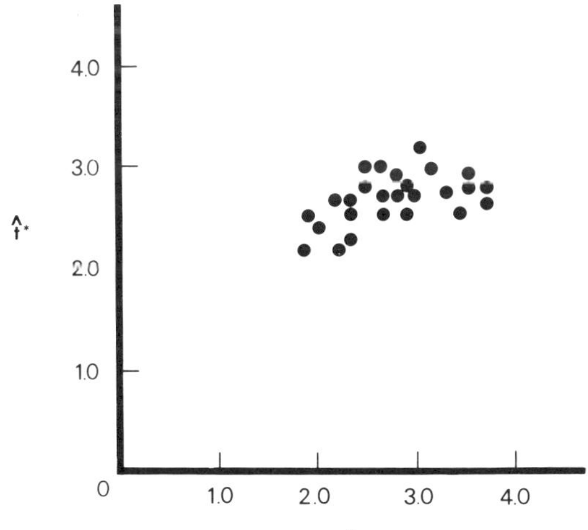

FIGURE 9.3. Scatterplot of the observed log response times t^* and the fitted log response times $\hat{t}^*$ for one examinee for the analogies.

Figure 9.4 is a similar scatterplot of the log latencies t^* and the predicted latencies $\hat{t}^*$ for all 78 examinees on one of the analogy items, "cent : dime : : ______ : dollar." This analogy has near-average values for c_j and u_j but a high estimated slope. The correlation between the observed and fitted log latencies for this item was .65. Some 35 to 50% of the variance in log latency within an individual or an item was predicted by the model. For these data, the logit regression parameter b had a value of .20. The quantitative meaning of the regression of latency on effective ability varies along the time scale because of the log transformation. For an average item, a person of average slowness and effective ability ($\theta = 0$) should respond in about 12.5 sec, whereas a less able individual ($\theta = -1$) is expected to respond in about 15.2 sec.

Since some of the item parameters in the timed testing model are the same as those used in conventional item response analysis, the timed

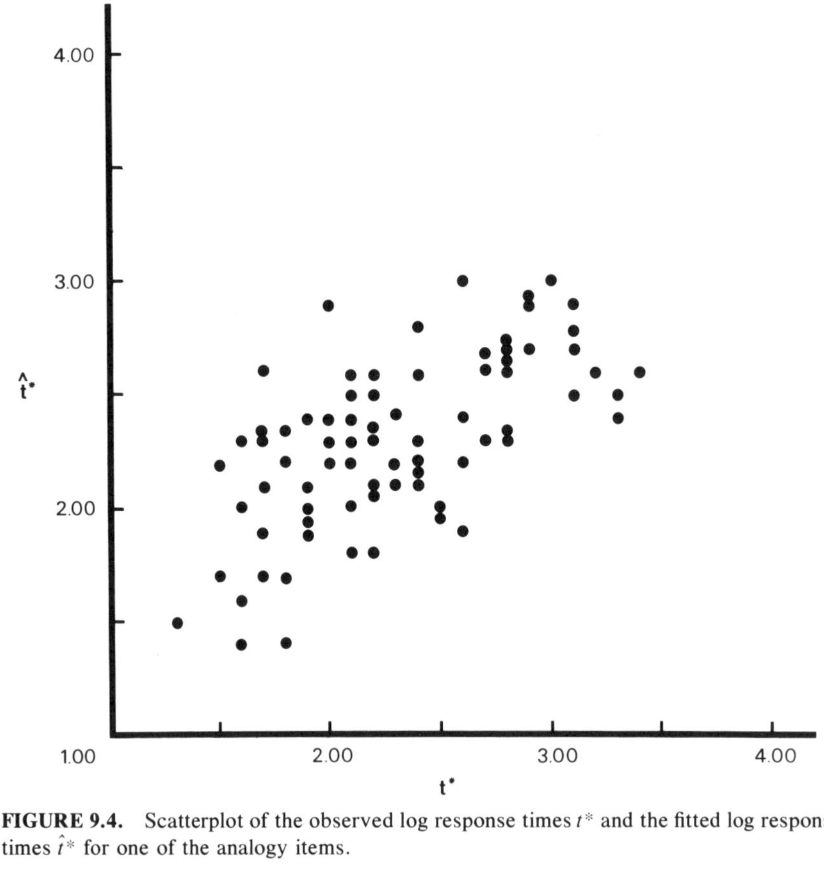

FIGURE 9.4. Scatterplot of the observed log response times t^* and the fitted log response times $\hat{t}^*$ for one of the analogy items.

testing model may be evaluated as an item calibration scheme by comparing the item parameters estimated using the timed testing model with item parameters estimated using large-sample item response techniques. Item easiness was estimated by Tinsley (1971) for the analogy items used here, employing the Rasch (1960) one-parameter logistic model; the data were from 641 examinees. The resulting item easiness parameters are plotted against c_j/a_j in Figure 9.5. The correlation between the two sets of easiness parameters was .80 (or .70 if the outlier in the upper right-hand corner is removed).

Slopes for each item, estimated by the unusual procedure of regressing the item responses on ability estimates derived from the one-parameter model, were available from the same large-sample item calibration study. They are plotted against the a_j estimated in the present study in Figure 9.6. The correlation was only .50, and only two-thirds of the points are within two of their own standard errors of the regression line. This may be because the large-sample slopes were estimated using ability derived from a one-parameter logistic model; and the standard errors of slopes esti-

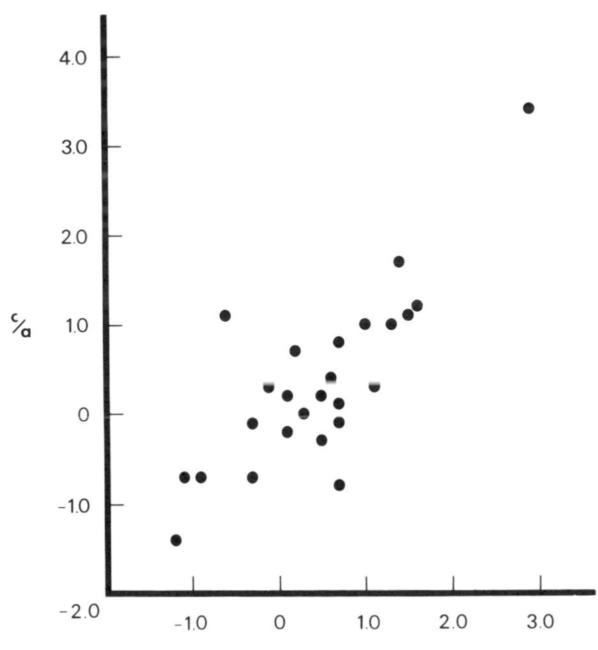

FIGURE 9.5. Timed testing easiness (actually, c/a) plotted against large sample easiness parameters for the analogies.

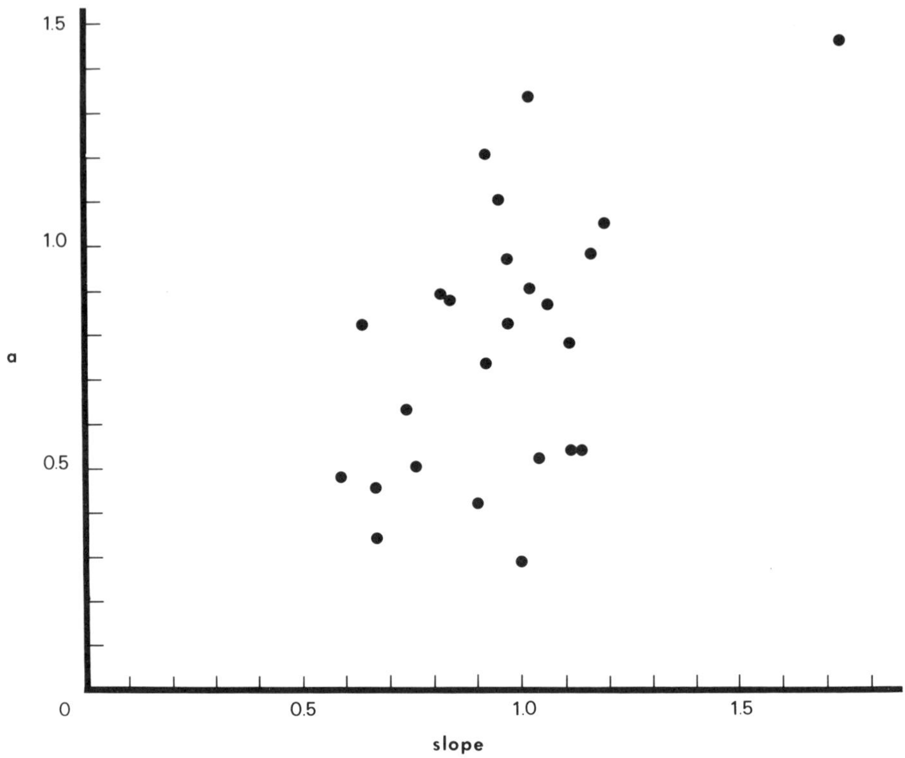

FIGURE 9.6. Timed testing discrimination parameters (a) plotted against large sample slope parameters for the analogies.

mated with only 78 examinees, even with the timed testing model, are liberally estimated to be about .12 for this test.

The item parameters estimated in this analysis showed some validity when they were related to individual item characteristics. Whitely (1977) placed these analogy items in eight categories defined by the types of relations among the elements. The average item parameters for the eight categories of analogies are given in Table 9.1. The easiness and item slowness parameters did not vary significantly among the categories, but the discrimination parameters were significantly related to analogy type ($F(7,18) = 3.16$, $p < .05$). Quantitative analogies (e.g., cent : dime : : ______ : dollar), word pattern analogies (e.g., owl : ant : : ______ : tan), and functional analogies (e.g., tree : man : : sap : ______) were strongly related to analogy-solving ability. Class-naming and similarity analogies (e.g., puzzle : ______ : : riddle : ocean), which are thinly disguised vocabulary items, were not strongly discriminating as analogies.

TABLE 9.1
Mean Item Parameters for Eight Categories of Analogies

Analogy category	Parameter a	c/a	u
Quantitative	1.16	.46	.08
Word pattern	.91	.21	.01
Functional	.90	.02	−.23
Opposites	.80	.73	.14
Conversion	.77	.59	.03
Class membership	.67	1.31	.12
Class naming	.53	−.47	−.26
Similarities	.42	.05	.09

Within sets of items of the same easiness, the u_j parameters describe other differences among the items. Some 58% of the examinees responded correctly to the quantitative analogy "cent : dime : : ______ : dollar" in a median time of 9.4 sec, with an estimated item slowness of −.13. Nearly the same number, 59%, of the subjects responded correctly to another of the quantitative analogies, "______ : yesterday : : tomorrow : today"; but they required an average of 15.5 sec, so u_j was .28. The consequences of this are that a person who responds to the first analogy correctly in 10 sec is average, but someone who responds to the second analogy correctly in 10 sec is either very fast or very able.

PROGRESSIVE MATRICES

The goodness of fit of the Progressive Matrices data to the timed testing model was not as good as for the Verbal Analogies. When the observed proportions of correct responses for a set of six ability groups were compared to the estimated logistic proportions of correct responses, 4 of the 34 individual item χ^2 tests were significant at the $p < .05$ level, and the total of the χ^2s was 201.0 with 136 *df* ($p < .01$), which indicates a significant lack of fit of the model. However, the problem seems to be poor test items, not widespread poor performance of the timed testing model, as the significant χ^2s were all due to items for which the observed proportions of correct responses did not form a strictly monotonic increasing function over estimated ability. For most of the items, the fit was acceptable.

The χ^2 test of the goodness of fit of the Gaussian distribution to the log latency residuals was highly significant: the likelihood ratio χ^2 was 58.5 with 7 *df* ($p < 0.1$). This indicates a substantial violation of the assumption of log-normal error for the response times. The problem arises because in this test, as in the Verbal Analogies, correct responses were faster than incorrect responses. The Progressive Matrices were much easier than the Verbal Analogies, so almost three-fourths of the responses were correct, and those were a little too fast. The total distribution of errors around the log latency model was somewhat skewed, and that is what the goodness-of-fit test detected.

Oddly enough, the prediction of the log latencies from the timed testing model was better for the Progressive Matrices than it was for the Verbal Analogies. Figure 9.7 is a scatterplot of observed (t^*) and fitted ($\hat{t}^*$) log latencies for the 34 Progressive Matrices items for the same examinee whose data are in Figure 9.3. The correlation was .84. There was substantial regression of log latency on effective ability and easiness for the Progressive Matrices data. The parameter b was estimated at .8, the location parameter for the log latencies was 3.10; that means that an item of easiness 0 should take an average examinee 22.2 sec, while an item of easiness 1 would be expected to take the same person only 10 sec.

Thissen (1976) estimated two-parameter logistic item parameters for 20 items drawn from the Progressive Matrices Sets A, B, and C administered to 570 junior high school students. Eighteen of those items form a subset of the 34 Progressive Matrices items included in the present analysis. Figure 9.8 is a plot of the corrected easiness parameters (c_j/a_j) from the present data against the same parameter for the same items from the junior high data; the correlation was .94. Figure 9.9 is a similar scatterplot for the slopes (a_j). The correlation in this case was .61, and there is some evidence of curvilinearity, mostly because there were four very high slopes in the earlier junior high calibration. These correlations, taken with those obtained for the Verbal Analogies, indicate that timed testing can yield reasonably effective item calibration with fewer than 100 subjects. It indicates that there are differences in discrimination between items like these, which are reliable across changing sets of items, and between tests administered to junior high students and college students.

CLOCKS

With only 17 usable items, the Clocks test of spatial visualization was the shortest of the three tests considered here. Four of the items had estimated discrimination parameters greater than 2.0, while three other items had slopes very near zero, indicating that the estimation procedure had taken a short test and made it shorter by allowing a fraction of the items to dominate the scoring. The high discriminations resulted in some outliers among the estimated values for θ; 4 of the 78 estimates were between 2.4 and 4.4. However, the high-scoring examinees did respond correctly to all items but one with near-zero discrimination and frequently

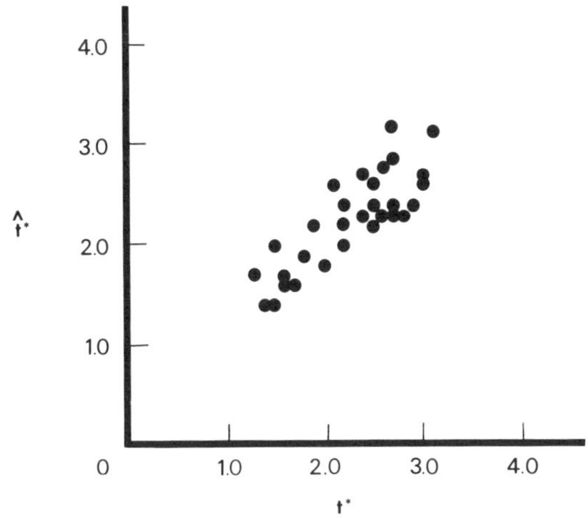

FIGURE 9.7. Scatterplot of observed log response times t^* and the fitted log response times $\hat{t}^*$ for the same examinee as in Figure 9.3 for the matrices.

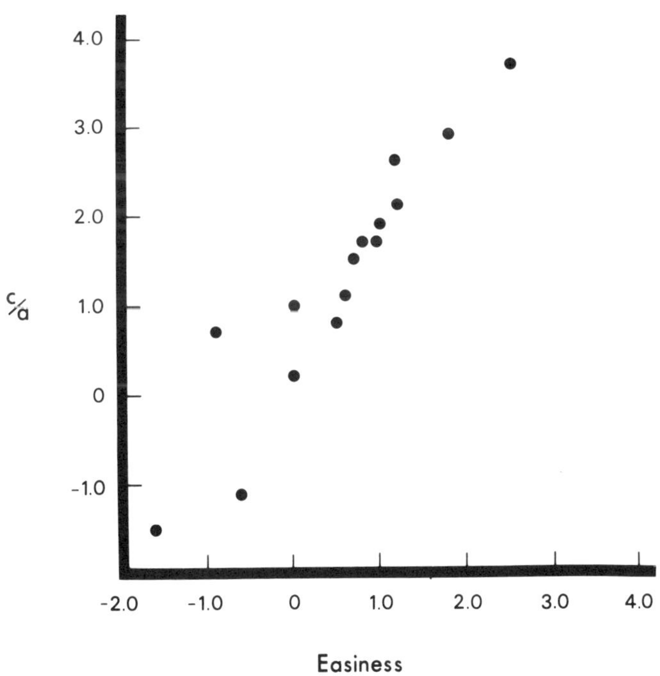

FIGURE 9.8. Timed testing easiness parameters (actually, c/a) plotted against large sample easiness parameters for the matrices.

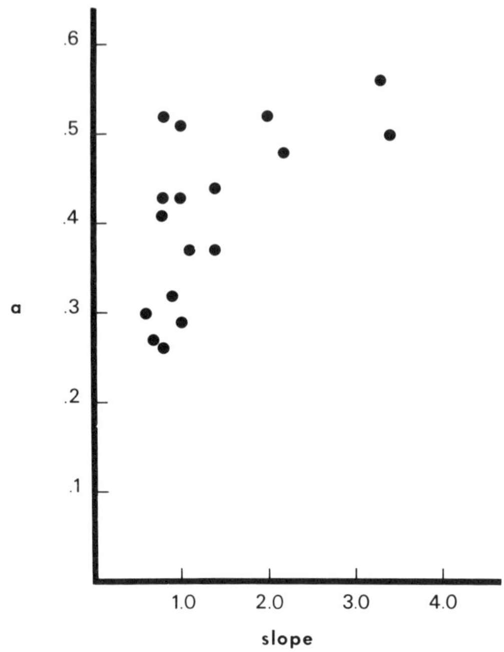

FIGURE 9.9. Timed testing discrimination parameters (*a*) plotted against large sample slopes for the matrices.

responded quite quickly, indicating that they were indeed individuals of very high spatial ability.

None of the item χ^2s was significant, and the total of 90.9 with 68 *df* was not highly significant either ($p = .05$). The probability of correct response as a function of effective ability seemed to be approximated well by the timed testing model.

The distribution of residuals from the log latency model showed approximately the same degree of nonnormality as was the case for the Progressive Matrices; the χ^2 was 39.4 with 7 *df* ($p < .01$). The pattern of nonnormality was the same as it was for the Progressive Matrices data. The estimated value for the regression parameter *b* for the Clocks test was only .025 (estimated SE = .003); although that indicates a significant relationship between response time and the probability of a correct response on the items of this test, the relationship was much smaller than for the other two tests. It seems that there were individual differences in the ability to respond correctly to these items, and individual differences in speed of response; but in striking contrast to the Progressive Matrices test, those two variables were unrelated. Indeed, the correlation between θ and s for the Clocks was $-.03$.

In the case of *this* spatial test, it was θ that showed a significant gender difference ($F(1,76) = 4.65$, $p < .05$). Even allowing for some moderate nonnormality, the male subjects scored substantially higher. There was no hint of a gender difference on s with the Clocks test. For this test, with three-dimensional rotations, it seems that the classical spatial trait simply determined whether the problem could be solved or not; using more or less time seemed to make little or no difference. This is in marked contrast to the results obtained earlier.

If effective ability is the spatial trait, what are the slowness parameters? The easiness of an item seems inversely related to the amount of rotation it requires, as in Shepard and Metzler (1971). Holding amount of rotation constant, the item slowness parameter is, in part, an increasing function of the number of symbolic "instructions" the item uses to achieve that rotation. The items in the test define the rotation to be applied by arrows on the surfaces of one, two, three, or four spheres; a rotation of 180° may be defined by any of those numbers of spheres, and in this subset of items, it was. The more spheres the item used to give the instruction, the longer it took to encode (presumably), and the longer it took for the examinee to respond. Regardless of the number of spheres involved, however, the 180° items had about equal easiness.

RELATIONSHIPS AMONG THE THREE TESTS

The correlations among the six scores are given in Table 9.2. The Progressive Matrices have been variously labeled as a test of abstract reasoning ability g and a number of other constructs. There is evidence in the present data that when given with no time limit, the Progressive Matrices became a test of slowness. Effective ability and slowness on the Progressive Matrices were correlated .94 in these data. Effective ability and slowness were not nearly so closely related for the Verbal Analogies ($r = .68$), and they were unrelated on the Clocks. One possible explanation is that some kinds of test items are affected more by slowness (and, possibly, by carefulness) on the part of the examinees than are others. This speculation has some support in these data, in the form of the relationship between effective ability and slowness on the Verbal Analogies and the parameters of the same names on the Progressive Matrices.

Both θ and s for the Progressive Matrices may be predicted from effective ability and slowness on the Verbal Analogies; the multiple Rs were .70 and .69. The standardized regression coefficients are given in Table 9.3. Effective ability on the Progressive Matrices was predicted by positively

TABLE 9.2
Estimated Correlations among the Person Parameters for the Three Tests

Test and parameter	Analogies θ	Analogies s	Matrices θ	Matrices s	Clocks θ	Clocks s
Analogies						
θ						
s	.68					
Matrices						
θ	.54	.69				
s	.39	.68	.94			
Clocks						
θ	.42	.28	.39	.29		
s	−.09	.40	.36	.55	−.03	

weighting effective ability on the Verbal Analogies and slowness on the Analogies. Individuals who went slowly on the Analogies responded correctly on the Matrices. Slowness on the Matrices was similarly predicted by slowness on the Analogies, with a small negative weight for Analogies effective ability: Individuals who went entirely too slowly on the Analogies, given their ability to respond correctly, would certainly go slowly on the Matrices.

It could be concluded that the three tests in this particular set represent three kinds of timed tests. On the Progressive Matrices, working slowly and carefully is strongly related to the probability of responding correctly, and what is measured is largely slowness. On the Verbal Analogies, working slowly and carefully is related to responding correctly but not so strongly; so two distinct traits are measured: analogical reasoning ability and slowness. On the Clocks, slowness did not seem to affect the probability of a correct response, and so spatial ability and slowness were essentially measured separately.

TABLE 9.3
Standardized Regression Coefficients: Prediction of the Matrices Scores from the Analogies Scores

Independent variable	Matrices θ	Matrices s
Analogies θ	.13	−.13
Analogies s	.60	.77

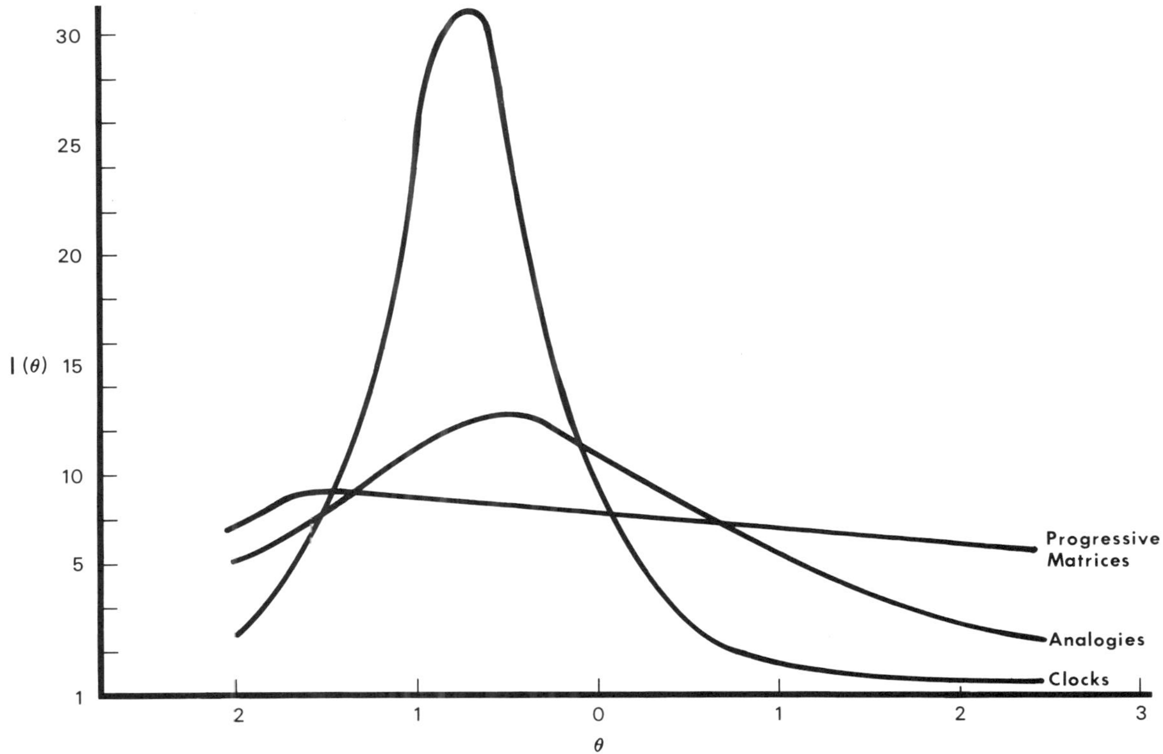

FIGURE 9.10. Test information functions for the three tests.

The differences among the three kinds of tests are graphically portrayed in the test information functions for θ for the three tests, shown in Figure 9.10. The latency data have almost no effect on the effective ability estimation in the Clocks; as a result, the test information function for the spatial test has a form frequently encountered in two-parameter logistic test information functions. All the information about ability comes from the item responses, and the information is only substantial near the difficulty level of the items, in this case between -1 and 0.

The Progressive Matrices, on the other hand, have a flat, regression-like test information curve. When b is estimated to be quite high, as it is in the case of the Progressive Matrices data in this example, effective ability is essentially estimated as though it were an element in a linear regression equation predicting log latency. The log latencies are assumed to provide the same amount of information about effective ability regardless of the value of θ; therefore the test information curve is nearly flat.

The Verbal Analogies test information function shows the situation in which information is being obtained both from the item response data (giving the curve its familiar peaked form) and the latencies. The information provided by the response times serves to raise the entire curve, so that there is *some* information available for estimating the effective ability of individuals with relatively extreme trait values. The test information function for the Verbal Analogies represents the desired outcome for the timed testing model; the other examples make it clear that we are only beginning to learn what can happen in a timed testing situation.

CONCLUSIONS

The latent trait model for timed ability testing described here is neither perfect nor complete; it still requires extensive (and expensive) computation to estimate the parameters of the model for fairly small samples, and it needs an additional parameter to absorb the relatively consistent difference between the residuals from the log latency model for correct and incorrect responses. Improved starting values for the maximum likelihood algorithm would go a long way toward solving the first problem.

The existence of the second problem is interesting. It is not too surprising that averaging over people or items, correct answers take less time than incorrect ones, because correct answers come from able people answering easy items and incorrect responses come from less able people responding to more difficult items. The timed testing model, however, corrects for the ability of the individuals and the easiness of the items, and it is still true that correct answers are associated with shorter latencies. This could be a real indication of some processing difference between correct and incorrect responses; future research could define the difference.

The timed testing model is essentially ready for use. However, it has become clear from repeated applications of this model that there is an essential ambiguity in the relationship of the two individual differences parameters, θ and s, to ability as it may have been classically defined with identical test items. For some tests, when examinees are permitted as much time as they want, many will respond correctly to most items and what had been individual differences in ability on the speeded test become individual differences in slowness; that seems to have happened on the Progressive Matrices. In other cases, both θ and s will absorb *part* of ability as defined by the speeded test, and analysis of the two-dimensional response space will be required to relocate the classical ability; this apparently happened with the Bock–Fitzgerald spatial test data. For other tests, ability and effective ability may be isomorphic; that seemed to be the case for the Clocks test data. Of course, the relationship between effective ability and slowness is simply another aspect of the so-called speed–accuracy trade-off, which has received so much attention in experimental psychology. This has implications for the use of the timed testing model.

Any researcher using timed testing must expect all analyses of the "test scores" to be two dimensional. Effective ability or slowness or both may be the valid response variable(s) in any given context. Univariate analyses of effective ability or slowness alone may be as misleading as analyses of speed or accuracy alone have been shown to be in experimental psychology.

The resulting increase in the complexity of the test scores is a high price to pay for this innovation, but it should be remembered that there is more detailed information about the form of the response when both effective ability and slowness are considered. At this time, only empirical analysis of any given measurement problem will determine if the added information is worth the added complexity.

ACKNOWLEDGMENTS

Some of the research reported here was completed in partial fulfillment of the requirements for the Ph.D. degree at the University of Chicago. I am deeply indebted to R. Darrell Bock, who supervised this project, and to Howard Wainer for their help, encouragement, and patience. I also thank Laura Baker for her assistance in the analysis of the University of Kansas data. This research was supported in part by grants from the National Science

Foundation (Grant No. S0C72-05576 and Grant No. GNS76-02849) and from the Foundations Fund for Psychiatry to R. Darrell Bock, and by grants from the University of Kansas General Research Fund to the author.

REFERENCES

Audley, R. J. A stochastic model for individual choice behavior. *Psychological Review,* 1960, *67,* 1–15.

Audley, R. J. Some observations of theories of choice reaction time. In S. Kornblum (Ed.), *Attention and performance IV.* New York: Academic Press, 1973.

Birnbaum, A. Some latent trait models and their use in inferring an examinee's ability. In F. M. Lord & M. Novick, *Statistical theories of mental test scores.* Reading, Massachusetts: Addison-Wesley, 1968.

Fletcher, R., & Reeves, C. M. Function minimization by conjugate gradients. *Computer Journal,* 1964, *7,* 149–154.

Furneaux, W. D. Intellectual abilities and problem solving behavior. In H. J. Eysenck (Ed.), *The handbook of abnormal psychology.* London: Pitman, 1961.

Guilford, J. P., & Zimmerman, W. F. *The Guilford–Zimmerman aptitude survey. IV. Spatial Visualization Form B.* Beverly Hills, California: Sheridan Supply Co., 1953.

Haberman, S. Subroutine MINIM. In R. D. Bock & B. H. Repp (Eds.), *MATCAL: Double precision matrix operations subroutines for the IBM System/360-370 computer.* Chicago, Illinois: National Educational Resources, 1974.

Laming, D. R. J. *Information theory of choice reaction times.* New York: Academic Press, 1968.

Luce, R. D. Response latencies and probabilities. In K. J. Arrow, S. Karlow, & P. Suppes (Eds.), *Mathematical methods in the social sciences, 1959.* Stanford, California: Stanford University Press, 1960.

Rasch, G. *Probabalistic models for some intelligence and attainment tests.* Copenhagen: Danmarks Paedagogiske Institut, 1960.

Raven, J. *Progressive matrices.* London: Lewis, 1956.

Scheiblechner, H. Specifically objective stochastic latency mechanisms. *Journal of Mathematical Psychology,* 1979, *19,* 18–38.

Shepard, R. N., & Metzler, J. Mental rotation of three-dimensional objects. *Science,* 1971, *171,* 701–703.

Sternberg, R. J. *Intelligence, information processing, and analogical reasoning: The componential analysis of human abilities.* Hillsdale, New Jersey: Erlbaum, 1977.

Tatsuoka, K., & Tatsuoka, M. *A model for incorporating response time data in scoring achievement tests* (CERL Report E-7). Urbana: University of Illinois, Computer-Based Education Research Laboratory, July 1979.

Thissen, D. Information in wrong responses to the Raven Progressive Matrices. *Journal of Educational Measurement,* 1976, *13,* 201–214.

Thissen, D. Incorporating item response latencies in latent trait estimation. (Doctoral dissertation, University of Chicago, 1976). *Dissertation Abstracts International,* 1977, *37,* 4658B.

Tinsley, H. E. A. *An investigation of the Rasch simple logistic model for tests of intelligence or attainment.* Unpublished doctoral dissertation, University of Minnesota, 1971.

White, P. O. Individual differences in speed, accuracy and persistence. In H. J. Eysenck (Ed.), *The measurement of intelligence*. Lancaster, England: Medical and Technical Publishing Co., 1973.

White, P. O. *A latent trait model for individual differences in speed, accuracy, and persistance*. Paper presented at the annual meeting of the Psychometric Society, Monterey, California, June 1979.

Whitely, S. E. Relationships in analogy items: A semantic component of a psychometric task. *Educational and Psychological Measurement*, 1977, *37*, 725–739.

Whitely, S. E. Multicomponent latent trait models for ability tests. *Psychometrika*, 1981, *45*, 479–494.

10

The Mental Growth Curve Reexamined

R. DARRELL BOCK

A study purporting to show the growth of mental ability, as measured by the Binet test, as a function of chronological age was published in 1929 by Thurstone and Ackerson. The curve was based on a rescaling of Binet mental ages (MA) in a cross-sectional sample of 4208 children age 3 through 17 seen at the Institute for Juvenile Research in Chicago. The shape of that curve, which is reproduced in Figure 10.1, is surprising in one respect: it shows an inflection point at about 10 years of age, where an initial positive acceleration switches to negative. There is no precedent for this type of growth curve in any other aspect of human growth. All other such curves—in particular, those for growth in stature (see Bock & Thissen, 1980)—show a rapid deceleration from birth through adolescence, followed by a brief period of acceleration during the adolescent growth spurt. (In longitudinal growth records of individual children, a slight middle-childhood spurt can sometimes also be seen between 6 and 7 yr, but it is not evident in cross-sectional data.)

Any discussion of the shape of such curves requires that the unit of scale be equal at all points throughout the range of measurement. Because there is no reason to suppose that MA scores for the Binet have this property, some method of scaling the test scores that will yield a uniform unit must be adopted. Thurstone (1925, 1927, 1928) formulated such a

NEW HORIZONS IN TESTING
Latent Trait Test Theory and
Computerized Adaptive Testing

ISBN 0-12-742780-5

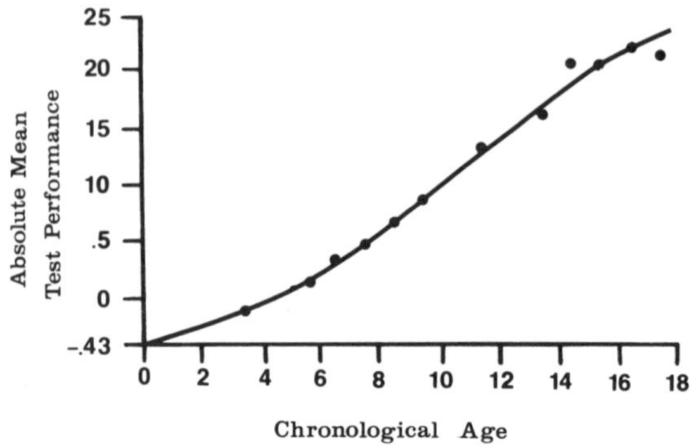

FIGURE 10.1. Thurstone's curve for Binet mental growth. (From Thurstone & Ackerson, 1929.)

method. It rests on two very general assumptions: (*a*) that the distributions of mental age (or attainment) conditional upon chronological age have the same (continuous) functional form at all age levels but different means and dispersions (standard deviation); (*b*) that the origin of measurement can be so assigned that the dispersion of the conditional distributions is directly proportional to the mean, that is, so that the coefficient of variation is constant.

Thurstone pointed out that if the functional form of the common distribution is known, these assumptions may be checked (*a*) by converting the observed proportions of persons at each age level who respond correctly to each test item to the corresponding percentage point of the normal distribution and (*b*) by plotting the resulting transformed proportions as a function of age. If the points tend to lie on straight lines and the slopes of the lines decrease with increasing age, the assumptions are justified. Thurstone (1925, 1927, 1928) exhibited numerous examples of data in which these assumptions seem reasonable. He also developed simple numerical methods for estimating the item means (thresholds) and the constants of proportionality for the item standard deviations. He called this procedure the "method of absolute scaling." Although the method is no longer used, it is important as a forerunner of modern item response theoretic (IRT) scaling procedures.

This method was not used directly on the item data by Thurstone and Ackerson (1929), however; rather, they obtained the means and constants of proportionality indirectly from the MA distributions of yearly age groups. (In their supplementary tables, the actual data distributions are given in 3-month intervals for boys and girls separately, with boys substantially outnumbering girls in the sample.) This labor-saving compromise of the absolute scaling method can be justified on the grounds that the mean and dispersion obtained from the average percentage correct for items represented in the MA score will be a good approximation to the average of the means and dispersions of the separate items. There is no reason to believe that the unusual characteristics of the Thurstone–Ackerson curve for mental growth curve are due to their scaling the Binet data at the score level rather than at the item level.

A more plausible explanation is that the shape of the curve is influenced by Thurstone's use of the observed ratios of MA dispersions in successive chronological age groups to determine the factor of proportionality (coefficient of variation) relative to the average scaled mental age. The growth curve thus obtained, although independent of the arbitrary Binet MA scale in the conditional means, is not independent of the scale in the calculation of the conditional dispersions. A solution independent of arbitrary scale artifacts in both item thresholds and dispersions was not practical with the hand methods of computation then available to Thurstone.

A SCALING PROCEDURE FULLY INDEPENDENT OF CHRONOLOGICAL AGE UNITS

With the aid of modern computers, however, the Binet test, or similar tests referenced to chronological age or to another external criterion, can be scaled on the single assumption that the underlying distributions for item attainments conditional on age have a known common functional form indexed by a threshold and a dispersion parameter. If this assumption is satisfied, scale values may be assigned to the chronological age groups so that with respect to the growth continuum, all the item response curves simultaneously fit the observed percentage-correct data for each item in each age group. On the further assumption that item responses within the age groups are locally independent, the goodness of fit of the solution can be tested by a large-sample statistical test.

A Biological Example

A maximum likelihood procedure for scaling by this method, when a normal ogive response curve is assumed, is presented in the appendix to Bock (1976). This procedure has been applied by Kolakowski and Bock (1981) to biological data consisting of counts of emerged permanent dentition in a large cross-sectional sample of Pima Indian children (Dahlberg & Menegaz-Bock, 1958). Reproduced in Figure 10.2 are the developmental age scale values obtained by Kolakowski and Bock (1981), plotted as a

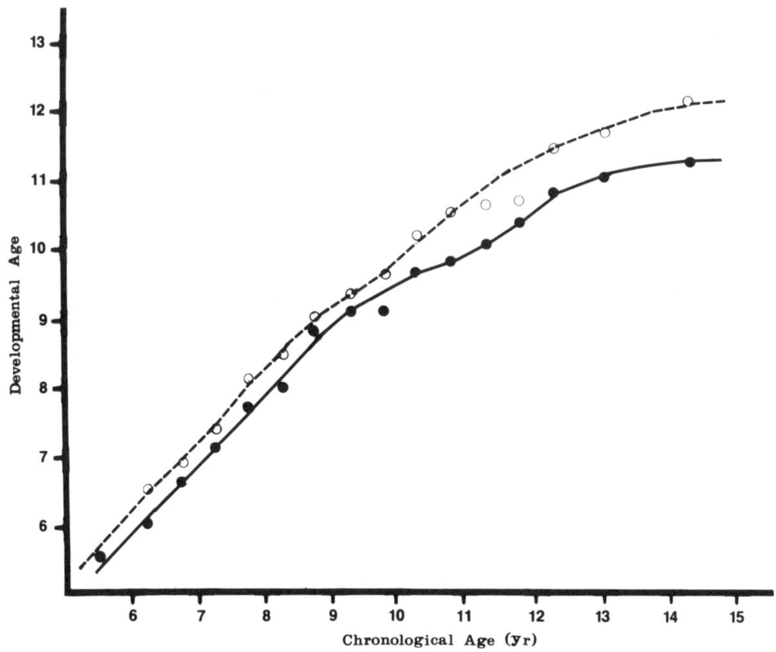

FIGURE 10.2. Developmental age curves inferred from emergence of permanent dentition in Pima Indian children. Open circles, girls; filled circles, boys. (From Kolakowski & Bock, 1981.)

function of chronological age. As can be seen, the curves based on incidence of emerged permanent teeth initially decelerate and show some suggestion of an adolescent growth spurt in both sexes. There is no evidence of the initial positive acceleration that was found in the mental growth curve by Thurstone and Ackerson (1929).

An unavoidable limitation of all such scaling methods is that the origin and unit of measurement of the scale is arbitrary in each sample analyzed. In the case of the tooth emergence data, Kolakowski and Bock (1981) adjusted the origin and unit so that the threshold and dispersion of one of the teeth that is known to show no sex difference in emergence time, an upper central incisor, had the same values as those in the literature based on probit analyses of tooth frequencies as a function of chronological age (Dahlberg & Menegaz-Bock, 1958). The curves for the two sexes in Figure 10.2 are based on this choice of origin and unit. Thurstone and Ackerson (1929) based the origin of their scale at an inferred point of zero variability (Thurstone's, 1982, "absolute zero" of intelligence) and set the unit so that the MA of the year group equaled chronological age (CA).

SCALING THE BINET TEST

Data and Method

Using data supplied by Reckase (1979), the Bock (1976) procedure was applied to 96 items of the current version of the Stanford–Binet test. These data, which are reproduced in Appendix Tables 10.A1 and 10.A2, are drawn from the full complement of 122 Binet tasks, with the first 13 omitted because all examinees responded correctly and the last 13 omitted because all examinees responded incorrectly.

The numbers and mean age of boys and girls in each CA group are shown in Table 10.1. In some instances, alternative forms of an item were treated in the scaling as if they were the same item. The data are strictly cross sectional and, like all such data, are not constrained to be increasing with chronological age (see Bock, 1979).

The scaling solutions based on Bock (1976) converged in 13 Newton–Raphson iterations. The computations required 63 sec of IBM 370/168 cpu time and 465K bytes of core storage. Scale values for successive 10-month chronological age groups were calculated. The origin and unit of the scale for boys were fixed so that the values for the 40-month and 160-month

TABLE 10.1
Mean Chronological Age (CA) and Sample Size for Each Age Group

		Boys (N = 342)		Girls (N = 81)	
Age group	CA interval in months	Mean	N	Mean	N
1	24–36	30.9	17	31.8	10
2	37–46	42.6	26	40.7	30
3	47–56	50.5	29	51.4	25
4	57–66	61.3	32	60.9	28
5	67–76	71.4	35	72.2	22
6	77–86	81.7	22	81.4	21
7	87–96	91.5	29	91.6	18
8	97–106	102.1	25	101.3	18
9	107–116	111.6	24	111.5	23
10	117–126	121.3	25	121.7	15
11	127–136	132.1	19	130.6	16
12	137–146	141.3	12	141.6	20
13	147–156	151.3	15	151.3	8
14	157–166	162.0	16	160.5	13
15	167–178	170.9	16	173.1	14

groups were 40 and 160, respectively. The unit of the girls' scale was then set so that the averages of the item dispersions for boys and girls were equal (to 19.55); and the origin of the girls' scale was set so that the scale value of the 160-month group was 160.

Results: The Revised Mental Growth Curve

The growth curves from this scaling solution are shown in Figure 10.3. The curve for boys is entirely plausible as a representation of growth. Unlike the Thurstone–Ackerson curve, it decelerates from the earliest age

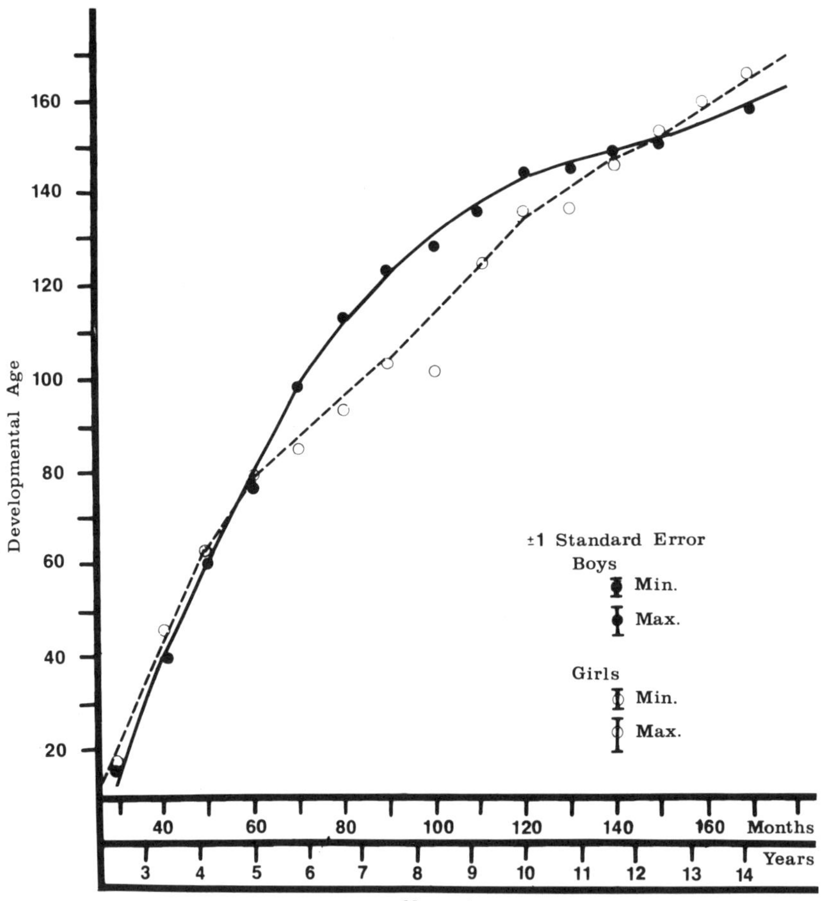

FIGURE 10.3. Proposed mental growth curve based on Binet item data collated by Reckase (1979).

until adolescence. The final two points suggest the possible upward inflection of a slight adolescent growth spurt in mental attainment. At age 14 the curve is still rising, and presumably would go higher if older age groups were included.

The curve for girls is less satisfactory. Initially, it resembles the curve for boys; but from years 6 through 11, the scale values for girls are irregular and considerably below those for boys of the same age. It is possible, of course, that the equating of boys and girls at 160 months is unfair to the girls. Perhaps they are actually 10 or 20 points higher at that age. If so, the points in the range 70–130 months would be more comparable in boys and girls. Such an adjustment, however, would make the girls' scores in the range 30–60 high relative to those of the boys. Inasmuch as the percentages correct for girls on items in this range, or indeed in the upper range of 140–170 months, were about the same as those for the boys, this interpretation does not seem plausible (compare Tables 10.A1 and 10.A2). The assumption that boys and girls have the same average Binet attainment at 160 months seems reasonable for these data.

The only explanation for the anomalous result for girls would seem to be that the samples for the two sexes were not comparable in some age groups. Some bias in selection of examinees or in administration of the tests must have operated against girls in the 70- to 130-month range. The irregularity of the girls' scale values in this range, especially the discrepant value at 100 months, suggests that the sample of girls may have been defective. Regrettably, no information is available on how the examinees were selected or how the tests were administered.

ADVANTAGES OF THE PRESENT SCALING PROCEDURE

Because the present scaling method does not force the dispersions of the conditional distributions to increase with age, the scale is not stretched to the left in order to make the conditional standard deviations small. It is this stretching of the scale that induces the initial positive acceleration in the Thurstone–Ackerson curve. When the dispersions were estimated without constraint, the more plausible initial negative acceleration seen in Figure 10.3 is obtained.

As discussed in Bock (1976) and in Kolakowski and Bock (1981), the item parameters estimated in the scaling solutions can also be used to assign developmental age scores to individuals by the method of maximum likelihood (see also Birnbaum, 1968; Samejima, 1969). In this role the present scaling solution has important methodological advantages. On the developmental-age scale, the item dispersions, rather than increasing as Thurstone had assumed, are relatively homogeneous. A solution with all item standard deviations set to their average value fit almost as well as the unconstrained solution. This implies that the maximum likelihood estimates of developmental age of individuals can be expressed with good accuracy as a function of the person's number-correct score. This is implied by the close similarity of a one-parameter normal ogive model with the one-parameter logistic model in which number correct is the sufficient statistic for the maximum likelihood estimate (Andersen, 1980).

Moreover, when the within-age-group standard deviations of the estimated developmental age scores were calculated (Table 10.2), they were also relatively homogeneous. This means that analysis of variance can be employed to investigate relations between developmental age and other age-structured data without violating the assumption of homoscedasticity. The conventional MA scores for the Binet test do not have this property.

The developmental-age scale may also have certain interpretational advantages whenever changes within examinees rather than normative comparisons with age-mates is at issue. Because the developmental age

TABLE 10.2
Developmental Age Means and Standard Deviations for Children in Successive Chronological Age (CA) Groups

Age group	Nominal CA	Boys		Girls	
		Mean	*SD*	Mean	*SD*
1	30	15.4	16.6	16.9	18.1
2	40	40.0	19.2	46.0	12.6
3	50	61.0	18.0	62.4	14.1
4	60	76.2	17.3	76.9	22.1
5	70	96.6	13.5	84.1	10.0
6	80	110.9	10.1	92.1	10.2
7	90	120.6	13.7	101.5	10.1
8	100	125.8	13.2	105.0	11.5
9	110	134.0	13.4	121.1	12.3
10	120	143.3	10.7	134.0	16.7
11	130	144.3	15.6	134.7	16.4
12	140	148.3	9.7	143.9	14.9
13	150	152.1	17.4	153.0	19.9
14	160	160.0	8.2	160.0	19.1
15	170	162.0	16.3	175.1	38.0

units are greater than chronological age units at young ages, the changes in scale values are more in accord with the rate of behavioral change (i.e., the surpassing of successive developmental tasks) than with changes in MA scores. Moreover, the growth of mental attainment in terms of scaled scores should parallel closely other quantitative indices of development, such as stature. Thus, the developmental age scores will tend to show simple linear relations with direct measures of development.

The present scale, however, does not exhibit increasing standard deviation with age and thus does not support Thurstone's definition of the absolute zero of intelligence. Provisionally, at least, it will be necessary to set the origin of the scale on some more arbitrary basis.

APPENDIX

TABLE 10.A1
Number of Correct Responses to Each Item for Boys in Successive Age Groups

	Age group														
Item	1	2	3	4	5	6	7	8	9	10	11	12	13	14	15
14	14	22	29	32	35	22	29	25	24	25	19	12	15	16	15
15	10	21	29	32	35	22	29	25	24	25	19	12	15	16	15
16	12	24	28	32	35	22	29	25	24	25	19	12	15	16	15
17	11	23	27	32	35	22	29	25	24	25	19	12	15	16	15
18	10	25	28	32	35	22	29	25	24	25	19	12	15	16	15
19	10	22	26	29	35	22	29	25	24	25	19	12	15	16	15
20	12	23	27	32	35	22	29	25	24	25	19	12	15	16	15
21	9	21	28	32	35	22	29	25	24	25	19	12	15	16	15
22	8	20	25	30	35	22	29	25	24	25	19	12	15	16	15
23	7	22	26	32	35	22	29	25	24	25	19	12	15	16	15
24	4	19	27	32	35	22	29	25	24	25	19	12	15	16	15
25	3	15	27	30	34	22	29	25	24	25	19	12	15	16	15
26	7	17	25	32	35	22	29	25	24	25	19	12	15	16	15
27	5	13	24	29	35	22	29	25	24	25	19	12	15	16	15
28	6	22	25	30	35	22	29	25	24	25	19	12	15	16	15
29	3	17	24	30	35	22	29	25	24	25	19	12	15	16	15
30	3	12	23	29	35	22	29	25	24	25	19	12	15	16	15
31	4	11	23	24	34	22	29	25	24	25	19	12	15	16	15
32	2	5	22	27	35	22	29	25	24	25	19	12	15	16	15
33	3	16	23	27	34	22	29	25	24	25	19	12	15	16	15
34	1	3	20	27	35	22	29	25	24	25	19	12	15	16	15
35	1	11	18	26	34	22	29	25	24	25	19	12	15	16	15
36	2	12	22	27	35	22	29	25	24	25	19	12	15	16	15
37	0	8	12	21	31	22	29	25	24	25	19	12	15	16	15

(*continued*)

TABLE 10.A1 (***Continued***)

	Age group														
Item	1	2	3	4	5	6	7	8	9	10	11	12	13	14	15
38	0	3	13	23	32	21	29	25	24	25	19	12	15	16	15
39	1	11	22	27	34	22	29	25	24	25	19	12	15	16	15
40	0	3	16	22	34	22	28	25	24	25	19	12	15	16	15
41	1	5	20	22	32	22	29	25	24	25	19	12	15	16	15
42	0	4	7	17	24	22	28	25	23	25	19	12	15	16	15
43	0	1	7	16	29	21	27	24	24	25	19	12	15	16	15
44	0	2	5	12	31	20	28	25	24	25	19	12	15	16	15
45	0	3	5	22	29	22	29	25	23	25	19	12	15	16	15
46	0	0	4	13	28	22	29	25	24	25	19	12	15	16	15
47	0	0	7	22	24	22	27	25	24	25	19	12	15	16	15
48	0	1	8	18	31	22	29	25	23	25	19	12	15	16	15
49	0	0	0	9	18	17	24	22	24	25	19	12	15	16	15
50	0	0	0	4	19	17	26	24	24	25	19	12	15	16	15
51	0	0	1	5	21	20	27	24	23	25	19	12	15	16	15
52	0	0	0	7	23	17	27	25	22	24	19	12	15	16	15
53	0	0	4	8	21	17	26	23	24	24	19	12	15	16	15
54	0	0	1	2	11	17	25	20	23	25	17	12	15	16	15
55	0	0	0	5	19	16	26	23	23	25	19	12	15	16	15
56	0	0	0	5	12	16	24	22	20	24	19	12	15	16	15
57	0	0	0	0	7	13	21	17	21	24	18	11	14	16	15
58	0	0	0	1	13	13	24	21	23	25	18	11	15	16	15
59	0	0	0	1	13	10	23	22	19	24	18	11	14	16	15
60	0	0	0	2	7	15	23	22	22	24	19	12	14	16	15
61	0	0	0	1	9	6	17	17	21	23	18	12	15	16	15
62	0	0	0	0	2	6	18	18	21	21	17	11	13	16	15
63	0	0	0	1	5	9	15	15	19	25	18	12	14	16	15
64	0	0	0	0	6	13	21	20	23	24	19	12	14	16	15
65	0	0	0	0	0	3	20	15	21	25	18	12	14	16	15
66	0	0	0	1	4	7	17	18	18	23	16	10	13	16	15
67	0	0	0	0	2	2	9	15	18	22	19	12	14	16	15
68	0	0	0	0	2	4	11	11	17	22	16	11	13	16	15
69	0	0	0	1	1	2	12	13	16	19	16	12	14	16	15
70	0	0	0	0	1	5	14	13	16	20	15	12	13	16	15
71	0	0	0	0	4	7	15	15	19	21	17	11	13	16	15
72	0	0	0	0	2	5	13	11	17	21	14	10	14	16	14
73	0	0	0	1	1	3	5	10	10	18	13	11	12	15	15
74	0	0	0	0	0	2	4	6	14	19	12	10	12	16	14
75	0	0	0	0	0	2	8	9	17	19	14	12	13	16	15
76	0	0	0	0	1	4	11	12	12	18	12	8	13	15	13
77	0	0	0	0	1	0	8	11	11	17	13	11	13	16	15
78	0	0	0	0	1	3	10	12	20	23	14	11	15	16	14
79	0	0	0	0	0	0	3	5	7	16	12	10	11	15	14
80	0	0	0	0	0	1	9	6	13	17	12	10	11	14	15
81	0	0	0	0	0	1	5	3	11	18	10	8	11	16	13

(*continued*)

TABLE 10.A1 (*Continued*)

Item	Age group 1	2	3	4	5	6	7	8	9	10	11	12	13	14	15
82	0	0	0	0	0	0	2	8	12	14	12	8	12	14	13
83	0	0	0	0	0	0	4	6	6	15	11	12	12	16	14
84	0	0	0	0	0	0	2	5	10	16	12	7	11	13	14
85	0	0	0	0	0	0	6	5	11	14	11	5	9	9	11
86	0	0	0	0	0	0	1	8	9	15	11	10	12	15	13
87	0	0	0	0	0	0	3	5	6	11	10	6	10	11	12
88	0	0	0	0	0	1	5	3	11	15	12	7	10	14	14
89	0	0	0	0	0	0	0	5	6	14	9	9	10	12	12
90	0	0	0	0	0	0	1	6	7	14	11	6	11	14	14
91	0	0	0	0	0	0	2	1	5	11	8	8	11	13	11
92	0	0	0	0	0	0	1	2	4	10	8	4	8	12	12
93	0	0	0	0	0	0	1	1	8	9	7	6	11	12	10
94	0	0	0	0	0	0	1	1	7	8	5	7	8	15	11
95	0	0	0	0	0	1	1	4	7	13	11	9	11	15	14
96	0	0	0	0	0	1	2	6	8	17	9	7	13	12	14
97	0	0	0	0	0	0	0	0	0	3	4	3	4	10	6
98	0	0	0	0	0	0	0	0	1	0	4	2	7	10	7
99	0	0	0	0	0	0	0	0	0	2	4	2	6	13	7
100	0	0	0	0	0	0	0	1	2	6	8	2	7	11	9
101	0	0	0	0	0	0	0	1	1	6	4	1	5	10	9
102	0	0	0	0	0	1	1	2	5	12	8	5	9	14	12
103	0	0	0	0	0	0	0	1	2	6	8	4	4	12	7
104	0	0	0	0	0	0	0	0	0	4	3	5	7	6	7
105	0	0	0	0	0	0	0	0	0	1	2	1	4	4	4
106	0	0	0	0	0	0	0	0	3	3	5	5	1	5	5
107	0	0	0	0	0	0	0	0	1	2	1	1	2	6	5
108	0	0	0	0	0	0	0	0	0	4	6	4	5	7	7
109	0	0	0	0	0	0	0	0	0	3	2	4	5	7	5
Attempts	17	26	29	32	35	22	29	25	24	25	19	12	15	16	15

TABLE 10.A2
Number of Correct Responses to Each Item for Girls in Successive Age Groups

Item	Age group 1	2	3	4	5	6	7	8	9	10	11	12	13	14	15
14	9	28	25	28	22	21	18	18	23	15	16	20	8	13	12
15	8	29	24	28	22	21	18	18	23	15	16	20	8	13	12
16	8	29	24	28	22	21	18	18	23	15	16	20	8	13	12
17	6	26	24	28	22	21	18	18	23	15	16	20	8	13	12

(*continued*)

TABLE 10.A2 (*Continued*)

Item	Age group 1	2	3	4	5	6	7	8	9	10	11	12	13	14	15
18	7	28	24	28	22	21	18	18	23	15	16	20	8	13	12
19	5	26	24	27	22	21	18	18	23	15	16	20	8	13	12
20	4	25	24	28	22	21	18	18	23	15	16	20	8	13	12
21	4	27	24	27	22	21	18	18	23	15	16	20	8	13	12
22	3	28	23	28	22	21	18	18	23	15	16	20	8	13	12
23	7	26	25	28	22	21	18	18	23	15	16	20	8	13	12
24	2	24	24	27	22	21	18	18	23	15	16	20	8	13	12
25	2	21	23	26	22	21	18	18	23	15	16	20	8	13	12
26	3	18	23	26	22	21	18	18	23	15	16	20	8	13	12
27	2	20	22	26	22	21	18	18	23	15	16	20	8	13	12
28	3	26	24	26	22	21	18	18	23	15	16	20	8	13	12
29	3	23	22	27	22	21	18	18	23	15	16	20	8	13	12
30	0	19	20	27	22	21	18	18	23	15	16	20	8	13	12
31	3	19	21	25	22	21	18	18	23	15	16	20	8	13	12
32	1	13	18	24	21	21	18	18	23	15	16	20	8	13	12
33	2	20	23	27	22	21	18	18	23	15	16	20	8	13	12
34	0	12	18	24	22	21	18	18	23	15	16	20	8	13	12
35	3	19	20	24	20	21	18	18	23	15	16	20	8	13	12
36	0	17	19	27	22	21	18	18	23	15	16	20	8	13	12
37	0	7	19	25	22	21	18	18	23	15	16	20	8	13	12
38	1	7	17	26	22	20	18	18	23	15	16	20	8	13	12
39	1	13	20	26	22	21	18	18	23	15	16	20	8	13	12
40	0	2	18	23	22	21	18	18	23	15	16	20	8	13	12
41	0	8	22	27	22	20	18	18	23	15	16	20	8	13	12
42	0	4	10	19	17	20	17	18	23	15	16	20	8	13	12
43	0	6	9	22	21	19	18	18	23	15	16	20	8	13	12
44	0	2	11	20	21	21	18	18	23	15	16	20	8	13	12
45	0	2	11	21	18	21	18	18	23	15	16	20	8	13	12
46	0	1	8	17	19	21	18	18	23	15	16	20	8	13	12
47	0	3	8	20	18	20	18	18	23	15	16	20	8	13	12
48	0	1	12	23	20	21	18	17	23	15	16	20	8	13	12
49	0	0	2	10	9	11	15	16	23	15	16	20	8	13	12
50	0	0	3	8	10	13	17	16	23	15	16	20	8	13	12
51	0	0	0	7	13	19	16	18	23	15	16	20	8	13	12
52	0	0	3	11	13	19	14	17	23	15	16	20	8	13	12
53	0	1	1	14	14	15	18	16	23	15	16	20	8	13	12
54	0	0	3	5	8	13	15	16	23	15	16	20	8	13	12
55	0	0	2	10	13	12	14	17	23	15	16	20	8	13	12
56	0	0	2	7	8	12	16	17	23	15	16	20	8	13	12
57	0	0	0	4	6	7	9	13	23	14	15	20	8	13	12
58	0	0	0	7	8	14	14	16	23	15	16	20	8	13	12
59	0	0	0	8	8	10	12	16	23	15	15	20	8	13	12
60	0	0	0	3	9	10	13	17	23	15	15	20	8	13	12
61	0	0	0	1	1	7	9	10	18	13	16	19	8	13	12

(*continued*)

TABLE 10.A2 (*Continued*)

Item	Age group														
	1	2	3	4	5	6	7	8	9	10	11	12	13	14	15
62	0	0	0	1	2	2	8	12	22	13	15	19	8	13	12
63	0	0	0	1	4	11	11	9	17	14	14	19	8	13	12
64	0	0	0	2	5	8	14	14	22	15	16	20	8	13	12
65	0	0	0	1	3	3	7	11	23	15	16	19	8	13	12
66	0	0	0	1	1	9	10	11	18	15	15	20	8	13	12
67	0	0	0	1	3	2	3	12	19	14	16	19	7	13	12
68	0	0	0	1	1	5	9	11	11	12	14	19	7	12	12
69	0	0	0	1	0	3	4	5	18	13	14	20	8	12	12
70	0	0	0	2	1	2	8	10	15	14	13	19	8	12	12
71	0	0	0	2	3	8	11	8	19	14	13	18	8	13	12
72	0	0	0	2	2	7	11	3	15	11	13	19	7	12	12
73	0	0	0	1	0	7	5	5	13	10	11	18	6	13	12
74	0	0	0	1	1	1	4	4	13	10	12	18	7	13	11
75	0	0	0	1	0	2	1	3	16	14	13	17	8	13	12
76	0	0	0	1	2	4	7	2	14	11	12	18	8	11	12
77	0	0	0	1	3	3	4	5	16	8	12	15	8	13	12
78	0	0	0	1	1	2	7	5	17	13	15	19	8	13	12
79	0	0	0	1	0	0	0	2	7	11	11	16	7	12	11
80	0	0	0	1	1	1	5	4	15	12	11	15	7	11	11
81	0	0	0	1	0	2	2	3	2	8	10	14	5	11	12
82	0	0	0	1	0	0	6	4	14	11	12	16	4	13	11
83	0	0	0	1	0	0	1	3	9	11	10	16	7	13	12
84	0	0	0	1	0	2	2	5	13	9	14	16	8	11	12
85	0	0	0	1	0	1	4	1	12	8	7	11	6	12	11
86	0	0	0	1	0	0	0	1	8	10	11	17	7	11	11
87	0	0	0	1	0	1	1	1	7	7	10	11	6	11	11
88	0	0	0	1	0	0	2	1	8	10	13	16	7	12	11
89	0	0	0	1	0	0	2	1	10	10	9	15	7	12	10
90	0	0	0	1	0	1	3	3	10	8	10	15	4	13	11
91	0	0	0	1	0	0	0	0	4	5	7	14	7	12	10
92	0	0	0	1	0	1	0	0	2	7	6	13	3	9	9
93	0	0	0	1	0	0	0	1	4	7	7	13	4	9	11
94	0	0	0	1	0	0	0	0	2	6	4	12	5	9	9
95	0	0	0	1	0	1	0	0	2	6	5	14	6	10	8
96	0	0	0	1	0	1	0	2	6	7	11	18	7	12	10
97	0	0	0	1	0	0	0	0	2	3	2	6	4	9	10
98	0	0	0	1	0	0	0	0	0	2	2	6	3	8	6
99	0	0	0	1	0	0	0	0	2	2	4	5	5	8	10
100	0	0	0	0	0	0	0	0	2	5	4	8	7	9	10
101	0	0	0	1	0	0	0	0	0	2	3	6	6	8	8
102	0	0	0	1	0	0	0	0	3	5	3	11	3	7	6
103	0	0	0	1	0	0	0	0	1	4	4	9	5	9	7
104	0	0	0	1	0	0	0	0	1	3	2	8	5	8	7
105	0	0	0	1	0	0	0	0	0	0	0	2	4	3	2

(continued)

TABLE 10.A2 (*Continued*)

Item	Age group 1	2	3	4	5	6	7	8	9	10	11	12	13	14	15
106	0	0	0	0	0	0	0	0	0	1	4	4	2	3	3
107	0	0	0	0	0	0	0	0	0	1	1	4	2	4	4
108	0	0	0	0	0	0	0	0	1	4	5	6	4	7	6
109	0	0	0	1	0	0	0	0	0	2	2	6	3	8	5
Attempts	10	30	25	28	22	21	18	18	23	15	16	20	8	13	12

ACKNOWLEDGMENT

Preparation of this chapter was supported in part by NSF Grants BNS-791247 and BNS 76-02849 to the University of Chicago.

REFERENCES

Andersen, E. B. *Discrete statistical models with social science applications*. Amsterdam: North-Holland Publishing Co., 1980.

Birnbaum, A. Some latent trait models and their use in inferring an examinee's ability. In F. M. Lord & M. R. Novick, *Statistical theories of mental test scores*. Reading, Massachusetts: Addison-Wesley, 1968.

Bock, R. D. Basic issues in the measurement of change. In D. N. M. de Gruijter & L. J. T. van der Kamp (Eds.), *Advances in psychological measurement*. New York: Wiley, 1976.

Bock, R. D. Univariate and multivariate analysis of time-structured data. In J. R. Nesselroade & P. B. Baltes (Eds.), *Longitudinal research in the study of behavior and development*. New York: Academic Press, 1979.

Bock, R. D., & Thissen, D. Statistical problems of fitting individual growth curves. In F. Johnston, A. F. Rocher, & C. Susanne (Eds.), *Methodologies for the analysis of human growth and development*. New York: Plenum, 1980.

Dahlberg, A. A., & Menegaz-Bock, R. M. Emergence of the permanent teeth in Pima Indian children. *Journal of Dental Research*, 1958, *37*, 1123–1140.

Kolakowski, D., & Bock, R. D. A multivariate generalization of probit analysis. *Biometrics*, 1981, *37*, 541–551.

Reckase, M. D. *Item response data for the Binet tests administered to a large sample of boys and girls age 30 to 170 months*. Personal communication, 1979.

Samejima, F. Estimation of latent ability using a response pattern of graded scores. *Psychometrika Monograph Supplement*, 1969, *34* (4, Pt. 2, Monograph No. 17).

Thurstone, L. L. A method of scaling psychological and educational tests. *Journal of Educational Psychology*, 1925, *16*, 433–451.

Thurstone, L. L. The unit of measurement in educational scales. *Journal of Educational Psychology*, 1927, *18*, 505–524.
Thurstone, L. L. The absolute zero in intelligence measurement. *Psychological Review*, 1928, *35*, 175–197.
Thurstone, L. L., & Ackerson, L. The mental growth curve for the Binet tests. *Journal of Educational Psychology*, 1929, *20*, 569–583.

IV

APPLICATION TO ADAPTIVE TESTING

As indicated earlier, a major use of adaptive testing is to improve the precision and efficiency of measurement, by selecting items from a larger item pool dynamically for each individual during the process of testing. Previous research on adaptive testing has focused on simulation studies that showed psychometric benefits, but few live-testing studies have verified the theoretical and simulation predictions. In a live-testing study, results in support of theoretical and simulation predictions verify both the mathematical model underlying the simulations and the utility of the methods in practical situations. In Chapter 11, McBride and Martin report live-testing data from the administration of IRT-based adaptive tests to Marine recruits. Their studies were designed to verify theoretical predictions of higher reliability and validity of adaptive tests in comparison to conventional tests.

In addition to efficiently measuring trait levels, adaptive testing can be used for making dichotomous classifications. Given a set of items to be used for making a dichotomous classification, it may be reasonable to select a set of items sequentially as opposed to adaptively. In a sequential test, items are ordered a priori for all individuals, but are administered sequentially in that order to each individual. Sequential tests simply have a variable termination rule, in which the length of the test is adapted to the

NEW HORIZONS IN TESTING
Latent Trait Test Theory and
Computerized Adaptive Testing

ISBN 0-12-742780-5

individuals, but the individuals who have taken tests of a given length have been administered the same items. In Chapter 12, Reckase investigates a sequential procedure based on item response theory, which is a modified version of a Waldian sequential probability ratio procedure, in terms of how well it classifies simulated examinees.

In Chapter 13, Kingsbury and Weiss compare an earlier version of Wald's sequential approach to a fully adaptive dichotomous classification procedure for making mastery decisions. In contrast to Reckase's Waldian-based procedure, which assumes that items vary according to the difficulty, discrimination, and pseudoguessing parameters, the sequential procedure studied by Kingsbury and Weiss assumes randomly selected items, an approach that is implicit in classical domain-sampling psychometrics or psychometrics based on generalizability theory. This procedure is compared to a fully adaptive procedure under a variety of circumstances, and circumstances are delineated where each of the procedures appears to be useful.

11

Reliability and Validity of Adaptive Ability Tests in a Military Setting

JAMES R. McBRIDE JOHN T. MARTIN

Since January 1976 all military services have used a common battery of mental tests for enlisted personnel selection and classification: the Armed Services Vocational Aptitude Battery (ASVAB). The current battery includes 10 subtests of cognitive aptitudes. These subtests are necessarily short; they are usually scored by hand; the raw scores are manually converted into service-specific scaled scores using conversion tables; and the scale scores are manually recorded and manually transcribed into permanent individual personnel records.

The U.S. Marine Corps has identified some difficulties with the ASVAB testing program. Now that the ASVAB has supplanted service-specific classification test batteries, a single test battery must serve all the special testing needs of the four services. In many cases, ASVAB subtests are excessively difficult for Marine Corps selection and classification purposes; this can result in inefficient and inaccurate classification. There has been some compromise of ASVAB test security: Test booklets and answer keys have been stolen. This problem, if uncontrolled, could seriously degrade the validity of the tests for classification purposes. The manual nature of the test scoring, score conversion, and score recording procedures provides opportunity for clerical error, and it is believed that such errors may have resulted in numerous accession errors.

NEW HORIZONS IN TESTING
Latent Trait Test Theory and
Computerized Adaptive Testing

ISBN 0-12-742780-5

The Marine Corps formulated an operational requirement to lessen or eliminate the impact of these problems. Computer-administered adaptive testing (CAT) was identified as one potential solution to all these problems. In an adaptive test, test difficulty is tailored dynamically to the ability level of the individual examinee; in principle, then, CAT eliminates the problem of excessive test difficulty and should yield scores that promote accurate selection and classification decisions. CAT addresses the test security problem by eliminating printed booklets and scoring keys and by administering an individually tailored set of test items to each examinee. Additionally, since CAT automates test administration, test scoring and recording are automated as well, thereby eliminating human clerical error from the testing process.

Recognizing the potential of CAT for selection and classification testing, the Marine Corps tasked the Navy Personnel Research and Development Center (NPRDC) with investigating the feasibility of CAT as part of a program of phased research and development related to military personnel accessioning.

BACKGROUND AND PURPOSE

Group-administered paper-and-pencil "objective" ability tests date back to World War I, when the introduction of the Army Alpha test signaled an era of vast improvements in the administrative efficiency of psychological testing. The price paid for this efficiency was loss of flexibility, since all examinees had to answer a common set of test questions. The psychometric effect of this was not too serious, provided that a test was designed to have a difficulty level appropriate to its intended application or that a test was sufficiently long to overcome minor design deficiencies. For persons whose ability levels were not near the target difficulty level of the test, however, the paper-and-pencil test was not a particularly accurate or precise measuring instrument.

The psychological tests used by the armed services for selection and classification are group-administered paper-and-pencil tests. Such tests lack the flexibility to measure well over a wide range of ability. In order to achieve that flexibility, the difficulty level of the test would have to be chosen to fit individual ability levels. Since individual ability levels are not known prior to testing, this is not practical; however, it can be accomplished using an adaptive test in which test items are chosen sequentially on the basis of the examinee's performance. This sequential item choice can best be accomplished using automated test administration, for example, by having the test administered at an interactive computer terminal.

The historical development of computer-administered adaptive testing was reviewed by Weiss and Betz (1973) and by Wood (1973). Weiss surveyed a variety of alternative adaptive testing methods (1974) and summarized a number of potential advantages of CAT over conventional paper-and-pencil tests (1975).

Research by Urry (1977) was based on a Bayesian sequential adaptive testing procedure proposed by Owen (1969, 1975) and demonstrated that the procedure could achieve satisfactory levels of measurement reliability in substantially less than half the number of items required of a conventional test; in one instance Urry (1977) estimated that an adaptive test was equivalent in reliability to a conventional test five times as long. It is this *efficiency* of measurement that has motivated most psychometric interest in adaptive testing, although test users have often been more attracted by its practical advantages.

The purpose of the research reported in this chapter was to report the results of the first in a series of live-testing studies investigating the psychometric characteristics of CAT in comparison with a conventional test. A pilot study based on a small sample of Marine recruits was followed by a larger study based on the same research design. The two studies were designed in part to address two psychometric research questions:

1. Is a computer-administered adaptive test more reliable than a conventional test, holding test length constant?
2. Is a computer-administered test more valid than a comparable conventional test with test length constant?

These questions were motivated by the results of previous research. The question of the advantages of adaptive tests over conventional ones in terms of reliability has a clear and positive theoretical answer: Holding test length and all else constant, a good adaptive test is superior, provided that highly discriminating test items are available (Urry, 1970).

This theoretical advantage is not always corroborated in empirical investigations. For instance, Bryson (1971) questioned the advantage of tailored testing over certain methods of conventional test design; Olivier (1974) failed to find an advantage for the flexilevel tests he used; and the results reported by Weiss and his colleagues have been less than unanimous in favor of adaptive tests. All these results are in contrast with those of Urry (1977), who reported that for his sample of 57 Civil Service job applicants an adaptive verbal ability test achieved an 80% reduction (compared to a conventional test) in the test length required to attain any of several specified levels of reliability. Urry's result, however, was based on indirect evidence: The conventional test reliabilities were based on Spearman–Brown equation adjustments to the reliability obtained in an independent sample, and the adaptive test reliability was merely assumed, not empirically verified.

Previous research into the reliability, validity, and efficiency of adaptive tests has often been inconclusive because of design flaws or nuisance factors. The major problem has been the lack of suitable means for estimating the adaptive test's reliability without making dubious assumptions. Another problem has been the general failure to match adaptive and counterpart conventional tests in item quality, with an unfair advantage usually in favor of the adaptive test. The research reported here was intentionally designed to remove those two problems—to provide credible indices of reliability that are appropriate for both test types and to provide a fair comparison by matching item quality across the test types. With those two problem sources eliminated, the research was designed to provide a clear comparison between adaptive and conventional test designs.

METHOD

The general method used in both studies was that of equivalent tests administered to independent examinee groups. One group took two equivalent computer-administered adaptive tests. The other group took two equivalent conventional tests, also administered by computer. In order to control for item quality, both test types were made up of items from the same source—a common pool of 150 verbal ability items, which had previously been calibrated using IRT methods in large samples of Marine recruits.

Research Design

Each examinee was randomly assigned to one of the two treatment groups—Group A (adaptive) or C (conventional). Group A took two 30-

TABLE 11.1
The Research Design for Administration of the Experimental and Criterion Tests

	Tests				
	Adaptive		Conventional		
Treatment group	Form 1	Form 2	Form 1	Form 2	Criterion
A	X	X			X
C			X	X	X

item adaptive verbal ability tests, followed by a 50-item criterion test of word knowledge. Group C took two 30-item conventional verbal ability tests, followed by the same criterion test. All tests were administered at a computer terminal. Table 11.1 shows the research design.

OBSERVATIONS

For each examinee who completed the tests, the following data were observed and automatically recorded:

1. For the conventional tests, the scored item response after each item was administered
2. For the adaptive tests, Owen's (1969) IRT Bayesian ability estimates and the Bayesian posterior variance of the ability estimate after each item
3. The number-correct score on the criterion test

Following administration of the conventional tests, item responses were rescored using Owen's Bayesian ability estimation procedure.

INDEPENDENT VARIABLES

For the comparisons between the adaptive and conventional testing methods, there were two independent variables: (*a*) test type (adaptive versus conventional) and (*b*) test length (5, 10, 15, 20, 25, 30 items for Study I, and 1, 2, 3, . . . , 30 items for Study II).

DEPENDENT VARIABLES

The dependent variables were alternate forms reliability coefficients after 5, 10, . . . , 30 items in Study I and 1, 2, 3, . . . , 30 items in Study II; and correlations of ability estimates with the criterion test scores after 5, 10, . . . , 30 items in Study I and after each item was administered in Study II.

Procedure

ITEMS

The 150 items in the pool were calibrated using Urry's (1977) ancillary estimation method and were selected according to his prescriptions: All item characteristic curve slope parameters a exceeded .80. The average value of the a parameter was 1.24; item difficulty b parameters ranged from -2.0 to $+2.0$; and there were no items with a pseudoguessing c parameter greater than .30.

EXAMINEES

Male Marine recruits reporting for duty at the Marine Corps Recruit Depot, San Diego, were the examinees. Examinees in Study I were tested one at a time at a Burroughs TD832 terminal controlled by a Burroughs B1717 time-sharing minicomputer system. Assignment to groups (Group A or C) was randomized. One hundred ninety-six examinees completed the tests—92 of these took the adaptive tests and 104 took conventional tests. In Study II, examinees were tested at cathode-ray terminals (CRTs) connected by a dedicated telephone line to a Hewlett–Packard minicomputer at the University of Minnesota in Minneapolis. The CRTs presented items at 120 characters/sec. Examinees were sequentially assigned by the computer to the adaptive and conventional groups. After eliminating three examinees with incomplete tests, there were 263 recruits in the adaptive group and 267 in the conventional test group.

TESTS

The conventional tests administered to Group C were rectangular tests spanning the difficulty range of the item pool. This broad range of difficulty was chosen in order to simulate the psychometric design of the verbal tests used in the ASVAB. Two 30-item equivalent forms—Form 1 and 2—were constructed from the 150-item pool. Items were chosen to be as highly discriminating as possible, consistent with the broad difficulty range. The two forms were constructed to be "weakly parallel" (Samejima, 1977), that is, to have approximately equal test information functions. Within each form, the 30 items were sorted into five difficulty levels, and then arranged in descending order of discrimination within each level. The first five items in each form were the most discriminating items at their respective difficulty levels; items 6–10 were the second most discriminating items at each level; and so on. This arrangement resulted in two 30-item tests consisting of a sequence of six 5-item subsets each. This design was intended to permit meaningful analysis in Study I of the psychometric properties of rectangular conventional tests of lengths of 5, 10, 15, 20, 25, and 30 items. In order to equalize any effects due to test length, fatigue, or other extraneous factors, the two conventional tests were administered in counterbalanced item order (i.e., the two 30-item tests were administered as one 60-item test in a repeating 12212112 order).

The two 30-item adaptive tests were based on Owen's (1969, 1975) Bayesian sequential testing procedure. For each examinee and each test form, an initial normal prior distribution of ability was assumed, with mean 0 and variance 1.0. The test form (either 1 or 2) was counterbalanced for each examinee during the administration of each test in a manner identical to that of the conventional tests: 12212112. . . . Both forms of the Bayesian test (Forms 1 and 2) drew items from the same 150-item pool; counterbalancing the order of administration here served the added purpose of equalizing item quality across the two forms. The two adaptive tests were independent of each other except for their use of a common item pool.

The criterion test was formed by concatenating two obsolete operational test forms measuring word knowledge. This resulted in a 50-item test expected to be a highly reliable and fairly broad-range test of an important facet of verbal ability.

RESULTS

Study I

RELIABILITY

Table 11.2 summarizes reliability and concurrent validity data for both the adaptive and conventional alternate forms tests at lengths of 5, 10, 15, 20, 25, and 30 items. Reliability was operationalized as the correlation between the Bayesian ability estimates on alternate forms at a given test length. From the table it is clear that the adaptive tests had substantially higher reliability coefficients than the conventional tests for any given test length. Viewing these data another way, it can be seen that the adaptive test reliability at a 5-item test length was almost equivalent to the conventional test's reliability at 15 items; similarly, the adaptive test's reliability at a length of 10 was superior to that of the conventional test at a length of 25.

TABLE 11.2
Psychometric Characteristics of the Computer-Administered Verbal Ability Tests as a Function of Test Type and Test Length

Psychometric characteristic and test	*N*	Test length					
		5	10	15	20	25	30
Reliability							
Adaptive	92	.78	.87	.89	.91	.92	.92
Conventional	104	.58	.73	.80	.83	.86	.89
Relative efficiency		2.57	2.48	2.02	2.07	1.87	1.42
Validity							
Adaptive	92	.76	.81	.82	.83	.84	.84
Conventional	104	.73	.80	.85	.86	.86	.87

Thus the adaptive tests achieved specific levels of reliability more efficiently (i.e., with fewer items) than the conventional tests. How much more efficiently is indicated by the "relative efficiency" data in Table 11.2. These data, based on the Spearman–Brown formula, estimate for each test length how much the conventional tests would have to be lengthened to attain the reliability of the adaptive tests. For example, the adaptive test reliability at 5 items, .78, was estimated to be equivalent to that of a conventional test 2.57 times as long, or 12.85 items in length. Note that the relative efficiency of the adaptive tests always exceeds unity but diminishes as test length increases. Thus the adaptive tests are more advantageous, at least in terms of relative efficiency, at fairly short test lengths. At lengths of 10 or fewer items, the adaptive tests were about 2.5 times as efficient as the conventional tests. At lengths of 15 and more items, however, the advantage, although still appreciable, is not quite so striking.

VALIDITY

The advantage of adaptive tests was not so clear when the validity of the two test types is compared. Validity was operationalized as the correlation between scores on the experimental tests and the examinee's raw score on the concurrently administered 50-item criterion test. From their superior reliability, it would be expected that the adaptive tests would also be superior in validity at any constant test length. As Table 11.2 indicates, the adaptive tests had higher validities at test lengths up to 10 items; at lengths of 15 and up, however, the conventional tests had slightly higher validity. None of the validity differences was statistically significant at the .05 level.

Study II

RELIABILITY

Alternate forms correlations for the two forms of the Bayesian adaptive tests and for the two forms of the conventional tests as a function of test length are plotted in Figure 11.1 (numerical values are shown in Table 11.3). As the data show, the Bayesian scores for the two adaptive tests correlated .45 after one item, increased rapidly to .78 after 7 items, then increased more slowly to .90 after all 30 items. The scores on the two forms of the conventional test correlated .16 after one item, dropped to .13 after the second item, increased to .76 after 12 items and then more slowly to .89 after all 30 items were administered. After using Fisher's z transformation, t tests for differences between the reliability correlations were computed. These t tests showed that for each test length up to 19 items

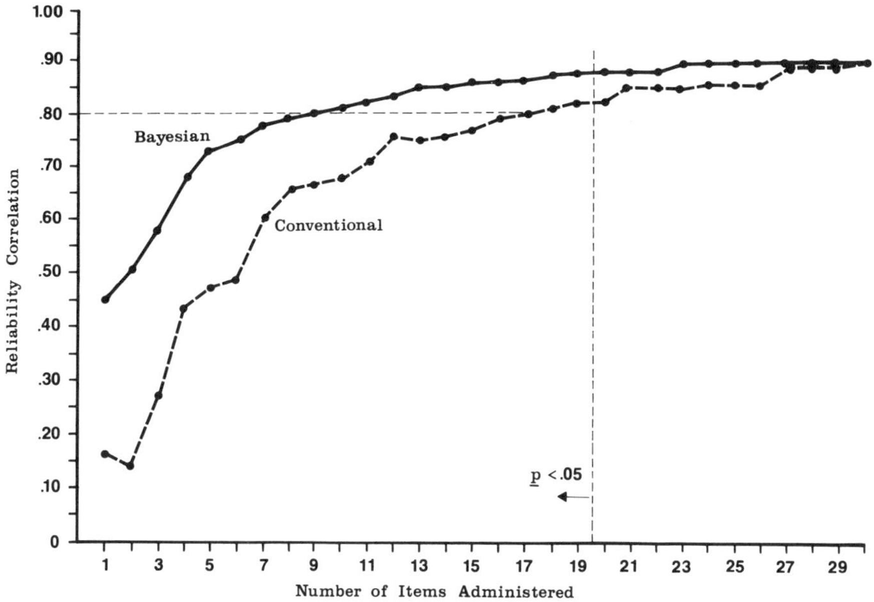

FIGURE 11.1. Alternate forms reliability correlations for the adaptive (N = 263) and conventional (N = 267) tests, as a function of the number of items administered.

TABLE 11.3
Alternate Forms Reliability Correlations as a Function of Test Length for Adaptive (N = 263) and Conventional (N = 267) Test Groups

Test length	Adaptive	Conven-tional	Test length	Adaptive	Conven-tional
1	.451	.162	16	.860	.789
2	.505	.130	17	.864	.798
3	.571	.271	18	.869	.816
4	.672	.427	19	.871	.824
5	.733	.463	20	.875	.831
6	.751	.487	21	.877	.846
7	.777	.607	22	.884	.850
8	.787	.659	23	.887	.853
9	.800	.675	24	.887	.857
10	.808	.694	25	.888	.857
11	.822	.709	26	.889	.862
12	.833	.756	27	.891	.869
13	.845	.748	28	.893	.875
14	.853	.760	29	.894	.881
15	.858	.771	30	.897	.886

(i.e., values to the left of the vertical dashed line in Figure 11.1), the adaptive forms correlated significantly higher ($p \leq .05$) with each other than did the conventional forms. Also, for all test lengths, the alternate forms reliabilities of the adaptive tests were higher than the reliabilities of the conventional tests. The horizontal dashed line in Figure 11.1 also shows that the adaptive test required only 9 items to achieve the same alternate forms reliability (.80) as a 19-item conventional test.

VALIDITY

Scores from each form of both the adaptive and conventional tests at lengths from 1 to 30 items were correlated with number-correct scores on the 50-item criterion test. These correlations are plotted in Figure 11.2; numerical data are in Table 11.4. Both adaptive forms correlated .39 with the criterion test after one item was administered, rising smoothly to .84 after all 30 items were administered. Scores on the conventional tests correlated .28 and .31, respectively, with the criterion test after one item and .80 and .81 after all 30 items were administered. The horizontal dashed line in Figure 11.2 shows that scores on Forms 1 and 2 of the

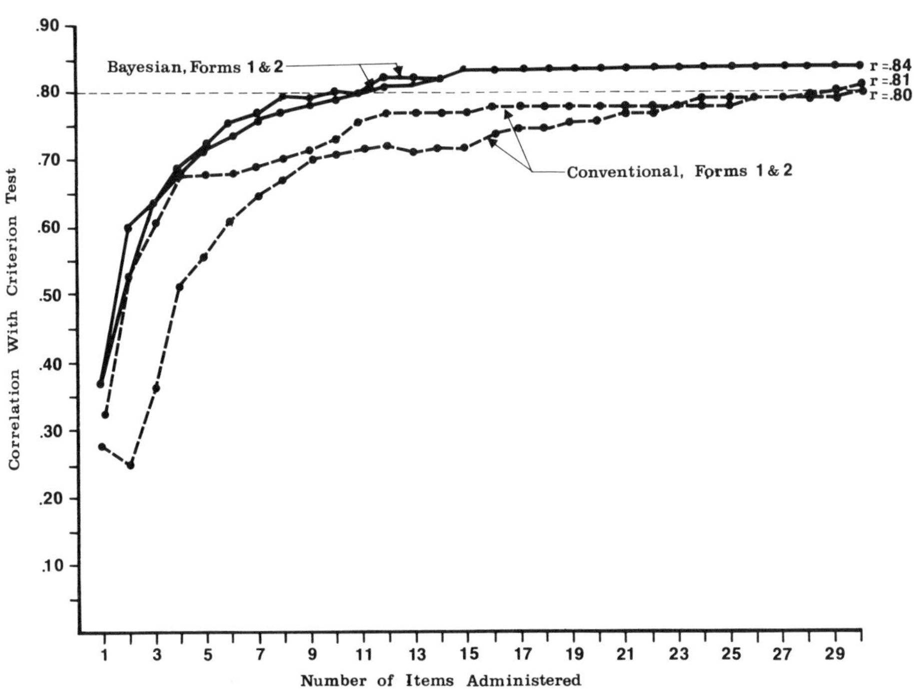

FIGURE 11.2. Validity correlations with the criterion test for two forms of the adaptive test and two forms of the conventional test, as a function of number of items administered.

TABLE 11.4
Pearson Product-Moment Correlations of Score on Forms 1 and 2 of the Adaptive Test (A1, A2) and the Conventional Test (C1, C2) with Number-Correct Scores on the Fifty-Item Criterion Test, and Results of Tests of the Significance of Differences in Pairs of Correlations

					Significant differences			
	Adaptive		Conventional		A1 vs.		A2 vs.	
Test length	Form 1	Form 2	Form 1	Form 2	C1	C2	C1	C2
1	.39	.39	.28	.31				
2	.60	.53	.25	.52	*		*	
3	.64	.63	.37	.61	*		*	
4	.69	.68	.51	.67	*		*	
5	.73	.72	.56	.67	*		*	
6	.76	.74	.61	.67	*	*	*	
7	.77	.76	.65	.68	*	*	*	
8	.79	.77	.67	.70	*	*	*	
9	.79	.78	.70	.71	*	*	*	
10	.80	.79	.71	.73	*		*	
11	.80	.80	.71	.76	*		*	
12	.82	.81	.72	.77	*		*	
13	.82	.81	.71	.77	*		*	
14	.82	.82	.72	.77	*		*	
15	.83	.83	.72	.77	*		*	
16	.83	.82	.74	.78	*		*	
17	.83	.82	.75	.78	*		*	
18	.83	.82	.75	.78	*		*	
19	.83	.83	.76	.78	*		*	
20	.83	.83	.77	.78				
21	.83	.83	.77	.78				
22	.84	.83	.77	.78	*			
23	.84	.83	.78	.78	*			
24	.84	.83	.78	.79				
25	.84	.84	.78	.79	*			
26	.84	.84	.79	.79				
27	.84	.84	.79	.79				
28	.84	.84	.79	.80				
29	.84	.84	.79	.80				
30	.84	.84	.80	.81				

* $p \leq .05$.

adaptive tests correlated .80 with scores on the criterion test after 10 items and 11 items, respectively, whereas scores on two forms of the conventional test required 30 items and 28 items, respectively, to achieve the same validity.

Table 11.4 also shows results of the pairwise comparisons (between forms of the adaptive and conventional tests) of the correlations with the criterion test. In all 120 comparisons, scores on the adaptive tests correlated more highly with scores on the criterion test than did scores on the conventional test. Although some of the differences were slight, 43 of them were sufficiently large to be statistically significant at the .05 level. Most of the significant differences occurred at test lengths of 18 items or less, but validity correlations of Form 1 of both the adaptive and the conventional tests were statistically different at test lengths of 22, 23, and 25 items.

Since the differences in the validity correlations might have been partially a function of the differences in reliabilities of the two testing strategies, Table 11.5 shows validity correlations from Table 11.4 for tests of length 5, 10, 15, 20, 25, and 30 items that have been corrected for attenuation caused by imperfect reliability in both the experimental tests and the criterion test. Alpha reliability for the 50-item criterion test was .85 in both the adaptive and conventional test groups; for experimental tests of a given length, the alternate forms reliabilities were used for these computations. Overall, scores on the Bayesian adaptive tests showed higher attenuation-corrected validity correlations than did scores on the conventional tests. For example, at the 15-item test length, average corrected validities for the adaptive tests were .97; those for the conventional tests averaged .92; at 25 items, average validities were .97 for the adaptive tests and .915 for the conventional tests. (The corrected correlation of 1.07 for Form 2 of the conventional test at five items was a result of sampling artifacts.) The implication of these corrected correlation coefficients seems to be that the ability dimension that was measured by the criterion test was more nearly identical to that measured by the Bayesian adaptive tests than by number-correct scores on the conventional test (i.e., Bayes-

TABLE 11.5
Validity Correlations Corrected for Attenuation for Forms 1 and 2 of the Adaptive and Conventional Tests, as a Function of Test Length

Test length	Adaptive		Conventional	
	Form 1	Form 2	Form 1	Form 2
5	.93	.91	.90	1.07
10	.96	.95	.93	.95
15	.97	.97	.89	.95
20	.97	.97	.92	.93
25	.97	.97	.91	.92
30	.96	.96	.92	.93

ian adaptive scores contained less error and specific variance than did number-correct scores on conventional tests of the same length).

CONCLUSIONS

Based on the data reported above, several conclusions can be drawn:

1. Striking psychometric efficiency was demonstrated for the adaptive tests of verbal ability used in these studies. It appears that in military personnel testing applications, well-constructed short adaptive tests can achieve high levels of measurement reliability with less than half the number of items required using conventional testing procedures.
2. Short adaptive tests of about 15 items per examinee seem to be sufficiently reliable for personnel testing purposes. The adaptive tests achieved a minimally satisfactory reliability level (.80) in just 5 items in Study I and 9 items in Study II; additional test lengths beyond about 10 items in Study I and about 13 items in Study II did not yield psychometric returns proportional to the added administration time required.
3. Although the validity results in Study I did not show a clear advantage for the adaptive tests, the data of Study II based on larger sample sizes and permitting statistical tests of significance showed validity differences in favor of the adaptive test for test lengths up to 19 items. The data also supported the use of short adaptive tests since there was no change in adaptive test validities for tests longer than 15 items.

Thus, based on these data, adaptive tests of about 15 items in length appear to provide verbal ability measurements that are both more reliable and more highly correlated with scores on criterion tests than do comparable conventional tests of equal length.

ACKNOWLEDGMENTS

Any opinions expressed in this chapter are those of the author and not necessarily those of the Department of Defense or the Department of the Navy.

REFERENCES

Bryson, R. *A comparison of four methods of selecting items for computer-assisted testing* (Technical Bulletin STB 72-8). San Diego, California: Naval Personnel and Training Research Laboratory, December 1971.

Olivier, P. *An evaluation of the self-scoring flexilevel testing model.* Unpublished doctoral dissertation, Florida State University, 1974.

Owen, R. J. *A Bayesian approach to tailored testing* (Research Bulletin 69-92). Princeton, New Jersey: Educational Testing Service, 1969.

Owen, R. J. A Bayesian sequential procedure for quantal response in the context of adaptive mental testing. *Journal of the American Statistical Association,* 1975, *70,* 351–356.

Samejima, F. Weakly parallel tests in latent trait theory with some criticisms of classical test theory. *Psychometrika,* 1977, *42,* 193–198.

Urry, V. W. *A Monte Carlo investigation of logistic test models.* Unpublished doctoral dissertation, Purdue University, 1970.

Urry, V. W. Tailored testing: A successful application of latent trait theory. *Journal of Educational Measurement,* 1977, *14,* 181–196.

Weiss, D. J. *Strategies of adaptive ability measurement* (Research Report 74-5). Minneapolis: University of Minnesota, Department of Psychology, Psychometric Methods Program, December 1974.

Weiss, D. J. Computerized adaptive ability measurement. *Naval Research Reviews,* 1975, *28,* 1–18.

Weiss, D. J., & Betz, N. E. *Ability measurement: Conventional or adaptive?* (Research Report 73-1). Minneapolis: University of Minnesota, Department of Psychology, Psychometric Methods Program, February 1973.

Wood, R. L. Response-contingent testing. *Review of Educational Research,* 1973, *43,* 529–544.

12

A Procedure for Decision Making Using Tailored Testing

MARK D. RECKASE

There are many applications of testing technology that require decisions about whether a person is above or below a criterion score. Criterion-referenced testing and its special case, mastery testing, are examples of such a decision. In the criterion-referenced testing application, it would be especially useful if decisions could be made quickly and conveniently for each student in an individualized instruction program. The recently developed technology of tailored–adaptive testing has the potential to fulfill the requirements of such a testing system. There is no generally accepted procedure for making classification decisions using tailored testing, however, probably because these testing techniques are still relatively new. The few procedures that do exist are either based on randomly sampling items (Epstein, 1978; Sixtl, 1974), which does not take advantage of the power of tailored testing, or on heuristic techniques (e.g., see Chapter 13), which do not have a sound theoretical base. The purpose of this chapter is to present a decision procedure that operates sequentially and can easily be applied to tailored testing without loss of any of the elegance and mathematical sophistication of the examination procedures.

NEW HORIZONS IN TESTING
Latent Trait Test Theory and
Computerized Adaptive Testing

ISBN 0-12-742780-5

TAILORED TESTING PROCEDURES

Numerous tailored (i.e., adaptive, response contingent, sequential) testing procedures now exist in the research literature, ranging from simple two-stage procedures (Betz & Weiss, 1973) to complex Bayesian procedures (Owen, 1969; see Weiss, 1974, for a good review of the tailored testing procedures that were developed prior to 1974). Although many procedures exist, only tailored testing procedures using item response theory (IRT) and maximum likelihood ability estimation will be considered in this chapter. It will also be assumed that the tests are administered to the examinees on a computer terminal and that the items are selected to maximize the value of the information function at the previous ability estimate. Despite the narrow definition of tailored testing used in this chapter, the results should generalize to any procedure based on IRT.

In applying the decision procedure discussed in this chapter, two specific IRT models will be used: the one- and three-parameter logistic models. Although any other IRT model could just as easily have been used, these models were selected because of their frequent appearance in the research literature and because of the existence of readily available calibration programs (e.g., LOGIST and BICAL) and tailored testing programs (Reckase, 1974).

SEQUENTIAL DECISION PROCEDURES

A cursory review of the statistical literature indicates that much has been written about sequential estimation and classification procedures. Although somewhat more obscure than ANOVA and regression procedures, most intermediate-level mathematical statistics books include at least one chapter on sequential analysis (for example, see Brunk, 1965, chap. 16). In a review of the extensive literature on this topic, it has been found that most procedures fall into one of three categories: (*a*) sequential probability ratio tests (SPRT; Wald, 1947); (*b*) Bayesian sequential procedures (e.g., DeGroot, 1970); and (*c*) curtailed single sampling plans (Dodge & Romig, 1929). Of these procedures, only the SPRT is narrowly specified; the other two refer to families of procedures rather than a single technique.

Although these statistical procedures are widely applied for quality control, little use has been made of them in the area of mental testing, probably because operable sequential testing procedures did not exist until recently. To date, most references in the testing literature to sequential decisions have used the SPRT (Epstein, 1978; Reckase, 1978; Sixtl, 1974). The SPRT will therefore be described and studied here. Since the Bayesian procedures have not been fully developed for practical implementation, and the curtailed sampling plans cannot readily be applied to the commonly used tailored testing procedures, they will not be discussed in this chapter.

THE SEQUENTIAL PROBABILITY RATIO TEST

The sequential probability ratio test (SPRT) was initially developed by Wald (1947) as a quality control device for use by the Armed Forces during World War II. In addition to Wald's (1947) excellent book on the subject, Epstein (1978) also clearly described this procedure. The procedure will therefore be only briefly described here in order to generalize it so that it will more directly apply to tailored testing.

Application to Mastery Decisions

Wald originally developed the SPRT as a statistical test to decide which of two simple hypotheses is more correct. For example, it may be interesting to determine whether a student can answer correctly 60% or 80% of the items in an item pool. The basic philosophy behind the procedure used to decide between these two alternatives is to determine the likelihood of an observed response to an item under the two alternative hypotheses. If the likelihood is sufficiently larger for one hypothesis than for the other, that hypothesis is accepted. If the two likelihoods are similar, another observation is taken. Wald (1947) has shown that one hypothesis will always be selected over another after the administration of a finite set of items.

To demonstrate this procedure, suppose an item is randomly selected from an item pool and administered to a student. If a correct response were obtained, the likelihood under H_1 (80% knowledge) would be .80, and the likelihood under H_0 (60% knowledge) would be .60. To evaluate these likelihoods, Wald takes the ratio of the two,

$$\frac{L(x = 1 | H_1)}{L(x = 1 | H_0)} = \frac{.80}{.60} = 1.67. \tag{12.1}$$

If the ratio is sufficiently large, H_1 is accepted; if it is sufficiently small, H_0 is accepted; and if it is near 1.0, another observation is taken. The values of this ratio that are considered sufficiently large or small depend on what is considered acceptable for the two possible decision errors: (*a*) accepting H_1 when H_0 is true (α error) and (*b*) accepting H_0 when H_1 is true (β error).

Although Wald (1947) developed a procedure for determining the exact values of these decision points, the procedure is very complex and is seldom used. Instead, good approximations can be determined using the following formulas:

$$\text{lower decision point} = B = \beta/(1 - \alpha), \tag{12.2}$$

$$\text{upper decision point} = A = (1 - \beta)/\alpha. \tag{12.3}$$

Thus, if the likelihood ratio is less than or equal to B, H_0 is accepted with error probability approximately β. If the likelihood ratio is greater than or equal to A, H_1 is accepted with error probability approximately α. If the ratio is between B and A, another item should be randomly sampled and administered and the decision rule implemented again. If $\alpha = .05$ and $\beta = .10$, for example, the decision points would be at $B = .105$ and $A = 18$. Since the likelihood ratio (1.67) in the above example is between these two values, no decision would be made, and another item would be selected and administered.

Since the responses to the items follow a binomial distribution in this example, a general expression for the likelihood ratio can be developed for the administration of n items:

$$\frac{L(x_1, x_2, \ldots, x_n | H_1)}{L(x_1, x_2, \ldots, x_n | H_0)} = \frac{p_1^{\Sigma x_i}(1 - p_1)^{n - \Sigma x_i}}{p_0^{\Sigma x_i}(1 - p_0)^{n - \Sigma x_i}}$$

$$= \left(\frac{p_1}{p_0}\right)^{\Sigma x_i} \left(\frac{1 - p_1}{1 - p_0}\right)^{n - \Sigma x_i}, \tag{12.4}$$

where x_i is the score on item i (0 or 1), p_1 is the proportion of items in the item pool known by the student under H_1, and p_0 is the proportion in the item pool known under H_0. If

$$\frac{L(x_1, \ldots, x_n | H_1)}{L(x_1, \ldots, x_n | H_0)} \geq A, \tag{12.5}$$

accept H_1. If

$$\frac{L(x_1, \ldots, x_n | H_1)}{L(x_1, \ldots, x_n | H_0)} \leq B, \tag{12.6}$$

accept H_0. Otherwise, continue administering items.

This procedure was originally developed to test simple hypotheses, but Wald (1947) has shown that the procedure operates in the same way for composite hypotheses. For example, suppose it is desired to know whether a student knew more than some proportion p of the items in an item pool. In order to use the SPRT to make this decision, a region for which it does not matter which decision is made must first be selected around p, say, $p_0 < p < p_1$. If p_0 is close to p_1, a very precise decision is required. If p_0 and p_1 define a wide indifference region around p, a rather gross decision rule is all that is needed. The SPRT is then carried out in the same fashion as above, using p_0 and p_1 as the values for hypotheses H_0 and H_1, respectively. When the decision points A and B are computed as above, the error rates α and β hold for true values of p at p_0 and p_1. For true values of p more extreme than p_0 or p_1, the error rates are lower.

Evaluating Outcomes

In order to evaluate the properties of the SPRT, two functions have been derived: the operating characteristic (OC) function and the average sample number (ASN) function. The OC function is defined as the probability of accepting hypothesis H_0 as a function of the true proportion of the item pool known by the student. Although the derivation of the OC function is somewhat complex, the function can be approximated by the following two formulas:

$$p = \frac{1 - [(1 - p_1)/(1 - p_0)]^h}{(p_1/p_0)^h - [(1 - p_1)/(1 - p_0)]^h} \qquad (12.7)$$

and

$$L(p) \simeq \frac{[(1 - \beta)/\alpha]^h - 1}{[(1 - \beta)/\alpha]^h - [\beta/(1 - \alpha)]^h}. \qquad (12.8)$$

These equations are used by substituting various arbitrary values of h and solving for p and $L(p)$. $L(p)$, the probability of accepting H_0, is then plotted against p to describe the OC function. Figure 12.1 shows an OC function for $\alpha = .05$, $\beta = .10$, $p_0 = .6$, and $p_1 = .8$. Note that at $p = p_0$ the height of the curve is equal to $1 - \alpha$, and at $p = p_1$, the height of the curve is equal to β. Note that the OC function is dependent only on α, β, p_0, and p_1. Also, the steeper the curve, the more accurate the SPRT decision rule.

The ASN function is defined as the expected number of items required to make a decision at the various values of the true proportion of known items, $E(n|p)$. The formula for the ASN function for the binomial case described above is

$$E(n|p) = \frac{L(p) \ln B + [1 - L(p)] \ln A}{p \ln(p_1/p_0) + (1 - p) \ln[(1 - p_1)/(1 - p_0)]}, \qquad (12.9)$$

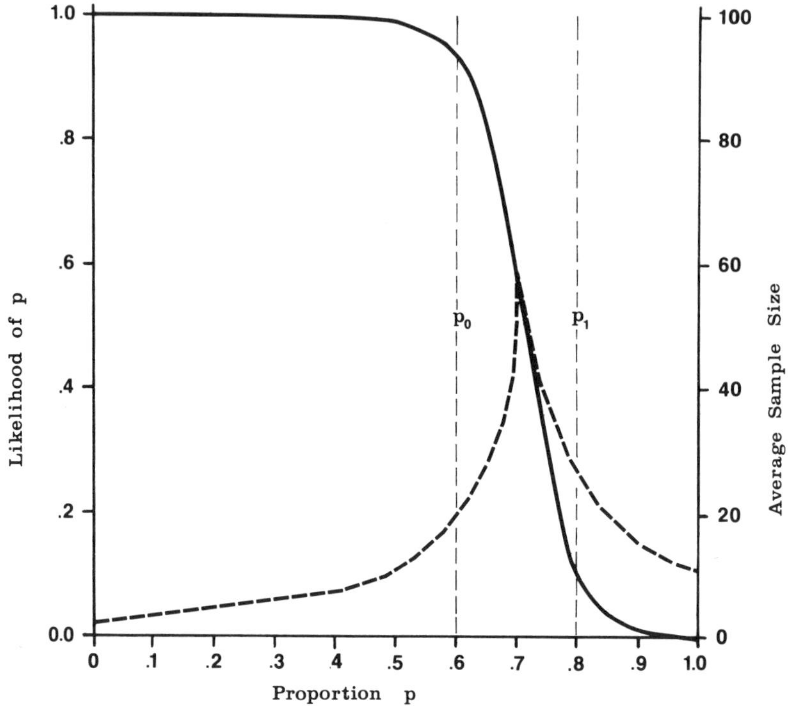

FIGURE 12.1. Example of the OC (solid line) and ASN (dashed line) functions.

where all of the symbols are as described above and the logarithms are to the base e. Figure 12.1 also shows the ASN function for the example presented above. Note that the ASN function is highest between the points p_0 and p_1 and that the closer together the values of p_0 and p_1 are, the higher the curve in that region. In general, the lower the ASN curve, the more efficient the decision rule since fewer observations are required, on the average, to make a decision.

Application to Tailored Testing

Although the SPRT as defined above is a valuable technique for decision making in many situations, it makes an implicit assumption that limits its usefulness for tailored testing. The technique as presented assumes that the probability of a correct response is the same for all items in the pool. This assumption is reasonable if items are randomly selected and p is the proportion of the items that a student can answer correctly, but it is not reasonable if items are selected to maximize information at an ability level. Under the tailored testing procedure assumed in this chapter, the probability of a correct response changes with each item, requiring a modification of the SPRT technique.

Fortunately, a detailed analysis of Wald's (1947) work indicates that the sequential random sampling assumption is not necessary for the application of the SPRT but is needed only for the derivation of the OC and ASN functions. The SPRT can be directly applied to tailored testing, but the OC and ASN functions must be determined in a different manner. One approach to determining these functions will be presented below.

To demonstrate the application of the SPRT to tailored testing as defined by this chapter, suppose that a tailored test is being used to determine whether a student has exceeded the criterion score specified for a criterion-referenced test. Although the method for selecting this criterion score is currently not well specified, assume that a value θ_c has been determined and that students above this value on the latent achievement scale pass the unit, while those below θ_c are given more instruction.

In order to use the SPRT, a region must be specified around θ_c for which it does not matter whether a pass or a fail decision is made. If high accuracy is desired for the decision rule, a narrow indifference region must be specified, but more items will be required to make the decision. As the region gets wider, the decision accuracy declines, but fewer items are required. Values on the ability scale, θ_0 and θ_1, mark the boundaries of this indifference region ($\theta_0 < \theta_c < \theta_1$). Once these values have been selected, the likelihood ratio can be defined as

$$\frac{L(x_1, \ldots, x_n \mid \theta_1)}{L(x_1, \ldots, x_n \mid \theta_0)} = \left(\prod_{i=1}^{n} P_i(\theta_1)^{x_i} Q_i(\theta_1)^{1-x_i}\right) \Big/ \left(\prod_{i=1}^{n} P_i(\theta_0)^{x_i} Q_i(\theta_0)^{1-x_i}\right), \quad (12.10)$$

where $L(x_1, \ldots, x_n \mid \theta_k)$, $k = 0, 1$, is the likelihood of the student's response string of n items administered so far; x_i is the 0, 1 score on item i; $P_i(\theta_k)$ is the probability of a correct response to item i assuming ability θ_k and the appropriate latent trait model; and $Q_i(\theta_k) = 1 - P_i(\theta_k)$.

If the one-parameter logistic model is used as a basis for the tailored testing procedure, Eq. (12.10) becomes

$$\frac{L(x_1, \ldots, x_n \mid \theta_1)}{L(x_1, \ldots, x_n \mid \theta_0)} = \left(\prod_{i=1}^{n} \frac{\exp[x_i(\theta_1 - b_i)]}{1 + \exp(\theta_1 - b_i)}\right) \Big/ \left(\prod_{i=1}^{n} \frac{\exp[x_i(\theta_0 - b_i)]}{1 + \exp(\theta_0 - b_i)}\right), \quad (12.11)$$

where b_i is the difficulty parameter for item i. Equation (12.11) can be simplified to

$$\frac{L(x_1, \ldots, x_n \mid \theta_1)}{L(x_1, \ldots, x_n \mid \theta_0)} = \left(\exp \sum_{i=1}^{n} x_i(\theta_1 - \theta_0)\right) \prod_{i=1}^{n} \frac{1 + \exp(\theta_0 - b_i)}{1 + \exp(\theta_1 - b_i)}. \quad (12.12)$$

The values of this likelihood ratio can then be used to test whether the student is above or below θ_c using the same method presented earlier. If the ratio is greater than $A = (1 - \beta)/\alpha$, the student is classified as being above θ_c; if it is below $B = \beta/(1 - \alpha)$, the student is classified below the criterion score; otherwise, another item is administered. If the three-parameter logistic model is the basis for the tailored testing procedure, the SPRT procedure is applied in the same manner as above, except that

$$P_i(\theta_k) = c_i + (1 - c_i) \frac{\exp[Da_i(\theta_k - b_i)]}{1 + \exp[Da_i(\theta_k - b_i)]} \quad (12.13)$$

is used in Eq. (12.10) instead of the simple logistic form.

The evaluation of the OC and ASN functions cannot be performed as easily as for the simple binomial model because of the presence of the item parameters in the formula for computing the probability of a correct response. Since the item parameters for the next item to be administered are dependent on the item pool used and on the responses to the previous items, the derivation of these functions depends on a complex string of conditional expectations. The conditional probabilities involved make the derivation of these functions, for all practical purposes, impossible. Therefore, the OC and ASN functions can only be approximated using simulation techniques, but these approximations should be adequate for most purposes. Note, however, that although the full OC function cannot be derived, the value of the function is equal to $1 - \alpha$ at θ_0 and to β at θ_1, assuming that the item parameters are known. In reality, these two values of the function are not known either, since in all cases except simulations the item parameters are only estimated.

RESEARCH DESIGN

The purposes of this research were (*a*) to obtain information on how the SPRT procedure functioned when items were not randomly sampled from the item pool; (*b*) to gain experience in selecting the bounds of the indifference region, θ_0 and θ_1; and (*c*) to obtain information on the effects of guessing on the accuracy of classification when the one-parameter logistic model was used.

Tailored Testing Procedure

To determine the effects of these variables, the computation of the SPRT was programmed into both the one- and three-parameter logistic tailored testing procedures that were operational at the University of Missouri—Columbia. Since these procedures have been described in detail previously (Koch & Reckase, 1978), they will be merely summarized here. The programs implementing both models used a fixed stepsize method for branching through an item pool until both a correct and an incorrect response had been given. After that point, all ability estimates were obtained using an empirical maximum likelihood estimation procedure. Items were selected for both models to maximize the item information at the most recent ability estimate.

To evaluate the decision-making power of the SPRT, subjects with known ability were needed. Therefore, a simulation routine was built into the tailored testing program in place of the responding live examinee. At the beginning of each simulation run, the true ability of the simulated examinee was input into the program. This value was used to determine the true probability of a correct response to the administered items based on the model used (one- or three-parameter logistic) and the estimated item parameters. A number was then randomly selected from a uniform distribution in the range from 0 to 1. If the randomly selected number was less than or equal to the probability of a correct response, the item was scored as correct. If the randomly selected number was greater than the probability of a correct response, the item was scored as incorrect. This procedure continued for each item in the tailored test.

Tailored tests were simulated 25 times at each true ability level using different seed numbers for the random number generator. True abilities from -3 to $+3$ at .25 intervals were used for both the one- and three-parameter models to evaluate the performance of the SPRT. In addition, simulations were run on a composite procedure in which the tailored testing procedure and the probability ratio calculations [Eq. (12.11)] were based on the one-parameter model, but the item responses were determined by using the three-parameter model. This was done to determine the effects of guessing on correct classification using the one-parameter logistic model.

Criterion Values

In computing the probability ratios, three sets of limits of the indifference regions were used: $\pm.3$, $\pm.8$, ±1. A criterion of $\theta_c = 0$ was assumed in all cases. The ratios were computed after each item was administered, and the results were compared to an A value of 45 and a B value of .102. These were determined based on $\alpha = .02$ and $\beta = .10$. A classification was made the first time these limits were exceeded. If the limits were not exceeded before 20 items had been administered (an arbitrary upper limit on test length), the values above 1.0 were classified as above θ_c and the values below 1.0 were classified as below θ_c. This is called a truncated SPRT. At each true ability used for the simulation, the proportion of the 25 administrations classified below θ_c and the average number of items administered were computed. Plots of these values against the true abilities approximate the OC and ASN functions, respectively. These plots were made for each combination of indifference region and tailored testing method, yielding nine plots of the OC and ASN functions.

Item Pools

Two different item pools were used for this study. For the analyses using just the one-parameter or the three-parameter model, an existing pool of 72 vocabulary items was used. This item pool had an approximately normal distribution of difficulty parameters. For the one-parameter tailored test using three-parameter responses, an item pool with 181 items, rectangularly distributed between -3 and $+3$ in difficulty, was used. These simulated items had constant discrimination parameters of .588 (this value yields a discrimination of 1.0 when multiplied by $D = 1.7$ in the exponent of Eq. (12.13)) and a pseudoguessing parameter of .12. This simulated item pool was selected over the real vocabulary pool to have better control over the guessing parameters. The one-parameter tailored testing procedure used only the b values from the pool, whereas the item responses were determined on the basis of all three parameters.

RESULTS

One-Parameter Model

Figure 12.2 shows the OC functions for the one-parameter logistic model based on the vocabulary item pool. The figure shows three graphs, one for each of the $\pm.3$, $\pm.8$, and ±1 indifference regions. Note that the curves are similar regardless of the indifference region. The data indicate that in all three cases the classification accuracy was nearly the same.

The values of the curves at the limits of the indifference region give further evaluative information. At the lower point the OC function should

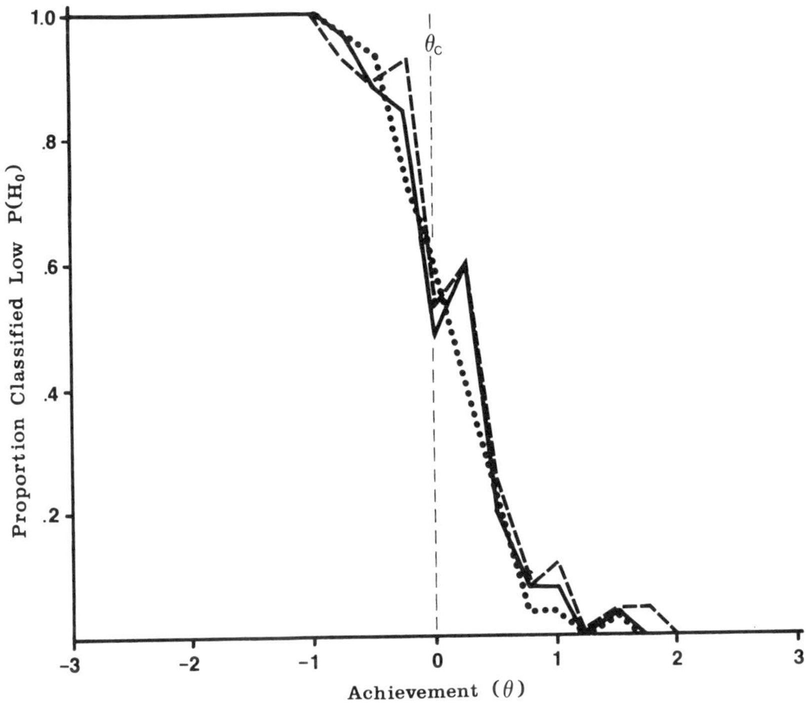

FIGURE 12.2. One-parameter OC functions for three indifference regions: solid line, ±.3; dashed line, ±.8; dotted line, ±1.0.

pass through $1 - \alpha$. At the −.3 value the curve is in fact .85 when it should be .98, showing the degrading effect of the restrictive stopping rule used by the tailored testing procedure. At the −.8 and −1 points for the corresponding curves, the results are about as expected, being .94 and 1.00 rather than .98.

At the upper limit of the indifference region, the OC function should have a value of .1. For the +.3 case it is in fact .5 rather than .1, again showing the effects of truncating the procedure. At the values of +.8 and +1 the values of the OC function were near or better than what they should have been based on the theoretically expected results.

The ASN functions for the one-parameter model are given in Figure 12.3. The curves plotted correspond to the ASN functions using indifference regions of ±.3, ±.8, and ±1. It can immediately be seen that there was substantial variability in the average number of items needed to reach a decision, with the greatest number required when the indifference region was narrowest. It can also be seen that the largest expected number of

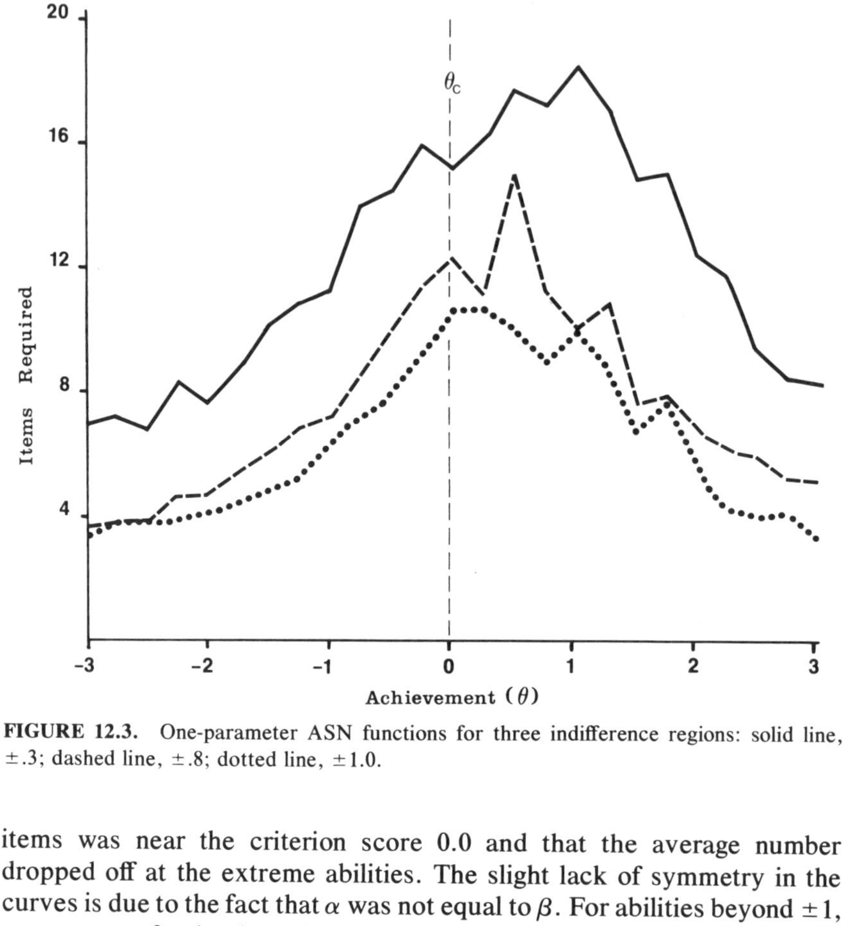

FIGURE 12.3. One-parameter ASN functions for three indifference regions: solid line, $\pm.3$; dashed line, $\pm.8$; dotted line, ±1.0.

items was near the criterion score 0.0 and that the average number dropped off at the extreme abilities. The slight lack of symmetry in the curves is due to the fact that α was not equal to β. For abilities beyond ±1, an average of only about 3–5 items was needed for classification for the wider regions, but 6–11 items were needed for the $\pm.3$ indifference region. Note that the $\pm.3$ curve approached the arbitrary 20-item limit for the tailored tests.

Figure 12.4 shows, for comparison purposes, the theoretical curves for the ASN and OC functions based on the $\pm.3$ indifference region. An infinite number of items with difficulty 0.0 was assumed for the theoretical functions, and the tests were assumed to have no upper limit on the number of items administered. A comparison of Figures 12.2 and 12.3 with Figure 12.4 shows that the OC curve for the theoretical function is steeper at the cutting point than the simulated curves, and that the ASN function is substantially higher. The difference in the theoretical and simulated OC

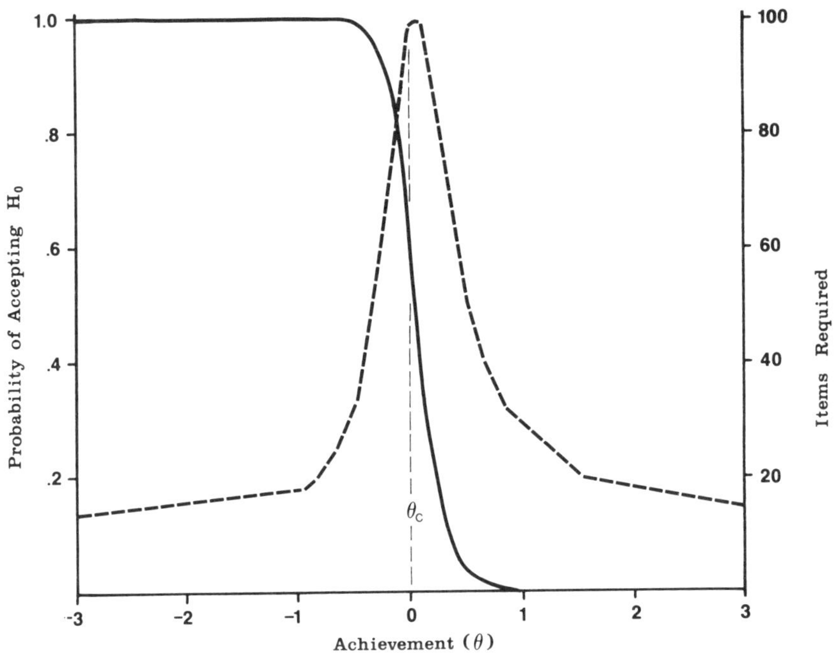

FIGURE 12.4. Theoretical OC (solid line) and ASN (dashed line) functions.

curves shows the effect of the 20-item stopping rule and the selection of items of differing difficulty.

Three-Parameter Model

The results of the simulation of the three-parameter logistic tailored test are given in Figures 12.5 and 12.6. Figure 12.5 presents the OC functions for the three-parameter model, again using the indifference regions of $\pm.3$, $+.8$, and ±1. Note that, as with the one-parameter model, the OC curves are fairly similar for the three indifference regions throughout most of the range of ability. However, there are discrepancies for the ±1 indifference region curve near the $+1$ and -1 points, indicating a decline in decision precision for that region. At the $-.3$ value for the $\pm.3$ indifference region, the value of the curve is .96, fairly close to the .98 theoretical value. At the upper end ($+.3$), however, the value is .2 instead of the .1 value that it should be. This may show the effects of guessing on the decision process. The $\pm.8$ and ±1 indifference regions again yield better error probabilities than would be expected from the theory.

The ASN function for the three-parameter model (Figure 12.6) also shows similar results to those obtained from the one-parameter model. The ±.3 indifference region required the greatest number of items, whereas ±.8 and ±1.0 required about the same number. As before, the largest number was required near the criterion score. However, with the three-parameter model far fewer items, on the average, were required to make a decision than for the one-parameter model. Of special note is the ASN value of about 1.0 in the −1 to −3 range on the ability scale. Decisions seem to be possible with very few items in that range.

Because of the guessing component of the three-parameter logistic model, the ASN function tended to yield more asymmetric results than the one-parameter model. More items were required when classifying high than when classifying low to compensate for the nonzero probability of a correct response. Also, the ASN curve for the ±.3 indifference region was much more peaked than its one-parameter counterpart. If the simulated

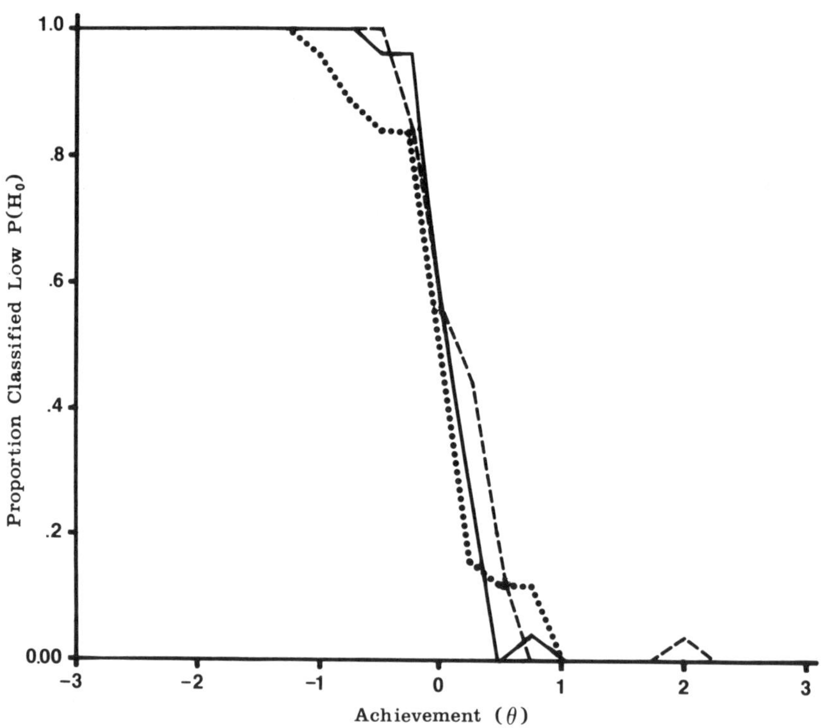

FIGURE 12.5. Three-parameter OC functions for three indifference regions: solid line, ±.3; dashed line, ±.8; dotted line, ±1.0.

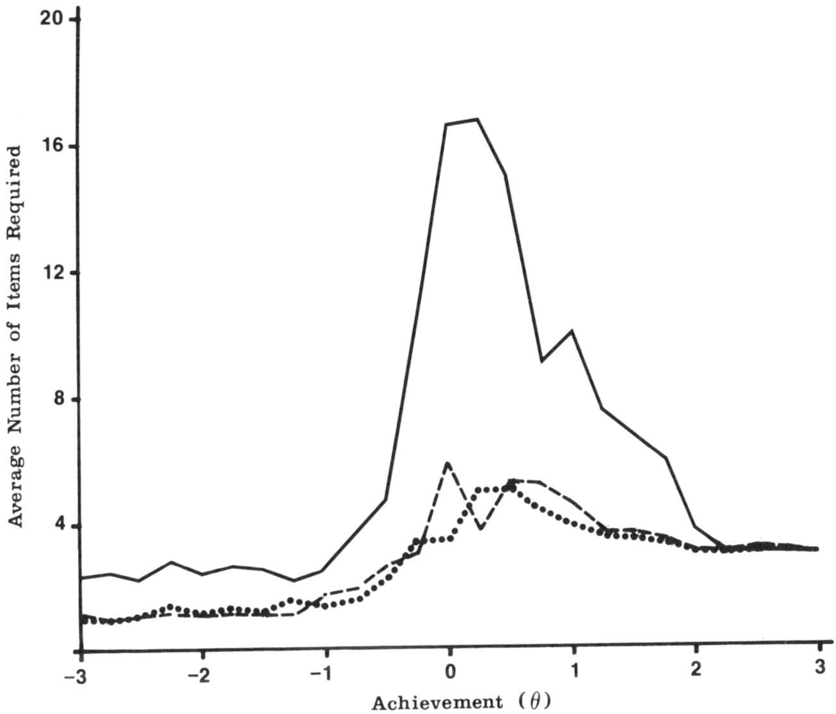

FIGURE 12.6. Three-parameter ASN functions for three indifference regions: solid line, ±.3; dashed line, ±.8; dotted line, ±1.0.

curves for the three-parameter model are compared to the theoretical curves presented in Figure 12.4, the OC functions can be seen to match the theoretical functions fairly closely, while the ASN functions show that substantially fewer items were required. Over much of the ability range, as many as 10 times more items were specified by the theoretical ASN curve when unlimited identical items were assumed. It should be noted, however, that the theoretical curves are based on the one-parameter model.

Effect of Guessing on the One-Parameter Model

Figure 12.7 shows the OC functions for the one-parameter model when the three-parameter model was used to determine the responses. The figure shows three graphs, one for each of the ±.3, ±.8, and ±1 indifference regions. Note that the curves are fairly similar regardless of the

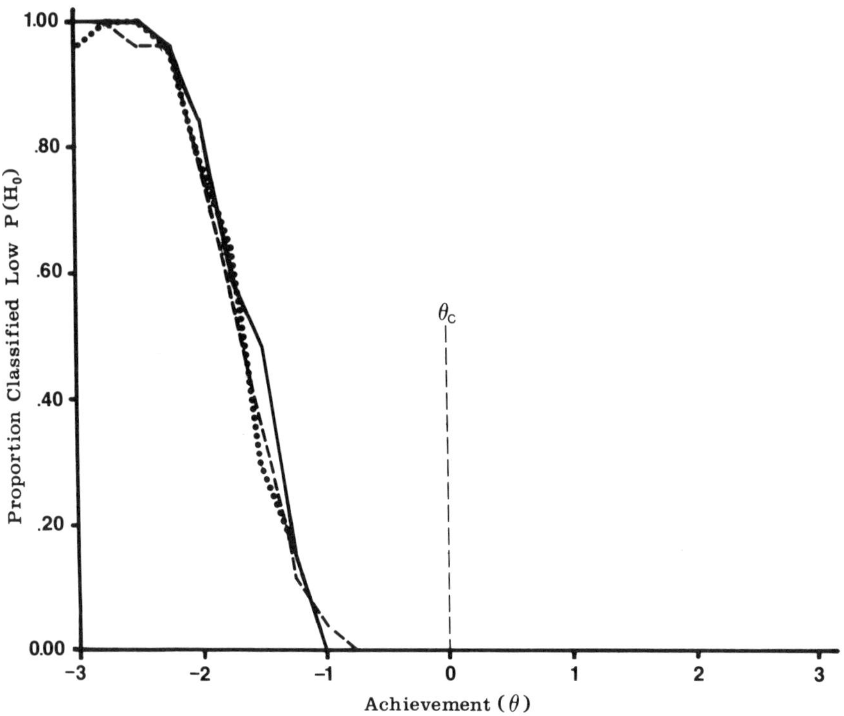

FIGURE 12.7. Composite OC functions for three indifference regions: solid line, ±.3, dashed line, ±.8; dotted line, ±1.0.

indifference region but that they are shifted substantially to the left compared to the previous OC curves. This indicates that the probability of classifying a person below θ_c has dropped off substantially until an ability of about -2 has been reached. In other words, it is much easier to be classified above the criterion score with this procedure than when guessing does not enter into the decision. Instead of being at zero, the effective criterion has been shifted down to -1.5. Clearly, the values of the OC function at the limits of the indifference region are entirely different from the theoretical values.

The ASN functions for the three indifference regions (±.3, ±.8, and ±1) are shown in Figure 12.8. The differences between these graphs and those presented in Figure 12.4 are that the curves are higher (more items were required) and the highest point of the curve is shifted to the steepest part of the OC curve. The relation between the height of the ASN function and the width of the indifference region still holds; however, as the region gets wider, the average number of items decreases.

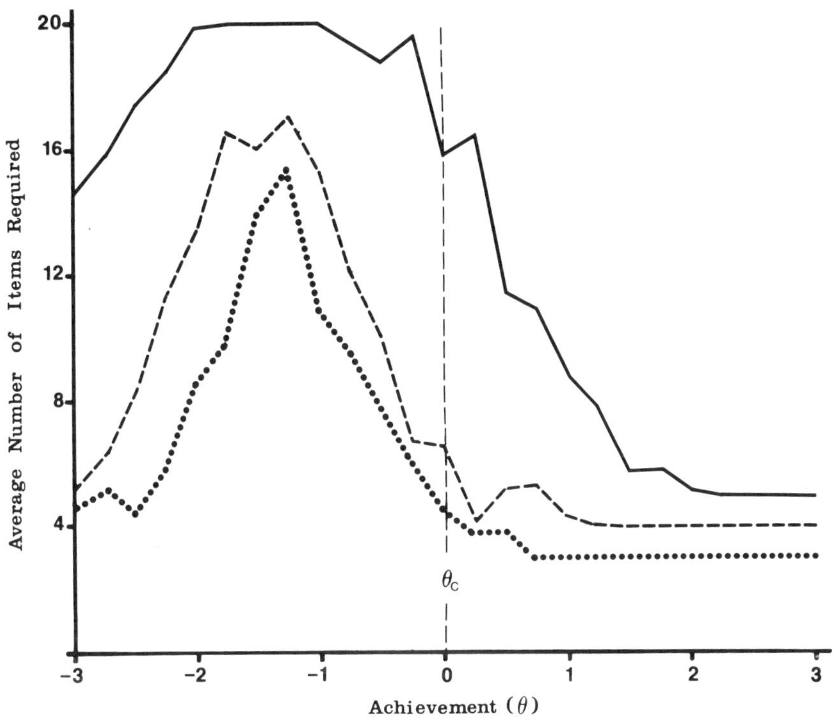

FIGURE 12.8. Composite ASN functions for three indifference regions: solid line, ±.3; dashed line, ±.8; dotted line, ±1.0.

SUMMARY AND CONCLUSIONS

The purpose of this study was to describe a procedure for making binary classification decisions using tailored testing in conjunction with the sequential probability ratio test (SPRT) and to present some simulation data showing the characteristics of the operation of the SPRT for two IRT models. The SPRT, which was developed by Wald for quality control work, has not been applied for tailored testing applications because the assumption of an equal probability of a correct response was made to facilitate the derivation of the operating characteristic (OC) and average sample number (ASN) functions. Since this assumption can be met for testing applications only by randomly sampling items for administration, the procedure has not been used with tailored testing. In the present research the probability of a correct response was allowed to vary from item to item, although it made the derivation of the OC and ASN functions impossible. Simulation procedures were used to estimate these functions.

The results of the application of the SPRT procedure in three simulation studies were described. The first study estimated the OC and ASN functions for a one-parameter logistic-based tailored testing procedure in which the size of the indifference region around the criterion score was varied. The results showed that the average number of items needed for classification was quite low when the true ability of a simulee was not too close to the criterion score and that the width of the indifference region did not greatly affect the OC function. The width of the indifference region did have a substantial effect on the ASN function. The accuracy of classification of the simulated tailored test was not quite as good as administering a large number of items with difficulty values equal to the criterion score. This result was explained by the arbitrary 20-item limit imposed on the tailored test and by the variation in the difficulty parameters of the items administered.

The second study estimated the OC and ASN functions for a three-parameter logistic tailored testing procedure, also varying the size of the indifference region. The results were similar to those for the one-parameter model, but even fewer items were generally needed for classification. The results of these first two studies both indicated that the SPRT could be successfully applied to tailored testing.

The third simulation study estimated the OC and ASN functions for the one-parameter model when guessing was allowed to enter into the responses to the items administered. The results showed that, in effect, guessing lowered the criterion score, making it easier to classify an examinee above the criterion and raising the average number of items needed for classification. This spurious shift in the effective criterion score greatly increased the error rates in classification. Although this effect was exaggerated in this study because the difficulty parameter estimates for the one- and three-parameter models were assumed to be equal except for the constant D when in reality there would be some difference in the parameter estimates, the effect was strong enough to raise serious questions concerning the use of the one-parameter model for making classification decisions when guessing is a factor in item responses.

On the basis of the results of these three simulation studies, several conclusions can be drawn. First, the combination of tailored testing and the SPRT results in a practical decision-making procedure. Second, accurate decisions can be made based on the administration of relatively few test items. Third, procedures based on the three-parameter model require fewer items for making decisions than do procedures based on the one-parameter model. Finally, guessing has a seriously detrimental effect on the decision making procedure when it is based on the one-parameter logistic model. Overall, the tailored testing–SPRT combination shows substantial promise for testing applications. Future research should concentrate on verifying the findings reported here in live-testing situations.

ACKNOWLEDGMENT

This research was supported by Contract No. N00014-77-C0097 from the Personnel and Training Research Programs of the Office of Naval Research.

REFERENCES

Betz, N. E., & Weiss, D. J. *An empirical study of computer-administered two-stage ability testing* (Research Report 73-4). Minneapolis: University of Minnesota, Department of Psychology, Psychometric Methods Program, October 1973.

Brunk, H. D. *An introduction to mathematical statistics* (2nd ed.). New York: Blaisdell, 1965.

DeGroot, M. H. *Optimal statistical decisions*. New York: McGraw-Hill, 1970.

Dodge, H. F., & Romig, H. G. A method of sampling inspection. *Bell System Technical Journal*, 1929, *8*, 613–631.

Epstein, K. Applications of sequential testing procedures to performance testing. In D. J. Weiss (Ed.), *Proceedings of the 1977 computerized adaptive testing conference*. Minneapolis: University of Minnesota, Department of Psychology, Psychometric Methods Program, 1978.

Koch, W. R., & Reckase, M. D. *A live tailored testing comparison study of the one- and three-parameter logistic models* (Research Report 78-1). Columbia: University of Missouri, Tailored Testing Research Laboratory, June 1978.

Owen, R. J. *A Bayesian approach to tailored testing* (Research Bulletin RB-69-92). Princeton, New Jersey: Educational Testing Service, 1969.

Reckase, M. D. An interactive computer program for tailored testing based on the one-parameter logistic model. *Behavior Research Methods and Instrumentation*, 1974, *6*, 208–212.

Reckase, M. D. *A generalization of sequential analysis to decision making with tailored testing*. Paper presented at the meeting of the Military Testing Association, Oklahoma City, November 1978.

Sixtl, F. Statistical foundations for a fully automated examiner. *Zeitschrift für Entwicklungspsychologie und Padagogische Psychologie*, 1974, *6*, 28–38.

Wald, A. *Sequential analysis*. New York: Wiley, 1947.

Weiss, D. J. *Strategies of adaptive ability measurement* (Research Report 74-5). Minneapolis: University of Minnesota, Department of Psychology, Psychometric Methods Program, December 1974.

13

A Comparison of IRT-Based Adaptive Mastery Testing and a Sequential Mastery Testing Procedure

G. GAGE KINGSBURY   DAVID J. WEISS

The use of criterion-referenced achievement test interpretation has gained great support within the educational measurement community since its introduction less than two decades ago (Glaser & Klaus, 1962). It is intuitively appealing to educators to be able to measure students' performances against absolute standards of behavior on prespecified learning objectives, and the use of criterion-referenced test interpretation gives educators this capability. One of the most basic forms of criterion-referenced test interpretation involves classifying students into two categories—one containing students who have achieved a sufficient command of the subject matter (mastery) and the other containing students who have not achieved a sufficient command of the subject matter (nonmastery). Traditionally, students are declared masters when their scores on a conventional classroom achievement test are as high as or higher than a prespecified cutoff level and are declared nonmasters if their scores on the test are lower than the cutoff level. This form of classroom testing has been called mastery testing and can be useful (*a*) in determining the degree of student proficiency within a classroom and (*b*) as a diagnostic tool to identify individuals who need further training in specific instructional areas (Nitko & Hsu, 1974).

As traditional mastery testing has been developing its own technology (see, e.g., Hambleton, 1980, for a review of technological developments)

NEW HORIZONS IN TESTING
Latent Trait Test Theory and
Computerized Adaptive Testing

ISBN 0-12-742780-5

adaptive testing technology has also developed to allow educators to make maximum use of classroom testing time while reducing to a minimum the amount of time spent on testing. The use of adaptive testing techniques has recently been shown to be effective in reducing test length while obtaining high-fidelity achievement level estimates in several instructional settings (e.g., Bejar, Weiss, & Gialluca, 1977; Brown & Weiss, 1977).

Mastery and adaptive testing technologies have each shown their usefulness in the academic setting for different, but compatible, reasons. It is, therefore, not surprising that a fusion of the two techniques should occur in order to allow mastery testing to be accomplished in the shortest possible class time while maintaining the accurate decisions necessary for correct diagnoses of student instructional problems.

ALTERNATIVES TO CONVENTIONAL MASTERY TESTS

Two alternatives have been proposed to conventional mastery tests in which a common set of test items is administered to every student: (*a*) applications of Wald's (1947) sequential probability ratio test (SPRT), and (*b*) Kingsbury and Weiss's (1979) formulation of an item response theory (IRT) adaptive mastery testing method. Both these testing procedures attempt to accomplish two common ends. First, the procedures seek to shorten the length of the test. Second, the procedures use statistical techniques designed to hold the number of misclassifications (i.e., individuals for whom the wrong decision is made) to some acceptable minimum. The methods by which these two procedures attempt to accomplish these ends are quite different, however.

The very fact that two procedures exist that attempt to accomplish the same basic ends through different techniques renders a comparison of the two methods desirable. The primary objective of this study was a comparison of the efficiency with which these two procedures for mastery testing achieved their goals of reducing test length while obtaining a high percentage of correct decisions.

SPRT Applied to Mastery Testing

THE SPRT

Wald's (1947) SPRT was originally designed as a quality control test for use in a manufacturing setting. It was designed to determine whether a large consignment of light bulbs (or any other product) contained a small enough proportion of defective bulbs to pass some prespecified quality criterion while testing only a small sample of the light bulbs in the consignment. Wald's solution to this problem was to draw light bulbs sequentially from the consignment, to test the light bulb drawn at each stage, and to determine at each stage the relative probabilities of the following two hypotheses:

$$H_0 : p = p_0, \tag{13.1}$$

$$H_1 : p = p_1, \tag{13.2}$$

where p is the proportion of defective elements (light bulbs) in the population (consignment); p_0 the proportion of defective elements in the population below which it is always desired to accept the quality of the population; and p_1 the proportion of defective elements in the population above which it is always desired to reject the quality of the population.

Since each stage of the sampling procedure may be viewed as a Bernoulli trial (given that each element is sampled at random without replacement from the population of equivalent elements and is assigned either nondefective or defective status), the probability of observing a certain number of defective elements in a sample of a certain size, given that either H_0 or H_1 is true, may be described with the binomial probability density function. Consequently, the probability of observing W defective elements in a sample of m elements (W_m), under $H_0 : p = p_0$ is

$$p_{0_m} = p_0^{(m-W_m)}(1 - p_0)^{W_m} \tag{13.3}$$

Under $H_1 : p = p_1$, the probability becomes

$$p_{1_m} = p_1^{(m-W_m)}(1 - p_1)^{W_m}. \tag{13.4}$$

The ratio of these two probabilities yields an index of the relative strengths of the two hypotheses such that at each stage in the sampling procedure the quality of the consignment may be either rejected or accepted, or sampling of elements may be continued. The stringency of the test is based (a) on the proportion α of errors one is willing to tolerate in rejecting the quality of consignments that actually do have the quality desired and (b) on the proportion β of errors one is willing to tolerate in accepting the quality of consignments that do not actually have the minimum acceptable quality.

In its final logarithmic form the test used by the SPRT at each stage of sampling specifies that if

$$\log(p_{1_m}/p_{0_m}) \geq \log[(1 - \beta)/\alpha], \tag{13.5}$$

the consignment is rejected; if

$$\log(p_{1_m}/p_{o_m}) \leq \log[\beta/(1 - \alpha)], \qquad (13.6)$$

the consignment is accepted; and if

$$\log[\beta/(1 - \alpha)] < \log(p_{1_m}/p_{o_m}) < \log[(1 - \beta)/\alpha], \qquad (13.7)$$

sampling continues.

Wald (1947) has shown that this testing procedure results in error levels approximating α and β across consignments. Further, it has been shown that the probability of not obtaining a decision for a consignment approaches zero as the sample size increases.

APPLICATION TO MASTERY TESTING

Ferguson (1969a,b) has applied the SPRT within a mastery testing situation using students' responses to test items in place of light bulbs and a domain of items that represents an instructional objective instead of a consignment. The quality that Ferguson evaluated was students' command of the content area being tested. Ferguson also branched through an instructional hierarchy, applying the SPRT to various objectives of instruction. The present study, however, considered only the application of the SPRT to a single instructional unit.

To employ the SPRT in a mastery testing situation, the educator must specify the following:

1. Two performance criterion levels: p_0, serving as the lowest level at which a mastery decision will be made, and p_1, serving as the highest level at which a nonmastery decision will be made. These two values bound the neutral region in which testing will continue.
2. Two levels of error acceptance (α and β), which determine the strictness of the decision test and should reflect the relative costs of the two error types.
3. A maximum test length to constrain the testing time for individuals who are very difficult to classify.

One characteristic of this form of mastery testing is that it is fairly simple to implement within a classroom situation. The decision rule is easily incorporated into a chart (e.g., Wald, 1947, sec. 5.3.2) showing the teacher or the student how many questions need to be answered correctly or incorrectly for each test length in order to terminate the test. Once the charts are made for various values of p_0, p_1, α, and β, the statistical work is completed and the SPRT procedure can be readily used by the educator. In this procedure items are presented in a single fixed sequence to all students; it is only the test length that varies for individuals.

IRT-Based Adaptive Mastery Testing

The paradigm for adaptive mastery testing (AMT) that Kingsbury and Weiss (1979) have proposed makes use of IRT and Bayesian statistical theory to adapt the mastery test to the individual's level of skill during the testing process. IRT is used (*a*) to estimate the parameters that most efficiently describe each of the items in the item pool; (*b*) to select the best items for each student; (*c*) to estimate students' achievement levels; (*d*) to determine whether those levels exceed the mastery level; and (*e*) to control the error rates for the mastery decisions. Given the IRT parameter estimates, it is possible to apply an adaptive item selection and test termination procedure that will allow mastery decisions that are quite accurate to be made while shortening the length of the test needed for most individuals.

The AMT procedure is based on three integrated procedures (see Kingsbury & Weiss, 1979):

1. A procedure for individualizing the administration of test items
2. A method for converting a traditional (proportion correct) mastery level to the latent achievement metric
3. A procedure for making mastery decisions using Bayesian confidence intervals

INDIVIDUALIZED ITEM SELECTION

To make mastery testing a more efficient process, it is desirable to reduce the length of each individual's test (*a*) by eliminating test items that provide little information concerning an individual's achievement level and (*b*) by terminating the AMT procedure after enough information has been gathered to ensure that the mastery decision can be made with a high degree of confidence. To operationalize this goal, the item to be administered to an individual at any point during the testing procedure is selected on the basis of the amount of information that the item provides concerning the individual's achievement level estimate at that point in the test; the item with the highest level of information is the item that most reduces the uncertainty in the person's achievement estimate. A procedure that selects and administers the most informative item at each point in an adaptive test—the maximum information search and selection (MISS) technique—has been described by Brown and Weiss (1977) and is used in the AMT procedure.

The information that an item provides at each point along the achievement continuum may be determined using the IRT model that is assumed to underlie the individual's responses to the test items. The AMT procedure assumes the three-parameter logistic item response function (IRF), or item characteristic curve model (Birnbaum, 1968). Using this model, the information available in any test item is (Birnbaum, 1968, Equation 20.4.16)

$$I_i(\theta) = (1 - c_i)D^2 a_i^2 \psi^2[DL_i(\theta)]/\{\psi[DL_i(\theta)] + c_i \Psi^2[-DL_i(\theta)]\}, \tag{13.8}$$

where $I_i(\theta)$ is the information available from item i at any achievement level θ; c_i the lower asymptote of the IRF for the item; $D = 1.7$, a scaling factor used to allow the logistic IRF to closely approximate a normal ogive; a_i the discriminatory power of the item at the inflection point of the IRF; ψ the logistic probability density function; $L_i(\theta) = a_i(\theta - b_i)$, where b_i is the difficulty of the item; and Ψ the cumulative logistic function.

If it is assumed that the achievement level estimate $\hat{\theta}$ is the best estimate of the actual achievement level θ, the item information of each of the items not yet administered may be evaluated at $\hat{\theta}$ at any point during the test by substituting $\hat{\theta}$ for θ in Eq. (13.8). The item that has the highest information value at the individual's current level of $\hat{\theta}$ is thus chosen to be administered next. For this study, a Bayesian estimator of the individual's achievement level, developed by Owen (1969), was used.

MASTERY LEVEL

The classical mastery testing procedure specifies a percentage of the items on a test that must be correctly answered by an individual in order for him/her to be declared a master. Using IRT, it is possible to generate an analog to the percentage-correct mastery level of classical theory for use in adaptive testing, even though the use of MISS will tend to result in each person answering about 50% of the items correctly given a large enough item pool (because items administered will most probably have difficulty levels very close to the individual's level of $\hat{\theta}$). The analog is based on the use of the test response function (TRF), or test characteristic curve (Lord & Novick, 1968). The TRF is the function that relates the achievement continuum to the expected proportion of correct answers that a person at any level of θ may be expected to obtain if all of the items on the test are administered.

For this procedure, the assumption was made that a three-parameter logistic ogive described the functional relation between the latent trait (achievement) and the probability of observing a correct response to any of the items on the test (i.e., the IRF). This assumption yields a TRF of the form

$$E(P|\theta) = \left(\sum_{i=1}^{n} c_i + (1 - c_i)\{1 + \exp[-1.7a_i(\theta - b_i)]\}^{-1} \right) \Big/ n, \tag{13.9}$$

where $E(P|\theta)$ is the expected value of the proportion of correct answers observed on the test, given any achievement level; n the number of items on the test; c_i the estimate of the lower asymptote for the IRF of item i; a_i the estimate of the item discrimination; b_i the estimate of the difficulty of the item; and θ any given achievement level. This monotonically increasing function enables one to express any given level of θ as its most likely proportion correct or, more important in this context, to determine the level of θ that will most probably result in any given proportion of correct answers (Kingsbury & Weiss, 1979, pp. 3–4).

MAKING THE MASTERY DECISION USING BAYESIAN CONFIDENCE INTERVALS

Although any achievement level estimate obtained using IRT-based scoring of any subset of the items from a test will be on the same metric as the TRF for the original test, two different subsets of items may result in achievement level estimates that are not equally informative. For example, if one test consisted of many items and the other used only a few items, the longer test would probably yield a more precise achievement level estimate if the items in the two tests had similar IRFs. Thus, IRT-based θ estimates that are on the same metric are comparable except for their differential precision. Comparisons of IRT-based θ estimates should therefore be based on confidence interval estimates instead of the point estimates.

For this reason, the AMT strategy makes mastery decisions with the use of Bayesian confidence intervals. Specifically, after each item is selected and administered to an individual (using MISS to choose the appropriate item at each point in the test), a point estimate of the individual's achievement level θ may be determined using Owen's Bayesian scoring algorithm, based on information gained from all items administered previously. Given this point estimate and the corresponding variance estimate, also obtained using Owen's procedure, a Bayesian confidence interval may be defined such that

$$\hat{\theta}_i - 1.96(\sigma_i^2)^{1/2} \leq \theta \leq \hat{\theta}_i + 1.96(\sigma_i^2)^{1/2} \qquad \text{with} \quad p = .95, \qquad (13.10)$$

where $\hat{\theta}_i$ is the Bayesian point estimate of achievement level, calculated following item i; σ_i^2 the Bayesian posterior variance following item i; and θ the true achievement level. Equation (13.10) may be interpreted as meaning that the probability is .95 that the true value of the achievement level parameter θ is within the bounds of the confidence interval. It may also be said that there is 95% confidence that the true parameter value lies within the confidence interval.

After this confidence interval has been generated, it is a simple matter to determine whether or not θ_m, the achievement level earlier designated as the mastery level on the θ metric, falls outside the limits of the confidence interval. If it does not, the AMT procedure administers another item to the individual and recalculates the confidence interval. This procedure continues until, after some item has been administered, the confidence interval calculated does not include θ_m, the mastery level on the achievement continuum. At this point the testing procedure terminates and a mastery decision is made. If the lower limit of the confidence interval falls above the specified mastery level θ_m, the individual is declared a master; if the upper limit of the confidence interval falls below θ_m, the individual is declared a nonmaster.

At any stage during the testing procedure, the individual's θ estimate will fall to one side or the other of the specified mastery cutoff, so for any person only half of the dual inequality in Eq. (13.10) needs to be evaluated at any time during the test. Thus, the decision rule implied by Eq. (13.10) is functionally equivalent to using a one-sided Bayesian confidence interval with $p = .975$. The error rate that the procedure actually produces will be examined further below.

Given a finite item pool size, the testing procedure may exhaust the pool before a decision can be made in this manner. It is possible to make a decision concerning mastery for any individuals for whom this occurs based on whether the Bayesian point estimate of their achievement level $\hat{\theta}$ is above or below the specified mastery level θ_m. These decisions, however, cannot be made with the same degree of confidence as those made with confidence intervals that do not contain the mastery level.

SPRT versus AMT

The two mastery testing strategies described above differ in a number of characteristics. The most salient of these differences are as follows:

1. Treatment of the items in the domain
2. Treatment of the uncertainty of decisions
3. Treatment of the mastery level
4. Treatment of the achievement metric

TREATMENT OF ITEMS

The SPRT, as outlined above, treats all the items in the mastery test as if they were perfect replicates of each other. Thus an individual's response to a particular item is viewed solely as a probabilistic function of the individual's true mastery status. This assumption is most appropriate in the production setting in which Wald originally designed his procedure—

that is, each light bulb can be expected to be like every other light bulb. This assumption may be less tenable in the mastery testing situation, where an individual's responses to test items may vary as a function of differential characteristics of the items themselves, as well as the individual's mastery status. However, this assumption may be tenable in a mastery testing environment in which items are randomly selected from a domain of items.

The AMT procedure assumes that if items differ, their individual characteristics may be described by a logistic ogive that varies as a function of the item's power to discriminate among individuals with different achievement levels (a), the item's difficulty (b), and the ease with which an individual may answer the item correctly with no knowledge of the subject matter (c). This assumption concerning the operating characteristics of the items is less restrictive than the assumption made in the SPRT; but to the extent that the items do not conform to the logistic form specified, the assumption might still restrict the efficiency of the AMT procedure.

Both mastery testing procedures, therefore, postulate some systematic similarities among the test items. To the extent that one of the postulations is closer to the actual state of the world than the other, it might be expected that the corresponding procedure would perform more efficiently. Thus the characteristics of the item pool itself are the first point at which the two testing strategies diverge.

TREATMENT OF UNCERTAINTY

The SPRT makes use of traditional hypothesis testing methods to determine the point at which an individual's item responses are sufficient evidence for making a decision concerning his/her mastery status. Here "sufficient" is defined in terms of the α and β error rates that are acceptable for the group of students being tested. α and β may be set independently to reflect the educator's concerns over the relative costs of the two error types.

The AMT procedure uses a symmetric Bayesian confidence interval to make the mastery decision. This functionally sets α equal to β and, by doing so, implies equal costs for the two error types. To the extent that the costs of the two error types are not equal, the SPRT procedure provides the educator with more flexibility than the AMT procedure as currently operationalized.

TREATMENT OF MASTERY LEVEL

The SPRT uses a neutral region, rather than a single mastery level, to define the mastery and nonmastery regions. The specification of this neutral region is based on a decision by the educator concerning the range that appropriately reflects uncertainty as to whether the student's performance is actually the performance of a master or a nonmaster. By contrast, the AMT procedure defines a single mastery level and determines whether an individual is significantly above or below the mastery level using a Bayesian confidence interval.

This difference between the two testing procedures renders tentative any comparison that might be made. The performance of the SPRT procedure will vary widely as a function of the uncertainty band chosen. For the AMT technique, this uncertainty is not directly taken into account. Any comparison between the two techniques is conditional upon the width and absolute bounds of the uncertainty region.

TREATMENT OF THE ACHIEVEMENT METRIC

The decisions made by the SPRT are dependent on the percentage of items that are correctly answered for any specific test length. Thus the metric of achievement assumed in this procedure is the proportion-correct metric. The AMT procedure assumes, because of the differential properties of the items in the item pool, that there is a nonlinear transformation of the proportion-correct metric that more accurately represents the achievement of the individuals taking the test. This latent continuum (the θ metric) serves as the achievement metric for the AMT procedure.

This difference in the achievement metric again renders comparisons between the two procedures somewhat difficult, since the "true" achievement levels of individuals must be postulated to fit one of these metrics. Any differences noted in the performance of the two procedures may be due to this difference in the achievement metrics assumed.

METHOD

Monte Carlo simulation was used to delineate circumstances in which one of the mastery testing procedures might have an advantage over the other. The method used to compare the two variable-length mastery testing procedures (AMT and SPRT) to one another, as well as to a conventional (fixed length) testing procedure, consisted of the following steps:

1. Four item pools were generated in which the items differed from one another to different degrees.
2. The desired mastery level on the proportion-correct metric was converted to the θ metric by means of the TRF from each item pool, as required by the AMT procedure.

3. Item responses were generated for 500 simulated subjects (simulees) for each of the items in the four item pools.
4. Conventional tests of three different lengths were drawn from the larger item pools; these conventional tests served as item pools from which the SPRT and AMT procedures drew items.
5. The AMT and SPRT procedures were simulated for each of the four different item pool types and the three conventional test lengths.
6. Comparisons were made among the three types of tests (AMT, SPRT, conventional) concerning the degree of correspondence between the decisions made by the three test types and the true mastery status. Further comparisons were made based on the average test length that each test type required to reach its decisions.

Item Pools

Four 100-item pools were generated to reflect different types of pools that might be used in a mastery test. Items were randomly ordered during the generation of each item pool.

UNIFORM POOL

The uniform pool consisted of 100 items that were perfect replications of one another. Each item had discrimination a of 1.0, difficulty b of 0.0, and lower asymptote (pseudoguessing level, or c) of .2. This pool was designed to correspond to the SPRT procedure's assumption that all items in the test are similar.

b-VARIABLE POOL

The b-variable pool varied from the uniform pool only in that the items had a range of difficulty levels. Eleven values of b were assigned to an approximately equal number of items in the pool. The values of b chosen were -2.5, -2.0, -1.5, -1.0 $-.5$, 0.0, .5, 1.0, 1.5, 2.0, and 2.5. Nine items at each level of difficulty were used in this pool, along with an additional item with $b = 0.0$ to bring the pool to 100 items (see Kingsbury & Weiss, 1980, for item parameters for items in this pool and the following pools).

a- AND b-VARIABLE POOL

The a- and b-variable pool differed from the b-variable pool in that the discriminations a of the items differed across a range of values. The a values used were .5, 1.0, 1.5, and 2.0. Each level of discrimination was equally represented in the item pool, and each level of item discrimination occurred with approximately the same frequency at each level of difficulty.

a-, *b*-, AND *c*-VARIABLE POOL

The *a*-, *b*-, and *c*-variable pool differed from the *a*- and *b*-variable pool in that the guessing levels of the items were allowed to spread across a range of values. The *c* values used were .1, .2, and .3. All *c* values were approximately equally represented. The parameter values were arranged such that each level of difficulty was represented by items that had approximately the same average *a* level and the same average *c* level.

CHARACTERISTICS OF THE ITEM POOLS

Each of the four item pools was designed so that a proportion-correct mastery level of .60 would correspond to an achievement level of 0.0 when conversion by means of the TRF was done. Although each of the item pools had the same mastery level, because of the way items were generated for each pool, they differed in their psychometric characteristics. The uniform item pool thus had all of the characteristics of a peaked test, concentrating its ability to differentiate contiguous θ levels immediately around $\theta = 0.0$. The other item pools had the characteristics of rectangular tests, with more widespread measurement capacity (see Kingsbury & Weiss, 1980, pp. 11–12, for the pool TRFs).

Item Response Generation

Achievement levels for 500 simulees were drawn from a normal distribution with a mean of 0.0 and a standard deviation of 1.0. Item responses for each of these simulees were then generated for each item in each of the four item pools using the three-parameter logistic IRT model. That is, knowing the θ level of the simulee and the parameters of the item in question, the expected probability of a correct response was calculated applying Eq. (13.9) to the item. A random number was then drawn from a uniform distribution ranging from 0 to 1. If this number was lower than the probability of a correct response, the simulee was given a correct response to the item. If the random number was higher than the correct response probability, the simulee was given an incorrect response. The "true" mastery level for each simulee was determined by comparing the θ level used to generate the item responses with the proportion-correct mastery level transformed to the θ metric.

Conventional Tests

Conventional tests (CTs) of three different lengths (10, 25, and 50 items) were drawn at random from each of the four item pools, with the stipulation that the 10-item test served as the first portion of the 25-item test and that the 25-item test in turn served as the first portion of the 50-item test. These 12 CTs served as subpools from which the AMT and SPRT procedures drew items during the simulations. This random sampling from a larger domain of items was designed to correspond to the traditional mastery testing paradigm and to the random sampling model underlying the SPRT procedure.

Simulation of the Testing Strategies

Using the item response data for the 500 simulees and the item parameters available for each of the items (for the AMT procedure), the three testing strategies (AMT, SPRT, CT) were employed to make mastery decisions for each individual. Each testing procedure was used with each of the 12 subpools.

CT PROCEDURE

Two different decision processes were used with the CTs to examine the effect of choosing an achievement metric. The first decision process used proportion-correct scores (CT/PC) and a mastery criterion of 60% correct responses. After all the items in the CT were administered, if the simulee answered 60% or more items correctly, the simulee was declared a master. If the simulee's score was less than 60% correct, the simulee was declared a nonmaster. This decision rule corresponded to the normal classroom use of conventional mastery tests with the proportion-correct achievement metric.

The second decision process used with the CTs employed the latent achievement continuum and Bayesian scoring (CT/B) to make mastery decisions. The response of a simulee to all of the items in the CT were scored using the known IRF parameters for the items and Owen's Bayesian scoring procedure, assuming a prior distribution having a mean of 0.0 and a variance of 1.0. If the final θ estimate was higher than the mastery level on the θ metric, the simulee was declared a master. If the final estimate was lower than the mastery level, the simulee was declared a nonmaster.

SPRT PROCEDURE

For the SPRT procedure the limits of the neutral region were set at proportion-correct values of .5 and .7. Values of α and β were each set to .10. For simulees for whom no decision was made by the SPRT procedure before the item pool was exhausted, the mastery decision was made in the same way as it was for the conventional testing procedure, using a mastery proportion-correct value of .6.

AMT PROCEDURE

For the AMT procedure, the mastery levels in each of the 100-item pools corresponding to 60% correct were determined from the TRF. This mastery level was used with each of the smaller item pools, even though they had not been designed to result in the same mastery level on the θ metric. This procedure added some sampling error to the AMT procedure in order to reflect more appropriately the error inherent when using estimated item parameters to determine the mastery level. As with the Bayesian scoring of the CT, each individual was given a prior distribution with a mean of 0.0 and a prior distribution with a variance of 1.0 for the Bayesian scoring of the adaptive test.

Comparison among the Testing Procedures

The AMT and SPRT procedures were compared in terms of the average reduction in the length of the test required to make mastery decisions across the entire group of individuals. All three procedures were also compared in terms of how well the decisions they made corresponded with the true mastery status of the simulees. Both comparisons were made across all 12 combinations of test lengths and item pool types. For the CT procedure, the correspondence of decisions with true mastery status was examined using both the CT/PC and CT/B scores.

RESULTS

Test Length

Table 13.1 shows the mean test length required by each of the testing procedures to make a decision concerning the mastery status of the simulees in the test group.

UNIFORM POOL

As can be seen from Table 13.1, the AMT procedure resulted in some test length reduction for each maximum test length (MTL), with the reduction in test length increasing as the MTL increased. For the 10-item MTL, the percentage by which the test length was reduced was 10%; for the 25-item MTL, the reduction was 36%; and for the 50-item MTL, the observed reduction was 54%.

For the SPRT procedure, increasing test length reduction was also noted as MTL increased; and some reduction was noted at each level of MTL. For the 10-item MTL, the reduction observed was 12%. The 25-

TABLE 13.1
Mean Number of Items Administered to Each Simulee for Three Mastery Testing Strategies Using Each Item Pool, at Three Maximum Test Lengths

Item pool and testing strategy	Maximum test length		
	10	25	50
Uniform pool			
Conventional	10.00	25.00	50.00
AMT	9.03	15.99	23.00
SPRT	8.75	13.12	15.39
b-variable pool			
Conventional	10.00	25.00	50.00
AMT	9.43	18.09	27.17
SPRT	9.62	16.79	21.41
a- and *b*-variable pool			
Conventional	10.00	25.00	50.00
AMT	8.55	15.78	24.07
SPRT	9.41	15.78	18.55
a-, *b*-, and *c*-variable pool			
Conventional	10.00	25.00	50.00
AMT	8.73	16.35	23.39
SPRT	8.62	13.42	15.70

item MTL resulted in a 48% reduction. For the 50-item MTL, the reduction was 69%. At all MTL levels the SPRT procedure resulted in a greater reduction of test length than the AMT procedure.

b-VARIABLE POOL

For the pool in which the difficulty levels of the items differed, the data in Table 13.1 show the same trends that were noted for the uniform pool. The AMT procedure reduced the test length at each MTL, and the reduction increased with the MTL level. For the 10-item, 25-item, and 50-item MTL levels, the AMT procedure reduced test length by 6, 28, and 46%, respectively.

The SPRT procedure also reduced test length at each MTL level; however, in this case, the reductions were larger for the longer MTL levels, relative to those obtained with AMT. At the 10-item, 25-item, and 50-item MTL levels the test length reductions observed were 4, 33, and 57%, respectively.

For this pool, the AMT procedure resulted in a slightly shorter mean test length (average of 9.43 items) than the SPRT procedure (average of 9.62 items) at the 10-item MTL level, whereas the SPRT procedure resulted in shorter mean test lengths for the longer MTL levels. Across all MTL levels, both procedures reduced test length somewhat less for this item pool than for the uniform item pool.

a- AND *b*-VARIABLE POOL

The data in Table 13.1 indicate that the AMT procedure again resulted in test length reduction at each MTL level with this item pool. Test length reduction was greater for the larger MTL levels. For the 10-item, 25-item, and 50-item MTL levels, the reductions in test lengths were, 14, 37, and 52%, respectively. For the SPRT procedure, test length reduction was again observed, increasing with MTL. The reductions in test length noted were 6, 37, and 63%, respectively, for the 10-item, 25-item, and 50-item MTL levels.

With this item pool, the AMT procedure resulted in a shorter mean test length (8.55 items) than the SPRT procedure (9.41 items) for the 10-item MTL. For the 25-item MTL level, both procedures produced the same mean test length (15.78 items). The SPRT procedure resulted in a shorter mean test length for the 50-item MTL. The two testing procedures resulted in mean test lengths which were, on the average, slightly longer than those observed for the uniform pool but shorter than those observed for the *b*-variable pool.

a-, *b*-, AND *c*-VARIABLE POOL

Table 13.1 shows that when the AMT procedure was used with this item pool, test length was again reduced at each MTL, with this reduction greater for the longer MTL levels. For the 10-item, 25-item, and 50-item MTL levels, the observed reductions in test length were 13, 35, and 53%, respectively.

For the SPRT procedure with this item pool, test length reduction was once more observed, with an increasing reduction as the MTL increased. The reductions noted were 14, 46, and 69%, respectively, for the 10-item, 25-item, and 50-item MTL levels.

For this item pool, the SPRT procedure terminated using a smaller average number of items than the AMT procedure for each MTL. Furthermore, the degree of test length reduction in this pool for both procedures at all MTL levels was quite similar to that observed for the uniform item pool.

Correspondence with True Mastery Status

Table 13.2 shows phi correlations between true mastery status and the mastery state determined by each of the testing procedures (correspondence coefficients) for each MTL level and pool type.

TABLE 13.2
Phi Correlations between Observed Mastery State and True Mastery State for Each Mastery Testing Strategy, Using Each Type of Item Pool, at Three Maximum Test Lengths

Item pool and testing strategy	Maximum test length		
	10	25	50
Uniform pool			
CT/PC	.771	.837	.875
CT/B	.706	.803	.863
AMT	.775	.840	.871
SPRT	.771	.837	.867
b-variable pool			
CT/PC	.541	.667	.783
CT/B	.533	.714	.791
AMT	.615	.715	.828
SPRT	.541	.656	.704
a- and *b*-variable pool			
CT/PC	.626	.719	.771
CT/B	.638	.763	.788
AMT	.638	.756	.778
SPRT	.626	.698	.720
a-, *b*-, and *c*-variable pool			
CT/PC	.290	.670	.735
CT/B	.485	.741	.804
AMT	.470	.733	.787
SPRT	.290	.592	.571

UNIFORM POOL

The major trend observed for the uniform pool was that for each testing procedure, an increase in the MTL level was accompanied by an increase in correspondence coefficients. In addition, it was observed that for the 10-item and 25-item MTL levels, the AMT procedure produced the highest correspondence coefficients observed (r = .775 and .840, respectively). For the 50-item MTL level, the CT/PC procedure resulted in the highest correspondence (r = .875). The CT/B procedure resulted in the lowest correspondence coefficient observed at each MTL level. However, the differences in correspondence between MTL levels within any testing procedure were generally larger than the differences noted between testing procedures within a single MTL level.

b-VARIABLE POOL

The same major trend that was found for the uniform pool was again observed in the *b*-variable pool: Each testing strategy resulted in higher correspondence as the MTL level increased. For this item pool, however, the AMT procedure resulted in the highest correspondence coefficient at each MTL level. The CT/B procedure resulted in the next highest correspondence coefficient for the two highest MTL levels, and the CT/PC procedure tied with the SPRT procedure for the next highest correspondence at the 10-item MTL level.

Differences in correspondence coefficients observed between testing procedures within an MTL level were larger in this pool than in the uniform pool but were still somewhat smaller than the differences noted between MTL levels, on the average. For this pool, each correspondence level observed was lower than for the uniform pool across all MTL levels and testing procedures. Both the AMT procedure and the CT/B resulted in higher correspondence coefficients at the 25-item MTL level (.715 and .714) than the SPRT procedure did at the 50-item MTL level (.704).

a- AND *b*-VARIABLE POOL

Again, each testing strategy resulted in higher correspondence coefficients as MTL increased. In this pool, however, CT/B resulted in the highest correspondence coefficients at each MTL level (tied with the AMT procedure at the 10-item MTL level). The AMT procedure resulted in the second highest correspondence coefficients at the 25-item and 50-item MTL levels.

Differences in correspondence coefficients among the various testing procedures within any MTL level still showed a tendency to be smaller than differences within any particular testing procedure across MTL levels. The correspondence coefficients observed for the CT/B and the AMT procedure at the 25-item MTL level (.763 and .756) were each higher than the correspondence coefficient observed for the SPRT procedure at the 50-item MTL level (.720).

a-, *b*-, AND *c*-VARIABLE POOL

The same pattern of increasing correspondence with increasing MTL was again noted for the CT/PC, CT/B, and AMT procedures. For the SPRT procedure, the correspondence peaked at $r = .592$ for the 25-item MTL and dropped to .571 for the 50-item MTL. The CT/B procedure resulted in the highest correspondence coefficients for all three MTL levels. The AMT procedure produced the next highest correspondence for all three MTL levels. The SPRT procedure resulted in the lowest level of correspondence at all MTL levels (tied with the CT/PC procedure at the 10-item MTL level).

Once again, the average difference in correspondence was much greater between MTL levels within testing strategies than between testing strategies within a single MTL level. Furthermore, on the average, the correspondence coefficients for this pool were lower than for the other pools, with rather large decreases at the 10-item MTL level, particularly for the CT/PC and SPRT strategies. With this item pool, the SPRT testing procedure produced a lower correspondence coefficient at the 50-item MTL level (.571) than any of the other three procedures produced at the 25-item MTL level (.670 for the CT/PC procedure, .741 for the CT/B procedure, and .733 for the AMT procedure).

Correspondence as a Function of Test Length

Figures 13.1a–d combine the observations made above concerning test length and correspondence. These figures show the correspondence coefficients (as reported in Table 13.2) as a function of the mean number of items administered by each testing strategy at each MTL level (as reported in Table 13.1) separately for each item pool. A testing strategy can be said to be most efficient to the extent that it results in the combination of highest correspondence level and shortest test length.

For the uniform item pool, Figure 13.1a indicates that the SPRT procedure was the most efficient. At all observed test lengths for which data were available, the SPRT procedure resulted in the highest correspondence coefficients. The correspondence coefficient observed for the SPRT at a mean test length of 15.39 items (the longest mean test length observed for the SPRT procedure) was .867. To achieve this correspondence level (interpolating from the data in Figure 13.1a), the AMT procedure would need to administer approximately 21 items, the CT/PC procedure would need about 45 items, and the CT/B procedure would need more than 50 items.

For the *b*-variable item pool, Figure 13.1b indicates that the AMT procedure was the most efficient testing procedure. The correspondence coefficient observed for the AMT procedure using an average test length of 27.17 items (the longest mean test length observed for the AMT procedure) was .828. As Figure 13.1b shows, this was the highest coefficient observed for any test strategy at any observed mean test length using this item pool. Each of the CT procedures would have required more than 50 items to achieve this level of correspondence. For the SPRT procedure, the longest mean test length observed was 21.41 items. At this mean test length the SPRT procedure resulted in a correspondence level of .704, whereas for the same test length the AMT procedure would have resulted in a correspondence level of approximately .745.

For the *a*- and *b*-variable item pool, Figure 13.1c shows that the AMT procedure was again the most efficient test procedure, achieving the high-

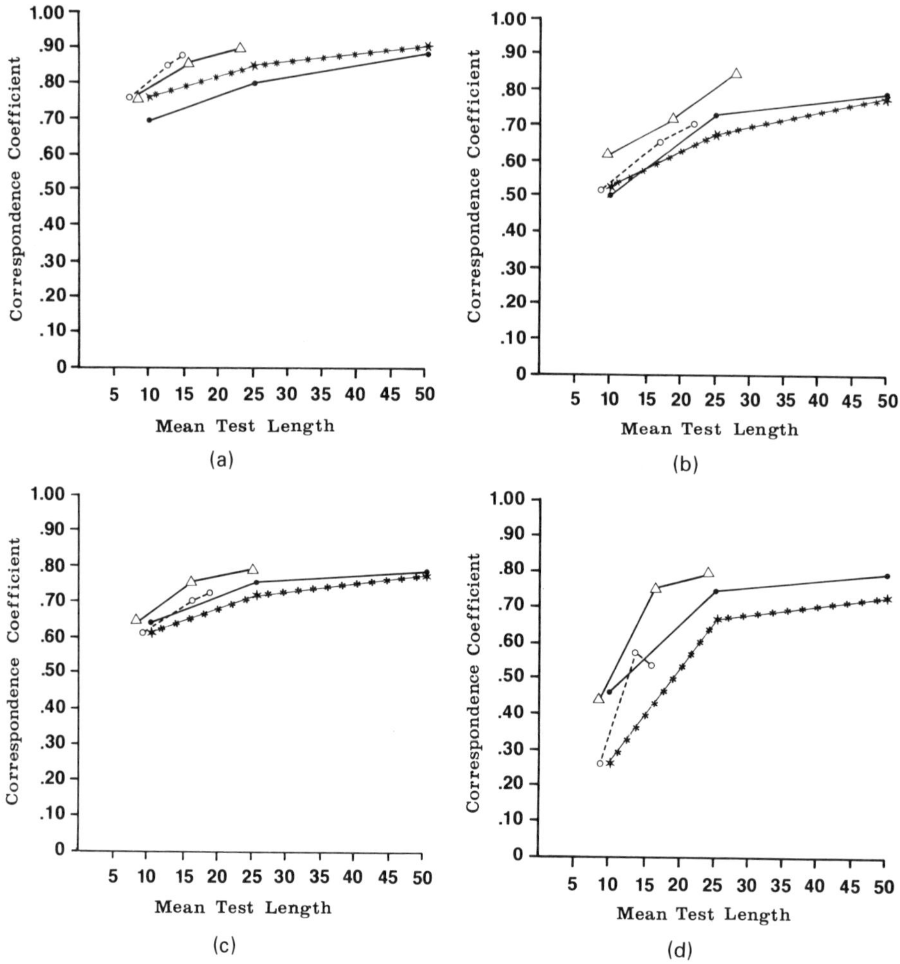

FIGURE 13.1. Correspondence coefficients as a function of the mean number of items administered by each testing strategy for each item pool: ○----○, SPRT; △——△, AMT; *-*-*, CT/PC; ●——●, CT/B. (a) Uniform item pool. (b) *b*-variable item pool. (c) *a*- and *b*-variable item pool. (d) *a*-, *b*-, and *c*-variable item pool.

est correspondence levels at each observed mean test length. At a mean test length of 24.07 items (the longest mean test length observed for the AMT procedure), the AMT procedure resulted in a correspondence coefficient of .778. The CT/PC procedure would require more than 50 items to produce as high a correspondence level, the CT/B procedure would require about 39 items. For the SPRT procedure, the longest mean test length observed was 18.55 items. At this test length the SPRT procedure resulted in a correspondence coefficient of .720, whereas interpolation shows that the AMT procedure at the same test length would yield a correspondence coefficient of about .765.

Figure 13.1d shows that the AMT procedure was also the most efficient testing procedure at all observed mean test lengths with the *a*-, *b*-, and *c*-variable item pool. At its longest observed mean test length, 23.39 items, the AMT procedure resulted in a correspondence coefficient of .787. The CT/PC procedure would need to administer more than 50 items to achieve this correspondence level, and the CT/B procedure would need about 46 items. The longest mean test length observed with the SPRT was 15.70 items, which resulted in a correspondence level of .571; interpolation of the data in Figure 13.1d shows that the AMT would probably result in a correspondence coefficient of about .710 at this test length.

The data in Figure 13.1 also show that the differences in the efficiencies of the testing strategies were much more pronounced in the *a*-, *b*-, and *c*-variable item pool than in any of the other three item pools. For instance, in the *a*-, *b*-, and *c*-variable item pool, the scoring system used with the conventional tests made a consistent difference in the correspondence coefficient observed at each test length. The magnitude of this difference was much greater with this item pool than with any other item pool.

Frequency and Type of Errors

To compare the performance of the mastery testing strategies further, the frequency with which each procedure made incorrect decisions of mastery and nonmastery was examined; the percentage of decision errors made by each of the testing strategies with each of the item pools at each MTL is shown in Table 13.3. The "Total" column in Table 13.3 reproduces in a different manner the information already reported from the correlational analysis. For each situation in which a high correlation was noted, a correspondingly low total error rate is noted in Table 13.3, as expected.

UNIFORM POOL

For the uniform pool, each of the testing strategies resulted in the same general pattern of errors across MTL levels. Each procedure resulted in fewer errors of each type with increased MTL. The difference in the frequencies of false mastery and false nonmastery decisions was smaller with higher MTL levels for all procedures except the CT/B procedure. The differences among the procedures in terms of the types of false decisions made were minimal, again except for the CT/B procedure, which

TABLE 13.3
Percentage of Incorrect Decisions by Type of Error Made by Each Testing Strategy Using Each Type of Item Pool, at Three Maximum Test Lengths

Item pool and testing strategy	10: False mastery	10: False nonmastery	10: Total	25: False mastery	25: False nonmastery	25: Total	50: False mastery	50: False nonmastery	50: Total
Uniform pool									
CT/PC	3.6	8.0	11.6	2.6	5.6	8.2	2.8	3.4	6.2
CT/B	5.4	9.4	14.8	5.0	4.8	9.8	3.8	3.0	6.8
AMT	3.6	7.8	11.4	3.0	5.0	8.0	3.0	3.4	6.4
SPRT	3.6	8.0	11.6	2.6	5.6	8.2	3.2	3.4	6.6
b-variable pool									
CT/PC	22.4	2.2	24.6	13.4	3.6	17.0	6.4	4.4	10.8
CT/B	14.0	9.2	23.2	7.8	6.4	14.2	4.8	5.6	10.4
AMT	12.2	7.0	19.2	6.6	7.6	14.2	3.4	5.2	8.6
SPRT	22.4	2.2	24.6	14.2	3.4	17.6	11.4	3.6	15.0
a- and *b*-variable pool									
CT/PC	13.4	5.4	18.8	6.8	7.2	14.0	7.2	4.2	11.4
CT/B	9.8	8.2	18.0	6.0	5.8	11.8	4.4	6.2	10.6
AMT	9.8	8.2	18.0	5.4	6.8	12.2	3.8	7.4	11.2
SPRT	13.4	5.4	18.8	8.4	6.6	15.0	9.2	4.8	14.0
a-, *b*-, and *c*-variable pool									
CT/PC	0.0	44.6	44.6	2.6	15.2	17.8	7.4	5.8	13.2
CT/B	8.4	18.0	26.4	5.0	8.0	13.0	3.8	6.0	9.8
AMT	8.0	19.4	27.4	5.2	8.2	13.4	5.0	5.6	10.6
SPRT	0.0	44.6	44.6	2.0	21.0	23.0	3.8	19.4	23.2

resulted in higher total error rates at each MTL level than any other testing strategy.

b-VARIABLE POOL

For this item pool, the patterns of errors made by the different testing strategies were less regular than in the uniform pool. The CT/PC and SPRT procedures produced more false mastery than false nonmastery decisions at all MTL levels. The AMT procedure produced more false mastery than false nonmastery decisions at the 10-item MTL level, but produced more false nonmastery than false mastery decisions at the two higher MTL levels. The CT/B procedure produced more false mastery than false nonmastery decisions for the two lowest MTL levels, but the trend was reversed for the highest MTL level. For the AMT procedure and the CT/B procedure, the discrepancy in the frequencies of the two types of errors was smaller than for the other two procedures at all three MTL levels and was quite small (less than 2%) for the two higher MTL levels. For the CT/PC procedure, the difference in the frequencies of the two types of errors was quite small at the highest MTL level; but for the SPRT procedure, a fairly large discrepancy between the two error rates (8% to 20%) was observed at each MTL.

In all testing conditions but one (AMT with a 25-item MTL), the use of the *b*-variable item pool resulted in higher discrepancies between the two observed error rates (as well as higher absolute error rates) than when the uniform pool was used.

a- AND *b*-VARIABLE POOL

Using this item pool, the SPRT procedure resulted in more false mastery decisions than false nonmastery decisions for all three MTL levels. The CT/B procedure resulted in a predominance of false mastery decisions for the two lowest MTL levels and a predominance of false nonmastery decisions at the 50-item MTL level. The CT/PC procedure produced a greater percentage of false mastery decisions at the 10-item and 50-item MTL levels, but a greater percentage of false nonmastery decisions at the 25-item MTL level. The AMT procedure resulted in a greater percentage of false mastery decisions at the 10-item MTL level, but a greater percentage of false nonmastery decisions at the 25-item and 50-item MTL levels.

Small discrepancies in error rates (less than 2%) were observed for the CT/B procedure at all MTL levels, for the AMT procedure at the 10-item and 25-item MTL levels, and for the SPRT procedure and the CT/PC procedure at the 25-item MTL level only. The SPRT procedure resulted in the largest discrepancy in error rates at all MTL levels (tied with the CT/PC procedure at the 10-item MTL).

a-, *b*-, AND *c*-VARIABLE POOL

For this item pool, each of the testing procedures resulted in higher frequencies of false nonmastery decisions than false mastery decisions for the 10-item and 25-item MTL. For the 50-item MTL, the CT/PC procedure resulted in a higher frequency of false mastery decisions, but the CT/B, AMT, and SPRT procedures still resulted in higher percentages of false nonmastery decisions.

The AMT procedure used with this item pool resulted in smaller differences in the frequencies of the two error types than any of the other testing procedures at the 50-item MTL level. The CT/B procedure resulted in the lowest discrepancy in error rates (9.6%) at the 10-item MTL level.

The CT/B and AMT procedures tied for the lowest discrepancy in error rates (3%) at the 25-item MTL level. For the 50-item MTL level, the AMT procedure produced a very small difference in the two error rates (.6%). The CT/PC procedure also produced a small difference in the two error rates for the 50-item MTL level (1.6%). The SPRT procedure resulted in the highest difference between the two error rates at all MTL levels (tied with the CT/PC procedure at the 10-item MTL).

One interesting result was observed when the errors made with the *b*-variable item pool were compared with those made using the *a*-, *b*-, and *c*-variable item pool. For the *b*-variable pool, each of the testing procedures was more likely to make false mastery decisions than false nonmastery decisions. This tendency was reversed for the *a*-, *b*-, and *c*-variable item pool, where each of the procedures made more false nonmastery decisions than false mastery decisions. These trends were most noticeable for each of the testing procedures at the 10-item MTL level and most noticeable for the SPRT procedure across all MTL levels.

It is probable that these trends were artifacts of the random sampling of items used to create the conventional tests, since the shorter conventional tests would be less representative of the item domain because of the small sample of items taken. The results obtained here would be explained by a very easy 10-item conventional test being drawn from the *b*-variable pool and a very difficult 10-item test being drawn from the *a*-, *b*-, and *c*-variable pool. In fact, the mean *b*-value for the 10-item conventional test drawn for the *b*-variable pool was −.80; for the *a*-, *b*-, and *c*-variable pool, it was 1.25. This would also explain the observation that the SPRT procedure most clearly showed these trends, since the SPRT procedure used shorter test lengths, on the average, than the other two procedures to make its final decisions and therefore was most prone to small-sample artifacts.

DISCUSSION AND CONCLUSIONS

Several trends were noted in the data concerning the performance of the different testing strategies with the four different item pools. In every instance the AMT and SPRT procedures produced reductions in the mean test length required to make mastery decisions. This reduction increased with the MTL in each item pool. The AMT procedure resulted in reductions of 6–54% from the length of the conventional test. The SPRT procedure resulted in reductions of 4–69%. On the average, the SPRT procedure required fewer items to make the mastery decision.

The correspondence between the estimated mastery status and the true mastery status systematically increased with MTL for all testing procedures in each item pool (with the exception of the SPRT procedure used with the *a*-, *b*-, and *c*-variable item pool). Either the AMT procedure or the CT/B procedure resulted in the highest level of correspondence in all circumstances but one (the CT/PC procedure performed best for the 50-item MTL with the uniform pool). On the average, though, the differences between different MTL levels were more pronounced than differences between testing procedures. Furthermore, the kind of item pool used had important effects on the correspondence obtained.

In the *b*-variable and the *a*- and *b*-variable item pools, the AMT and CT/B procedures resulted in higher levels of correspondence at the 25-item MTL level than did the SPRT procedure at the 50-item MTL level. For the *a*-, *b*-, and *c*-variable pool, the AMT, CT/B, and CT/PC procedures each resulted in higher correspondence at the 25-item MTL level than did the SPRT procedure at the 50-item MTL level.

When correspondence coefficients were examined as a function of the mean number of items administered by each of the testing strategies, the results differed depending on the type of item pool used. For the uniform item pool, the SPRT procedure was the most efficient testing procedure (i.e., it produced the highest correspondence level for any mean test length) for all observed mean test lengths. Using each of the other item pools, the AMT procedure was the most efficient testing strategy at all observed mean test lengths. Both the AMT and SPRT procedures were more efficient than either conventional testing procedure at all observed mean test lengths using the uniform and *b*-variable item pools. Using the *a*- and *b*-variable and *a*-, *b*-, and *c*-variable item pools, only the AMT procedure was consistently more efficient than the conventional testing procedures.

The AMT procedure resulted in the most even frequencies in the types of decision errors made across most MTL levels and item pools. This was desirable, since both error types were assumed to have the same relative cost. Furthermore, it was noted that the SPRT procedure was most susceptible to small-sample artifacts, resulting in the largest imbalance in the frequencies with which the two types of errors were made.

To prescribe the best testing strategy of those described here requires specification of priorities and conditionals. If testing time is at a premium in a course of instruction, then it might be important to shorten the length of tests in the course without much loss of decision accuracy. Both the AMT and SPRT procedures accomplish this. Results from this study indicate that the performance of these strategies is affected greatly by the characteristics of the items available for use. If a uniform item pool is available in which all items are equal in difficulty, discrimination, and guessing parameters, then the SPRT procedure as applied in this study will require the fewest items while resulting in decisions having correspondence coefficients that are quite comparable to the other three procedures. If, however, the item pool includes items with variable a, b, and c parameters, the SPRT procedure results in the shortest tests, but each of the other procedures will make more accurate classifications. Using a realistic item pool of this type, the AMT procedure provides the optimal combination of high decision correspondence and short test length. Combined with the results of Kingsbury and Weiss's (1981) live-testing application of the AMT strategy, which showed average test length reductions of 80% and higher criterion-related validities for the AMT strategy in comparison with "optimal" conventional mastery tests, the results of the present study support the use of the AMT strategy for making efficient and valid classifications.

ACKNOWLEDGMENTS

This research was supported by funds from the Air Force Human Resources Laboratory, the Air Force Office of Scientific Research, the Army Research Institute, and the Office of Naval Research, and monitored by the Office of Naval Research under Contract N00014-79-C-0172 NR 150-433.

REFERENCES

Bejar, I. I., Weiss, D. J., & Gialluca, K. A. *An information comparison of conventional and adaptive tests in the measurement of classroom achievement* (Research Report 77-7). Minneapolis: University of Minnesota, Department of Psychology, Psychometric Methods Program, October 1977.

Birnbaum, A. Some latent trait models and their use in inferring an examinee's ability. In F. M. Lord & M. R. Novick, *Statistical theories of mental test scores*. Reading, Massachusetts, Addison-Wesley, 1968.

Brown, J. M., & Weiss, D. J. *An adaptive testing strategy for achievement test batteries* (Research Report 77-6). Minneapolis: University of Minnesota, Department of Psychology, Psychometric Methods Program, October 1977.

Ferguson, R. L. *Computer-assisted criterion-referenced measurement* (Working Paper No. 41). University of Pittsburgh, Learning and Research Development Center, 1969. (ERIC Document Reproduction No. ED 037 089). (a)

Ferguson, R. L. The development, implementation, and evaluation of a computer-assisted branched test for a program of individually prescribed instruction. (Doctoral dissertation, University of Pittsburgh, 1969.) (University Microfilms No. 70-4530). (b)

Glaser, R., & Klaus, D. J. Proficiency measurement: Assessing human performance. In R. M. Gagne (Ed.), *Psychological principles in system development*. New York: Holt, 1962.

Hambleton, R. K. (Ed.). Contributions to criterion-referenced testing technology. *Applied Psychological Measurement*, 1980, *4* (4).

Kingsbury, G. G., & Weiss, D. J. *An adaptive testing strategy for mastery decisions* (Research Report 79-5). Minneapolis: University of Minnesota, Department of Psychology, Psychometric Methods Program, September 1979.

Kingsbury, G. G., & Weiss, D. J. *A validity comparison of adaptive and conventional strategies for mastery testing*. (Research Report 81-3). Minneapolis: University of Minnesota, Department of Psychology, Psychometric Methods Program, Computerized Adaptive Testing Laboratory, September 1981.

Lord, F. M., & Novick, M. R. *Statistical theories of mental test scores*. Reading, Massachusetts: Addison-Wesley, 1968.

Nitko, A., & Hsu, T. C. Using domain referenced tests for student placement, diagnosis, and attainment in a system of adaptive individualized instruction. *Educational Technology*, 1974, *14*, 48–53.

Owen, R. J. *A Bayesian approach to tailored testing* (Research Bulletin 69-92). Princeton, New Jersey: Educational Testing Service, 1969.

Wald, A. *Sequential analysis*. New York: Wiley, 1947.

V

MODELS FOR SPECIAL APPLICATIONS

Item information is an important concept in applications of item response theory. As indicated in Chapter 13, one way to select items in an adaptive test is by selecting items with highest levels of item information at a current estimated ability level. When item information is summed over a set of items, it becomes test information (or its reciprocal, the standard error of measurement or the standard error of estimate studied by Hambleton in Chapter 2), which is useful in test design.

For purposes of measuring the abilities of individuals when they are assumed to vary over a wide range, it is important that all ability estimates have the same standard error. One way of achieving this is through adaptive testing with a termination rule based on a value of the standard error of measurement. Since the standard error of measurement given a specified level of ability is inversely proportional to test information, this can be expressed as constant information over a defined range of θ. Another way to achieve constant information in the measurements is described by Samejima in Chapter 14. Samejima describes her potentially useful constant information model and explores some of its properties, such as the simplified ability and error variance estimation procedures which derive from the model.

Another variation of the general latent trait model is described by Fischer in Chapter 15. Unlike the other latent trait models described and applied in this book, Fischer's is based on qualitative variables. Also unlike the others, which are designed to measure status variables at one point in time, Fischer's model—based on the one-parameter logistic (Rasch) model—is explicitly designed to measure change. Fischer describes and explores a procedure for measuring change resulting from experimental treatments, with qualitative variables as dependent variables.

Both Chapters 14 and 15, as well as the other chapters in the book, demonstrate only two of the many possible variations of latent trait models for general measurement problems, both within and outside the field of psychological testing. These models and their applications emphasize the general power of this class of models for innovative solutions to a wide range of psychological measurement problems.

14

Constant Information Model on the Dichotomous Response Level

FUMIKO SAMEJIMA

Generally speaking, the fundamental role of the mathematical model in psychology is to simulate psychological reality following some sound rationale, using well-defined parameters. The normal ogive model in latent trait theory, for example, is one of such mathematical models. Another role of the mathematical model may be to provide some mathematical convenience, just as the logistic model does in its relationship with the normal ogive model.

Mathematical models of a third kind, which have their specific usefulness in the context of comprehensive theories and methods, can be conceived. The direct simulation of psychological reality is less important for this kind of model than it is for the first two kinds of models. The *constant information model* belongs to this new kind; its role is to be of assistance in the developmental stages of theories and methods, rather than to simulate psychological reality directly.

The constant information model (Samejima, 1979) is a new model on the dichotomous response level (Samejima, 1972). This model provides a constant item information for a finite interval of a latent trait. Although the usefulness of the model has not yet been fully investigated, an effective use of the model has been found in the process of estimating the operating characteristics of item response categories (Samejima, 1977b, 1977c, 1978a, 1978b, 1978c, 1978d, 1978e, 1978f).

NEW HORIZONS IN TESTING
Latent Trait Test Theory and
Computerized Adaptive Testing

ISBN 0-12-742780-5

Let θ be ability, or latent trait, which is assumed to be unidimensional. Let $g = 1, 2, \ldots, n$ be an item and $x_g = 0, 1, 2, \ldots, m_g$ denote an item response category, or item score. The operating characteristic $P_{x_g}(\theta)$ of the item score x_g is defined by

$$P_{x_g}(\theta) = \text{prob}[x_g | \theta]. \tag{14.1}$$

Let V be a response, pattern, such that

$$V = (x_1, x_2, \ldots, x_g, \ldots, x_n)', \tag{14.2}$$

and $P_V(\theta)$ be its operating characteristic. Because of the assumption of local independence,

$$P_V(\theta) = \prod_{x_g \in V} P_{x_g}(\theta) \tag{14.3}$$

can be written. The *item response information function* (Samejima, 1969) $I_{x_g}(\theta)$ is defined by

$$I_{x_g}(\theta) = -\frac{\partial^2}{\partial\theta^2} \log P_{x_g}(\theta) \tag{14.4}$$

and the *item information function* $I_g(\theta)$ is the conditional expectation of the item response information function, given θ, so that

$$\begin{aligned} I_g(\theta) = E[I_{x_g}(\theta)|\theta] &= \sum_{x_g=0}^{m_g} I_{x_g}(\theta) P_{x_g}(\theta) \\ &= \sum_{x_g=0}^{m_g} \left(\frac{\partial}{\partial\theta} P_{x_g}(\theta)\right)^2 [P_{x_g}(\theta)]^{-1} \end{aligned} \tag{14.5}$$

(cf. Samejima, 1969, chap. 6). The *response pattern information function* $I_V(\theta)$ can be written for a specified response pattern V such that

$$I_V(\theta) = -\frac{\partial^2}{\partial\theta^2} \log P_V(\theta) = \sum_{x_g \in V} I_{x_g}(\theta) \tag{14.6}$$

and the *test information function* $I(\theta)$ is defined as the conditional expectation, given θ, of the response pattern information function, which can be written

$$I(\theta) = \sum_V I_V(\theta) P_V(\theta) = \sum_{g=1}^{n} I_g(\theta). \tag{14.7}$$

When item g is binary (i.e., is scored either 0 or 1), u_g is used for the item score category instead of x_g. The item characteristic function $P_g(\theta)$

is defined by

$$P_g(\theta) = \text{prob}[u_g = 1|\theta], \tag{14.8}$$

and the other operating characteristic for $u_g = 0$, which is denoted by $Q_g(\theta)$, can be written

$$Q_g(\theta) = \text{prob}[u_g = 0|\theta] = 1 - P_g(\theta). \tag{14.9}$$

From Eqs. (14.5), (14.8), and (14.9)

$$I_g(\theta) = \left(\frac{\partial}{\partial\theta} P_g(\theta)\right)^2 [P_g(\theta)\, Q_g(\theta)]^{-1}, \tag{14.10}$$

which is identical with the item information function defined by Birnbaum (1968), can be obtained for the item information function.

THE CONSTANT INFORMATION MODEL

The item characteristic function of the new model, the constant information model (CIM), is given by

$$P_g(\theta) \begin{cases} = \sin^2[a_g(\theta - b_g) + \pi/4] & \text{for } \underline{\theta} < \theta < \bar{\theta} \\ = 0 & \text{otherwise,} \end{cases} \tag{14.11}$$

where

$$\underline{\theta} = -\pi a_g^{-1}/4 + b_g, \qquad \bar{\theta} = \pi a_g^{-1}/4 + b_g. \tag{14.12}$$

Figure 14.1 presents five examples of the item characteristic functions in the CIM with varieties of sets of parameters. From Eqs. (14.9) and (14.11), $Q_g(\theta)$ can be written for the other operating characteristic such that

$$Q_g(\theta) \begin{cases} = \cos^2[a_g(\theta - b_g) + \pi/4] & \text{for } \underline{\theta} > \theta > \bar{\theta} \\ = 1 & \text{otherwise.} \end{cases} \tag{14.13}$$

Since

$$\begin{aligned} \frac{\partial}{\partial\theta} P_g(\theta) &= 2 \sin[a_g(\theta - b_g) + \pi/4] \\ &\quad \times \cos[a_g(\theta - b_g) + \pi/4] \times a_g \\ &= 2a_g[P_g(\theta) Q_g(\theta)]^{1/2} \end{aligned} \tag{14.14}$$

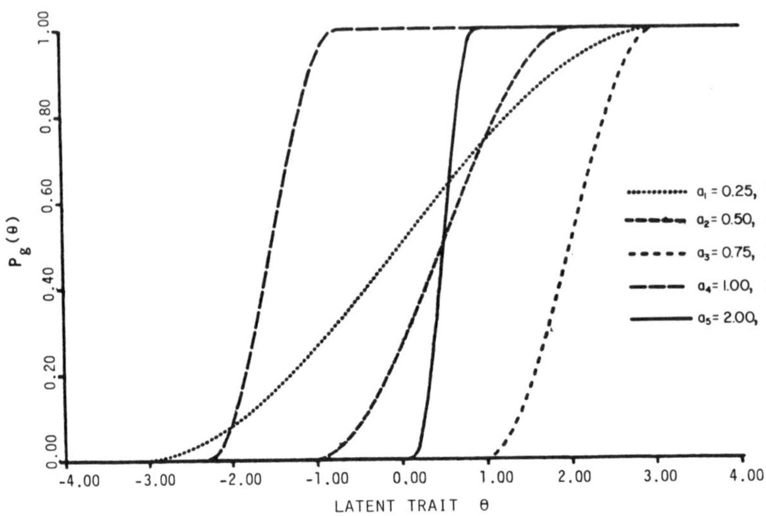

FIGURE 14.1. Item characteristic functions of five binary items following the constant information model.

can be written for the interval of θ such that

$$-\pi a_g^{-1}/4 + b_g < \theta < \pi a_g^{-1}/4 + b_g, \tag{14.15}$$

it is obvious that in this model (a) the item characteristic function is strictly increasing in θ in the above interval of θ, and (b)

$$\lim_{\theta \to \underline{\theta}} P_g(\theta) = 0, \qquad \lim_{\theta \to \bar{\theta}} P_g(\theta) = 1. \tag{14.16}$$

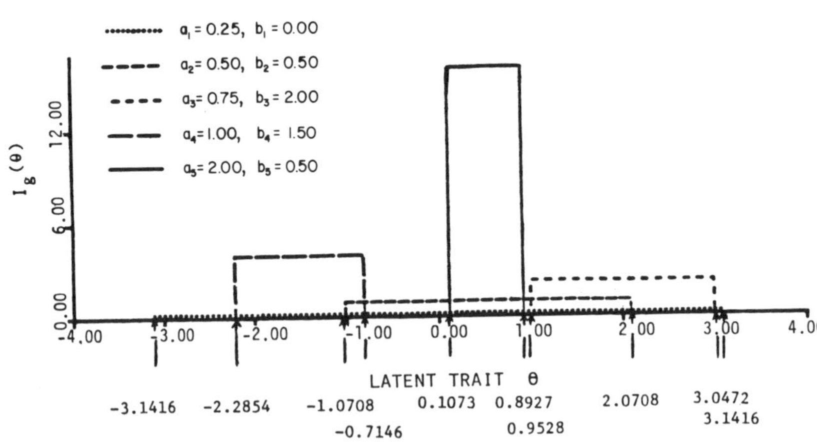

FIGURE 14.2. Item information functions of five binary items following the constant information model.

For convenience, hereafter, the CIM will be considered only for the range of θ given by Eq. (14.15), unless otherwise stated. The mathematical models that satisfy (*a*) and (*b*) will be called models of Type I.

It is obvious from Eqs. (14.10), (14.11), (14.13), and (14.14) that the model provides constant item information such that

$$I_g(\theta) = 4a_g^2 = C_g, \tag{14.17}$$

where C_g indicates the constant amount of information. Figure 14.2 presents the item information functions for the five items whose item characteristic functions are given in Figure 14.1. The length of the interval of θ for which the item information function equals C_g is given by

$$\bar{\theta} - \underline{\theta} = \pi C_g^{-1/2} = \pi(2a_g)^{-1}. \tag{14.18}$$

The basic function $A_{x_g}(\theta)$ (Samejima, 1969, 1972), which is defined by

$$A_{x_g}(\theta) = \frac{\partial}{\partial\theta} \log P_{x_g}(\theta) \tag{14.19}$$

for the item response category x_g, is obtained for the CIM from Eqs. (14.11), (14.13), (14.14), and (14.19) such that

$$A_{u_g}(\theta) \begin{cases} = -2a_g[P_g(\theta)]^{1/2}[Q_g(\theta)]^{-1/2} & \\ \quad = -2a_g \tan[a_g(\theta - b_g) + \pi/4] & \text{for} \quad u_g = 0 \\ = 2a_g[Q_g(\theta)]^{1/2}[P_g(\theta)]^{-1/2} & \\ \quad = 2a_g \cot[a_g(\theta - b_g) + \pi/4] & \text{for} \quad u_g = 1 \end{cases} \tag{14.20}$$

Figure 14.3 presents these basic functions for the five items shown earlier. It is clear from Eq. (14.20) that the basic function for $u_g = 1$ is a strictly decreasing function of θ with positive infinity and zero as its two asymptotes and that for $u_g = 0$ it is a strictly decreasing function of θ with zero and negative infinity as its two asymptotes, respectively. This is also confirmed visually by Figure 14.3.

The item response information functions for each of the binary scores, $u_g = 0$ and $u_g = 1$, are given from Eqs. (14.4), (14.11), (14.13), and (14.14) by

$$I_{u_g}(\theta) \begin{cases} = 2a_g^2 \sec^2[a_g(\theta - b_g) + \pi/4] & \\ \quad = 2a_g^2[Q_g(\theta)]^{-1} > 0 & \text{for} \quad u_g = 0 \\ = 2a_g^2 \csc^2[a_g(\theta - b_g) + \pi/4] & \\ \quad = 2a_g^2[P_g(\theta)]^{-1} > 0 & \text{for} \quad u_g = 1. \end{cases} \tag{14.21}$$

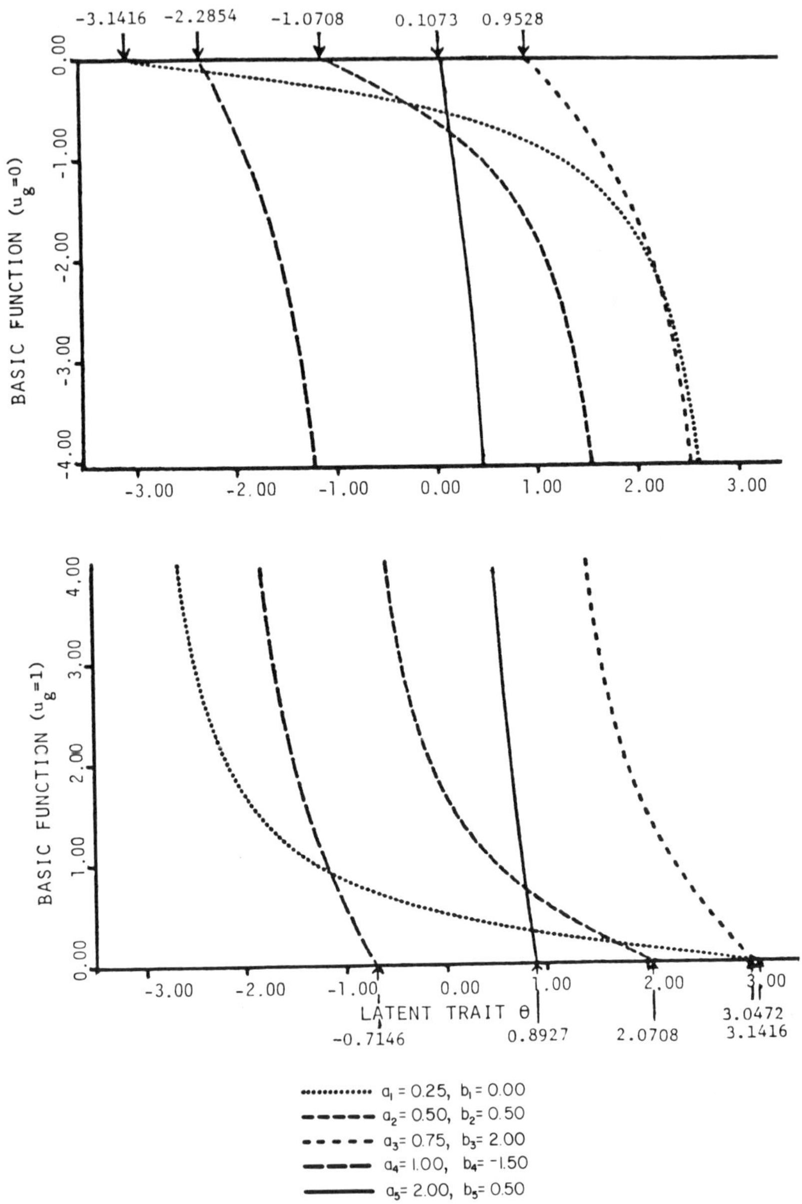

FIGURE 14.3. Basic functions of five items following the constant information model for $u_g = 0$ (upper graph) and for $u_g = 1$ (lower graph).

Figure 14.4 illustrates these two functions for the item with parameters $a_g = .25$ and $b_g = 0.00$, together with the constant item information (.25). The response pattern information function $I_V(\theta)$ is written from Eqs. (14.6) and (14.21) such that

$$I_V(\theta) = 2 \sum_{u_g \in V} a_g^2 [P_g(\theta)]^{-u_g} [Q_g(\theta)]^{u_g^{-1}} \tag{14.22}$$

and the test information function $I(\theta)$ is given from Eqs. (14.7) and (14.17) by

$$I(\theta) = 4 \sum_{g=1}^{n} a_g^2 . \tag{14.23}$$

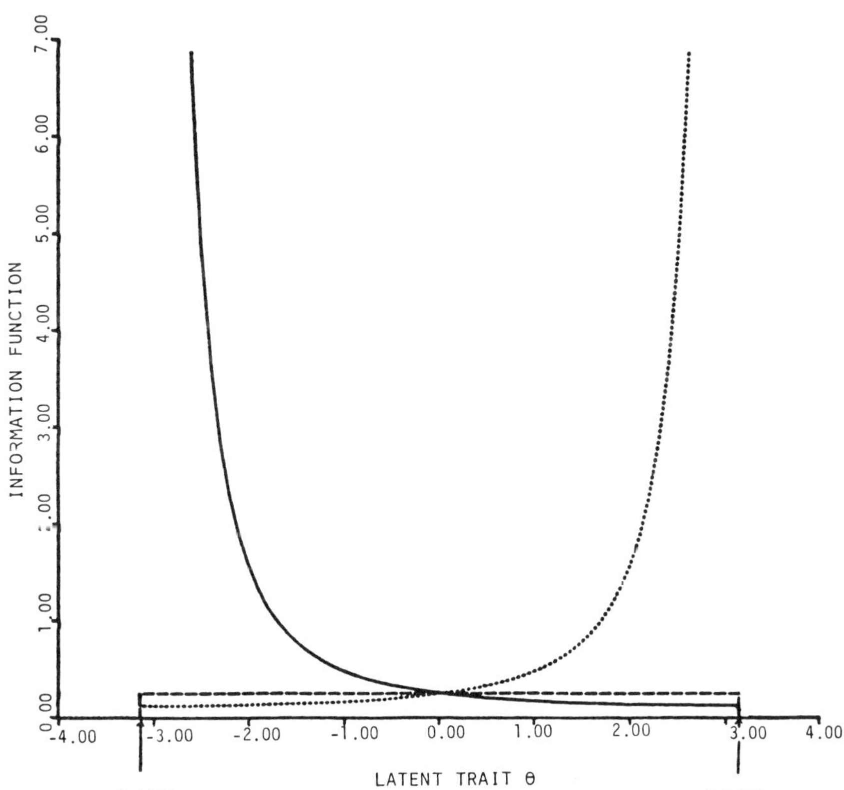

FIGURE 14.4. Item response information functions of an item following the constant information model, with the parameters $a_g = .25$ and $b_g = 0.0$ for $u_g = 0$ (dotted line) and for $u_g = 1$ (solid line), together with the constant item information function (dashed line).

Figure 14.5 presents both the set of four response pattern information functions and the test information function for a hypothetical test of two binary items, whose item parameters are $a_1 = .25$, $b_1 = 0.00$, $a_2 = .50$, and $b_2 = 1.00$, each of which follows the CIM.

It should be noted that the interval of θ for which the item information is a positive constant is always finite. This is related to the fact that

$$\int_{\underline{\theta}}^{\bar{\theta}} [I_g(\theta)]^{1/2} \, d\theta = \pi, \tag{14.24}$$

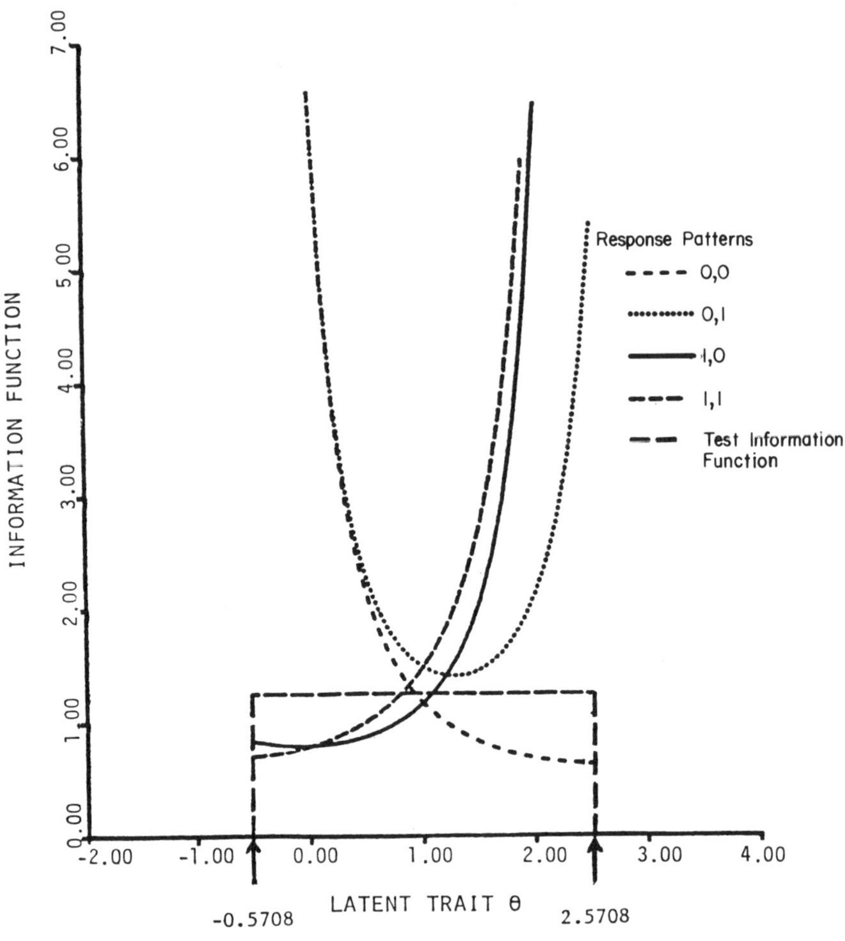

FIGURE 14.5. Response pattern information functions of the four possible response patterns following the constant information model with the parameters $a_1 = .25$, $b_1 = 0.0$, $a_2 = .50$, and $b_2 = 1.0$.

for any model of Type I, which includes the CIM (Samejima, 1979). Thus

$$C_g^{1/2}(\bar{\theta} - \underline{\theta}) = \pi, \tag{14.25}$$

and $\bar{\theta} - \underline{\theta}$ must therefore be a finite value.

USE OF THE CONSTANT INFORMATION MODEL IN THE ESTIMATION OF THE OPERATING CHARACTERISTICS OF ITEM RESPONSE CATEGORIES

The methods and approaches for estimating the operating characteristics of the item response categories of a test item developed so far (Samejima, 1977b, 1977c, 1978a, 1978b, 1978c, 1978d, 1978e, 1978f) have common characteristics such that (*a*) they are made without assuming any prior mathematical form and (*b*) they use a relatively small number of examinees in the process of estimation. One common restriction in these methods and approaches is that what is needed is an old test, consisting of the items whose operating characteristics are known, which provides a constant test information function for the interval of latent trait, or ability θ, of interest. With this setting, each examinee's ability level is estimated from his/her response pattern by maximum likelihood estimation; and the set of these maximum likelihood estimates is the main information source of the estimation procedures.

We note that these methods and approaches are useful in a situation when there already is an item pool and it is desired to add more test items to it. In a different setting where the starting point is the very beginning of item pool development, however, these methods appear to be useless, since there is no old test, and, consequently, the maximum likelihood estimate of each examinee's ability is not given a priori.

In the above setting, it is very common that a large number of test items, which include many items of intermediate difficulties, is administered to the group of examinees chosen for the purpose of developing the item pool. If, among the items administered, a substantially large subset of binary items can be found that can be regarded as equivalent items, then that subset of items can be used in place of the old test. Even though their item characteristic functions are not known, this can be done with the aid of the CIM. In other words, it is possible to expand the estimation methods developed so far to make them usable when there is no old test.

Estimation Procedures

In practice, it is necessary to identify these equivalent items without knowing their operating characteristics. This can be done as follows. First, the proportion correct must be obtained for each item. If a subset of items exists whose proportions correct are around .5 and close enough to one another, then these items make a good candidate for the subset of equivalent items. Second, a 2×2 contingency table, as exemplified in Table 14.1, must be made for each pair of items in the subset. In order for the items to be equivalent, it is necessary that within the range of sampling fluctuations, these 2×2 contingency tables of the bivariate frequency distributions should be symmetric and identical for all the pairs of binary items. This can be checked for every pair of binary items that have passed the first screening, and, very likely, more items must be eliminated through this second screening. Now there can be progression to the third screening of 2^3 contingency tables, to the fourth screening of 2^4 contingency tables, and so forth. Note, however, that an increasingly large number of more complicated contingency tables is usually encountered as progression is made to higher stages of screening. Therefore, some criterion must be set for terminating this process and the remaining items must be accepted by assuming their equivalence. Table 14.1 illustrates two typical 2×2 contingency tables—one is for a pair of equivalent binary items that has a common low discrimination parameter and the other is for a pair that has a common high discrimination parameter.

After the subset of equivalent binary items has been identified, the CIM

TABLE 14.1
Two Typical 2×2 Contingency Tables for a Pair of Equivalent Items with a Common Low Discrimination Parameter, and a Pair with a Common High Discrimination Parameter

	Item h		
Item g	$u_h = 0$	$u_h = 1$	Total
	Low discrimination parameter		
$u_g = 0$	110	243	353
$u_g = 1$	248	399	647
Total	358	642	1000
	High discrimination parameter		
$u_g = 0$	300	53	353
$u_g = 1$	58	589	647
Total	358	642	1000

is assumed for each equivalent item. This assumption is made without loss of generality, since the scale of the latent trait is subject to any strictly increasing transformation (Samejima, 1969, 1979).

The next step is to obtain the maximum likelihood estimate, $\hat{\theta}$, of θ on the response pattern of each of the N examinees, with respect to the above subset of equivalent items. Let V^* be the examinee's response pattern on the subset of k equivalent items. Since they are equivalent items, the simple test score t, where

$$t = \sum_{u_g \in V^*} u_g, \tag{14.26}$$

is a minimal sufficient statistic, regardless of the model that these item characteristic functions follow (cf. Birnbaum, 1968). Thus the procedure of maximum likelihood estimation becomes much simpler, using the test score t instead of the response pattern V^*. In general, for any model on the dichotomous response level,

$$P_{V^*}(\theta) = \prod_{u_g \in V^*} [P_g(\theta)]^{u_g}[Q_g(\theta)]^{1-u_g}. \tag{14.27}$$

Since this operating characteristic of the response pattern V^* is itself the likelihood function in estimating the examinee's ability, the symbol $L_{V^*}(\theta)$ will be used for this function in the present section. When all the items are equivalent, Eq. (14.27) can be rewritten in the form

$$L_{V^*}(\theta) = [P_g(\theta)]^t[Q_g(\theta)]^{k-t}. \tag{14.28}$$

Thus, for the likelihood equation we can write

$$\frac{\partial}{\partial\theta} \log L_{V^*}(\theta) = \left(\frac{\partial}{\partial\theta} P_g(\theta)\right) [t - kP_g(\theta)] \times [P_g(\theta)Q_g(\theta)]^{-1} = 0, \tag{14.29}$$

and the equation

$$t = kP_g(\theta) \tag{14.30}$$

is obtained. For the maximum likelihood estimate $\hat{\theta}$,

$$\hat{\theta} = P_g^{-1}(t/k) \tag{14.31}$$

Now, Eq. (14.11) must be substituted into Eq. (14.31), which results in

$$\hat{\theta} = a_g^{-1}[\sin^{-1}(t/k)^{1/2} - \pi/4] + b_g. \tag{14.32}$$

It is obvious from Eqs. (14.15) and (14.32) that the range of $\hat{\theta}$ is given by

$$-\pi a_g^{-1}/4 + b_g \leq \hat{\theta} \leq \pi a_g^{-1}/4 + b_g. \tag{14.33}$$

Thus the maximum likelihood estimate of each examinee's ability level is obtained, based on his/her test score for the subset of k equivalent items. This set of N maximum likelihood estimates can be used in place of the one obtained in the old test, and the process of operating characteristic estimation in any combination of a method and an approach can be followed. The error variance σ^2 is given by

$$\sigma^2 = [kC]^{-1} = [4ka^2]^{-1}, \tag{14.34}$$

where

$$C_1 = C_2 = \cdots = C_k = C \tag{14.35}$$

and

$$a_1 = a_2 = \cdots = a_k = a. \tag{14.36}$$

After the process of estimation has been completed, the latent trait θ can be transformed to another latent trait τ by any strictly increasing function $\tau(\theta)$. To give some examples, if θ is transformed to τ by

$$\tau = P_g^{*-1}(\{\sin[a_g(\theta - b_g) + \pi/4]\}^2), \tag{14.37}$$

where

$$P_g^*(\tau) = (2\pi)^{-1/2} \int_{-\infty}^{a_g^*(\tau - b_g^*)} \exp\left(\frac{-t^2}{2}\right) dt, \tag{14.38}$$

then all the equivalent items will follow the normal ogive model specified by Eq. (14.38), and

$$\underline{\theta} = -\infty, \qquad \bar{\theta} = \infty; \tag{14.39}$$

if the transformation is such that

$$\tau = (\beta_g - \alpha_g) \sin^2[a_g(\theta - b_g) + \pi/4] + \alpha_g, \tag{14.40}$$

then they will follow the linear model, whose item characteristic is given by

$$P_g(\theta) = (\theta - \alpha_g)(\beta_g - \alpha_g)^{-1} \tag{14.41}$$

with

$$\underline{\theta} = \alpha_g, \qquad \bar{\theta} = \beta_g \tag{14.42}$$

and if θ is transformed to τ by

$$\tau = [1/(Da_g^*)] \log\{\tan^2[a_g(\theta - b_g) + \pi/4]\} + b_g^*, \tag{14.43}$$

then they will follow the logistic model, which is characterized by

$$P_g^*(\tau) = \{1 + \exp[-Da_g^*(\tau - b_g^*)]\}^{-1} \tag{14.44}$$

with the $\underline{\theta}$ and $\bar{\theta}$ given in Eq. (14.39). Similar transformations can be made to change the item characteristic functions of k equivalent items from those of the CIM to those of any other models of Type I. In each case, the newly estimated operating characteristics of the other items will be transformed according to the specific transformation of the latent trait θ to τ.

Some Necessary Considerations

In using the generalized method of operating characteristic estimation, which was described in the preceding section, a few problems must be considered. First, the constant test information provided by the subset of equivalent binary items with the CIM should be substantially large so that the normal approximation for the conditional distribution of $\hat{\theta}$, given θ, will be acceptable. On the other hand, a substantially wide range of ability θ, for which the test information is constant, is needed in order to make the estimation of the operating characteristics of the other items in the item pool meaningful. These two are opposing factors, as is obvious from Eqs. (14.15) and (14.17). The solution for this problem is to use a substantially large number of equivalent binary items whose common discrimination parameter is low.

Another problem is the effect of the range of $\hat{\theta}$ on the speed of convergence of the conditional distribution of $\hat{\theta}$, given θ, to the normal distribution, $N(\theta,\{kC\}^{-1/2})$. Since the range of $\hat{\theta}$ is a finite interval, which is given by Eq. (14.33), it should be expected that the truncation of the conditional distribution makes the convergence slow around the values of θ close to $-\pi a_g^{-1}/4 + b_g$ and $\pi a_g^{-1}/4 + b_g$, as is illustrated in Figure 14.6. A solution for this problem is again to use a subset of equivalent binary items whose common discrimination parameter is low so that the range of θ is wide enough to include all the examinees sufficiently inside of the two endpoints of the interval of θ. An alternative for the above solution is to exclude examinees whose $\hat{\theta}$s are close to $-\pi a_g^{-1}/4 + b_g$ or $\pi a_g^{-1}/4 + b_g$. In the second solution, however, the number of examinees usable for the estimation will be decreased, and this may substantially affect the accuracy of the estimation of the operating characteristics. It is worth noting that the solution for the first problem also seems to be the best solution for the second problem.

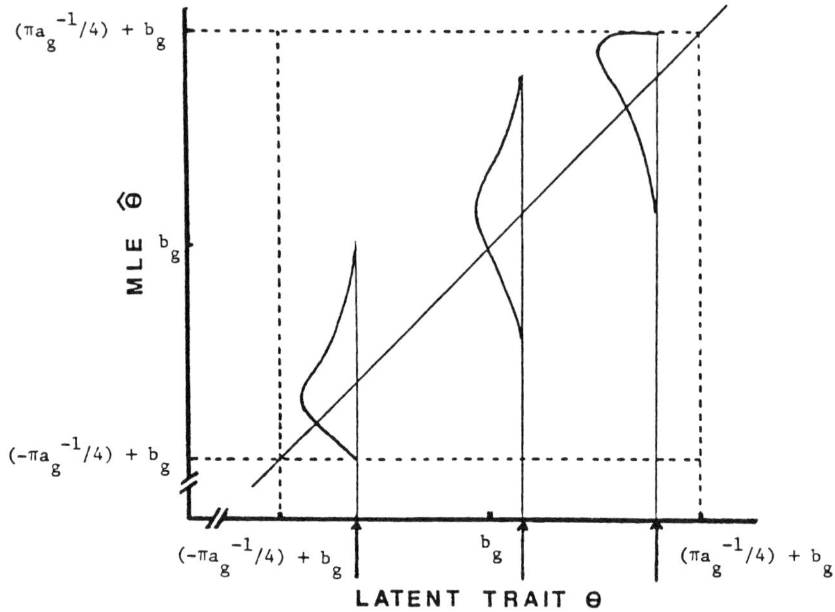

FIGURE 14.6. Graphical illustration of the conditional density functions of the maximum likelihood estimate $\hat{\theta}$, given the latent trait θ.

Monte Carlo Study

METHOD

To pursue the speed of convergence with which the conditional distribution of the maximum likelihood estimate $\hat{\theta}$ approaches normality, $N(\theta, I(\theta)^{-1/2})$, a Monte Carlo study was done. One hundred hypothetical examinees were assumed to exist at each of eight different levels of ability θ (i.e., -3.0, -2.2, -1.4, $-.6$, .2, 1.0, 1.8, and 2.6). It was assumed that each examinee had taken 20 sessions of equivalent tests, in each of which 10 equivalent binary items were given. Each binary item was assumed to follow the CIM with the parameters $a_g = .25$ and $b_g = 0.00$, and the response pattern was calibrated by the Monte Carlo method for each examinee in each of the 20 hypothetical sessions of testing. The cumulative test score was calculated after each session by summing all the binary item scores that the examinee had obtained. In this way, the number of binary items used for the computation of the cumulative test score was 10 after the first session, 20 after the second session, 30 after the third session, and 200 after the completion of the twentieth session. The maximum likelihood estimate of ability was obtained for each of the 800 examinees, based on each cumulative test score following the method described in the preceding section.

RESULTS

Figure 14.7 presents the resultant cumulative frequency distribution of the maximum likelihood estimates of the 100 examinees whose ability level was -2.2 (i.e., relatively close to the lower endpoint, $-\pi$, of the interval for which the item information $I_g(\theta)$ assumes the positive value .25), after the completion of each of Sessions 1, 2, 5, 10, and 20. In each graph, the normal distribution functions are also presented with $\theta = -2.2$ and $I(\theta)^{-1/2}$ as the two parameters and with the mean and the standard deviation of the observed 100 maximum likelihood estimates as the two parameters. As was expected, the cumulative frequency distribution shows a substantial skewness to the positive side when the number of binary items is as small as 10; and therefore the normal distribution function with the two empirically obtained parameters provides a curve that is located further to the right side of $N(\theta, I(\theta)^{-1/2})$. This tendency still holds, though slightly, even when the number of binary items is as large as 100.

For the purpose of comparison, Figure 14.8 presents a similar set of graphs for another group of 100 hypothetical examinees whose ability

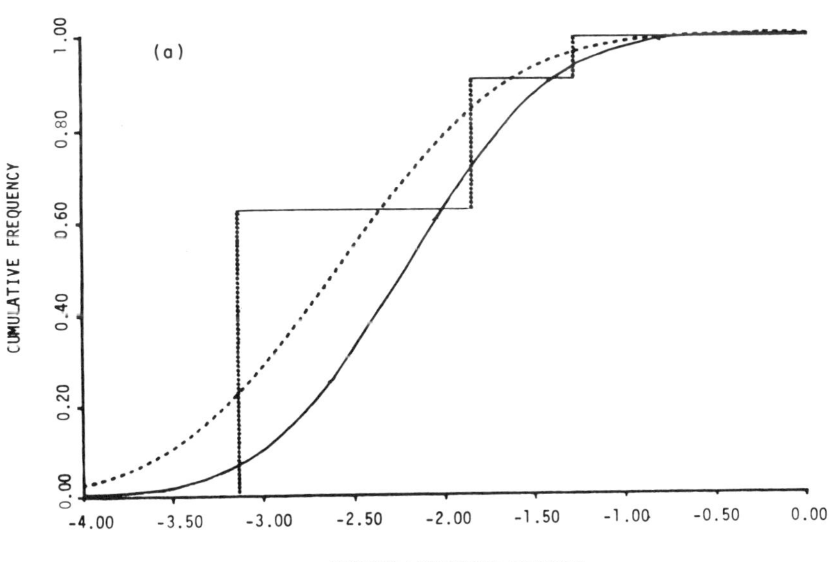

FIGURE 14.7. Cumulative frequency distributions of the 100 maximum likelihood estimates for $\theta = -2.2$ for Group 2, based on (a) 10, (b) 20, (c) 50, (d) 100, and (e) 200 items. (Parts (b) and (c) are on page 302; parts (d) and (e) on page 303.)

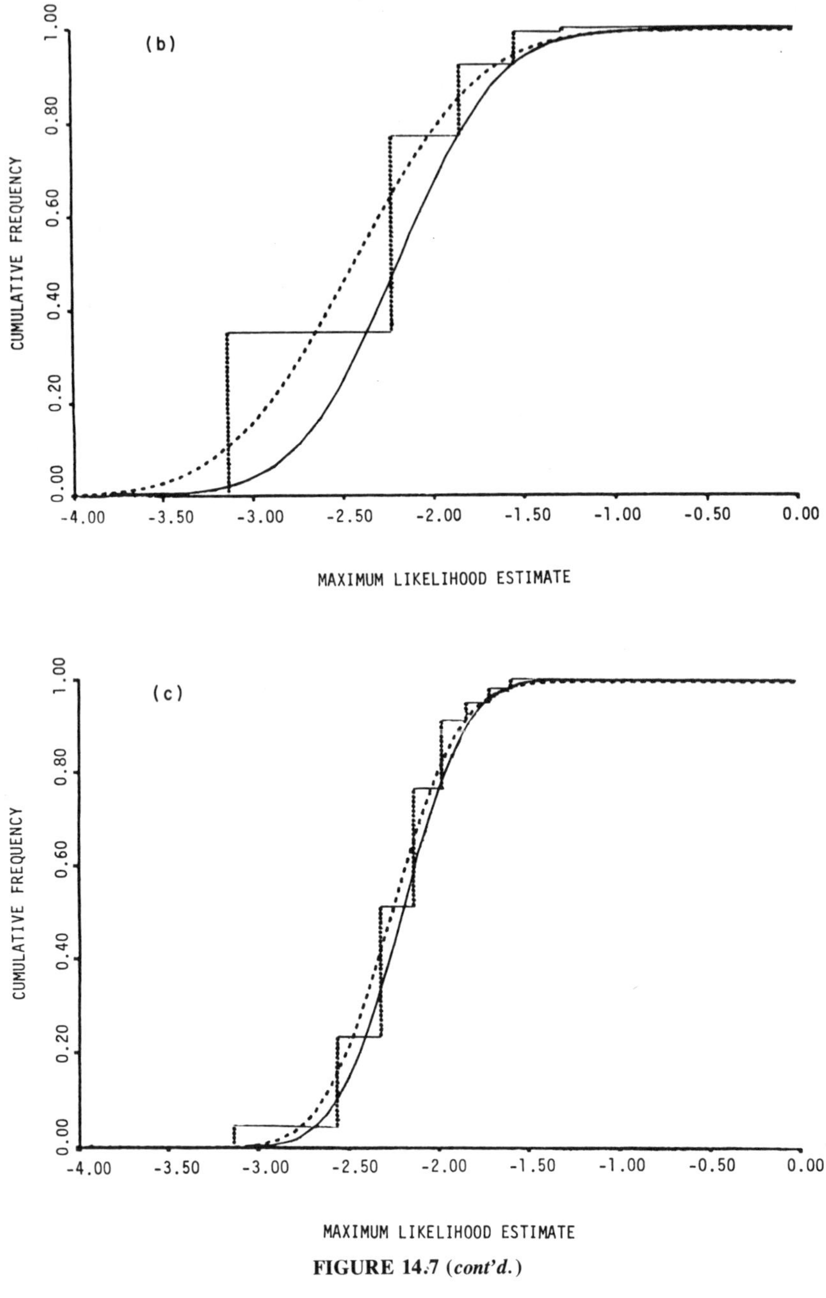

FIGURE 14.7 (*cont'd.*)

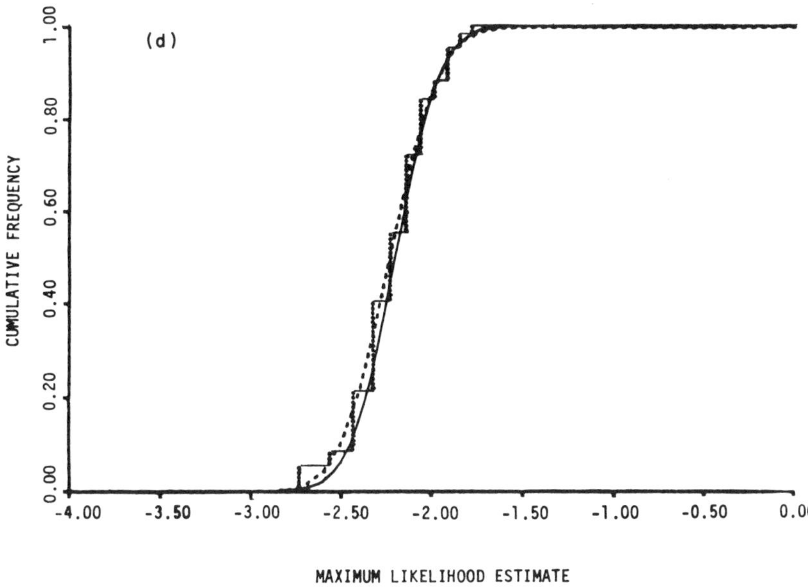

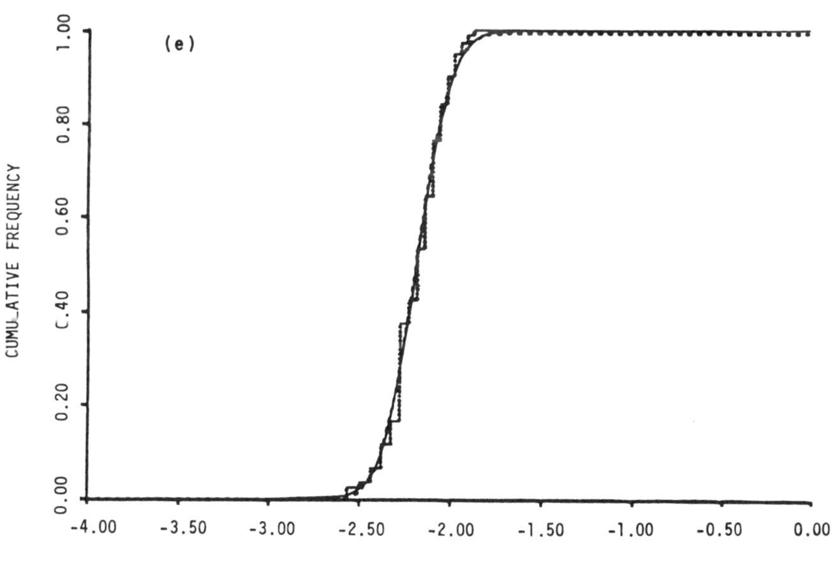

FIGURE 14.7 (*cont'd.*)

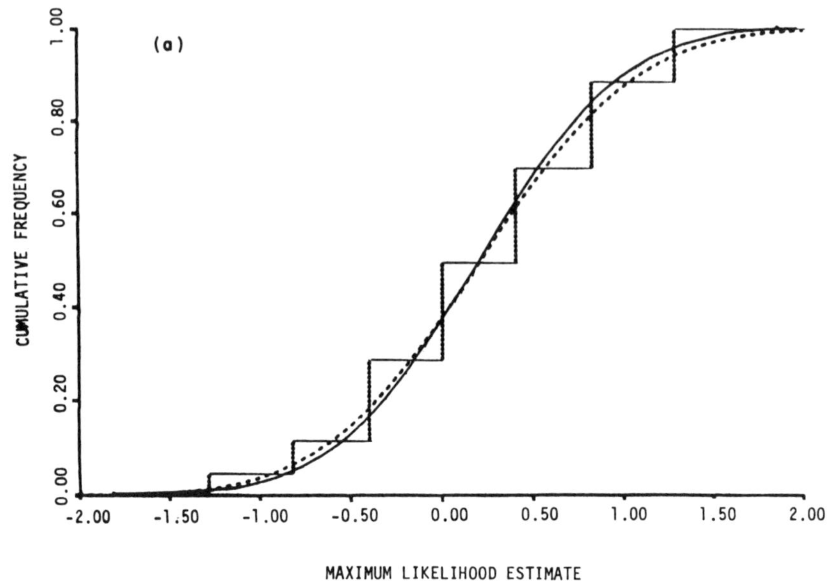

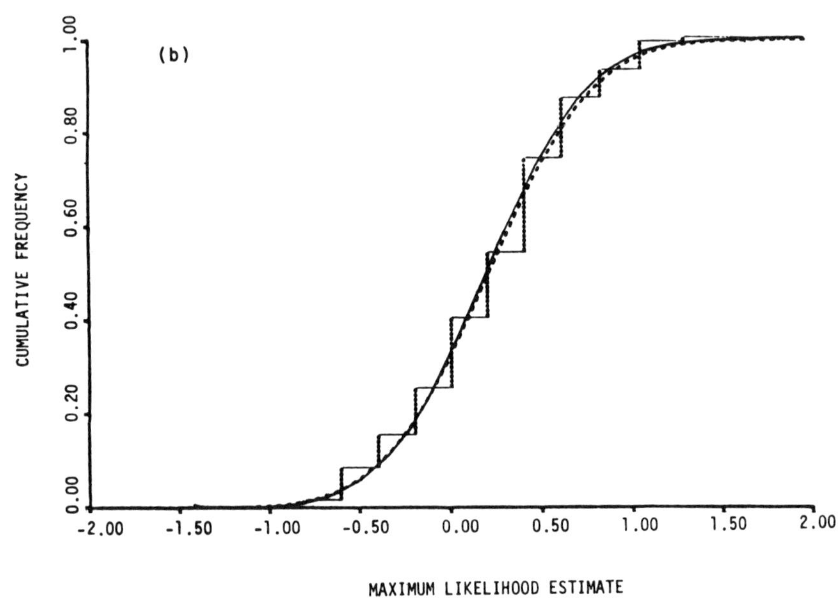

FIGURE 14.8. Cumulative frequency distributions of the 100 maximum likelihood estimates for $\theta = .2$ for Group 5, based on (a) 10, (b) 20, (c) 50, (d) 100, and (e) 200 items. (Parts (c) and (d) are on page 305; part (e) is on page 306.)

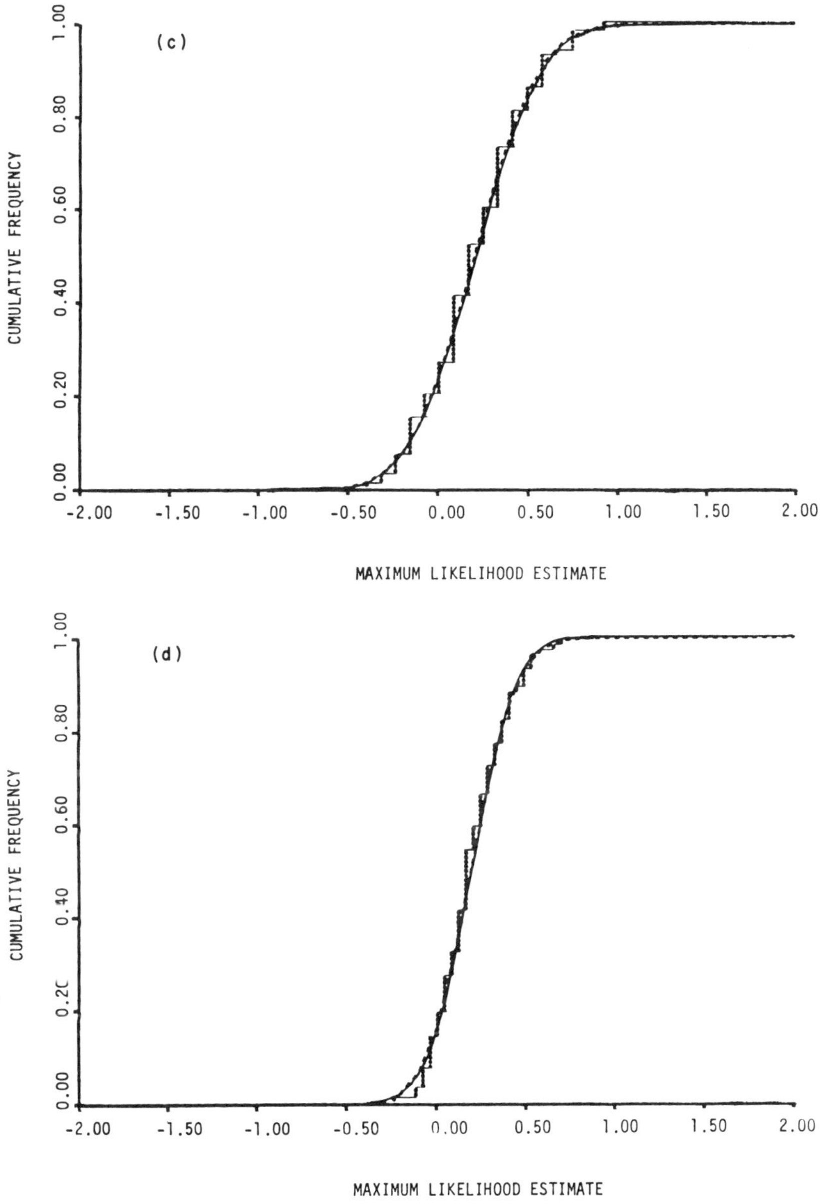

FIGURE 14.8 (*cont'd.*)

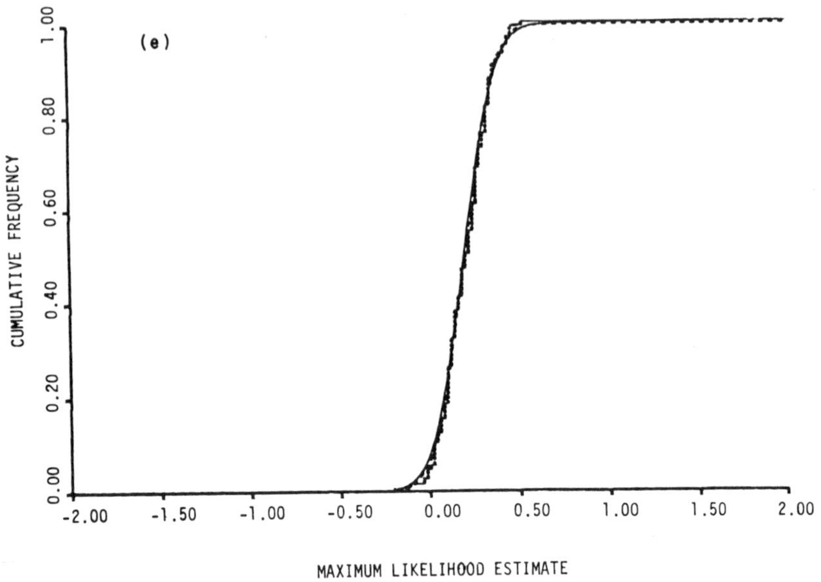

FIGURE 14.8 (*cont'd.*)

levels were uniformly .2 (i.e., a value which is far from the two endpoints, $-\pi$ and π, of the interval for which the item information function of each equivalent binary item assumes the positive constant .25). The results illustrated in Figure 14.8 make a good contrast with those in Figure 14.7, in which the two normal distribution curves almost overlap each other, even when the number of items used for obtaining the maximum likelihood estimate is as small as 20. This indicates a much faster convergence of the conditional distribution of the maximum likelihood estimate to normality with θ and $I(\theta)^{1/2}$ as the two parameters, in comparison with the case in which the ability level is closer to one of the two endpoints of the interval $(-\pi, \pi)$.

If the latent trait θ is transformed to τ through Eqs. (14.37) and (14.38) so that each equivalent binary item follows the normal ogive model with $a_g^* = 1.00$ and $b_g^* = 0.00$, then the values of τ corresponding to $\theta = -2.2$ and $\theta = .2$ are approximately -1.60 and .13. If Eq. (14.44) is used for the transformation so that each equivalent binary item follows the logistic model with the same parameters—a_g^* and b_g^*—with the scaling factor $D = 1.7$, then these corresponding values of τ are approximately -1.68 and .12, respectively. A similar transformation by means of Eq. (14.41), which provides the linear model with the parameters $\alpha_g = -2.5$ and $\beta_g = 2.5$ for each of the equivalent binary items, results in -2.23 and .25

as the approximate values of τ corresponding to $\theta = -2.2$ and $\theta = .2$, respectively.

DISCUSSION AND CONCLUSION

A new model for a binary item, the constant information model, has been introduced, and its characteristics and usefulness have been discussed. As was pointed out (Samejima, 1976, 1977a, 1977b, 1977d, 1978a, 1978b, 1978c, 1978d, 1978e, 1978f, 1978g), latent trait theory enlarges its horizon if full use is made of information functions, enabling types of research to be conducted which could not otherwise be done. For this reason, the constant information model will contribute to the productivity of research in the area of computerized adaptive testing as well as in other areas, as exemplified in this chapter.

REFERENCES

Birnbaum, A. Some latent trait models and their use in inferring an examinee's ability. In F. M. Lord & M. R. Novick, *Statistical theories of mental test scores*. Reading, Massachusetts: Addison-Wesley, 1968.

Samejima, F. Estimation of latent ability using a response pattern of graded scores. *Psychometrika Monograph Supplement*, 1969, *34* (4, Pt. 2, Monograph No. 17).

Samejima, F. A general model for free-response data. *Psychometrika Monograph Supplement*, 1972, *37* (1, Pt. 2, Monograph No. 18).

Samejima, F. Graded response model of the latent trait theory and tailored testing. In C. L. Clark (Ed.), *Proceedings of the first conference on computerized adaptive testing*. (U.S. Civil Service Commission, Personnel Research and Development Center PS-75-6). Washington, D.C.: U.S. Government Printing Office, 1976. (Superintendent of Documents Stock No. 006-00940-9)

Samejima, F. A use of the information function in tailored testing. *Applied Psychological Measurement*, 1977, *1*, 233–247. (a)

Samejima, F. A method of estimating item characteristic functions using the maximum likelihood estimate of ability. *Psychometrika*, 1977, *42*, 163–191. (b)

Samejima, F. *Estimation of the operating characteristics of item response categories I: Introduction to the two-parameter beta method* (Research Report 77-1). Knoxville: University of Tennessee, Department of Psychology, 1977. (c)

Samejima, F. *Estimation of the operating characteristics of item response categories II: Further development of the two-parameter beta method* (Research Report 78-1). Knoxville: University of Tennessee, Department of Psychology, 1978. (a)

Samejima, F. *Estimation of the operating characteristics of item response categories III: The normal approach method and the Pearson system method* (Research Report 78-2). Knoxville: University of Tennessee, Department of Psychology, 1978. (b)

Samejima, F. *Estimation of the operating characteristics of item response categories IV: Comparison of the different methods* (Research Report 78-3). Knoxville: University of Tennessee, Department of Psychology, 1978. (c)

Samejima, F. *Estimation of the operating characteristics of item response categories V: Weighted sum procedure in the conditional P.D.F. approach* (Research Report 78-4). Knoxville: University of Tennessee, Department of Psychology, 1978. (d)

Samejima, F. *Estimation of the operating characteristics of item response categories VI: Proportioned sum procedure in the conditional P.D.F. approach* (Research Report 78-5). Knoxville: University of Tennessee, Department of Psychology, 1978. (e)

Samejima, F. *Estimation of the operating characteristics of item response categories VII: Bivariate P.D.F. approach with normal approach method* (Research Report 78-6). Knoxville: University of Tennessee, Department of Psychology, 1978. (f)

Samejima, F. The application of graded response models: The promise of the future. In D. J. Weiss (Ed.), *Proceedings of the 1977 computerized adaptive testing conference.* Minneapolis: University of Minnesota, Department of Psychology, Psychometric Methods Program, 1978. (g)

Samejima, F. *Constant information model: A new, promising item characteristic function* (Research Report 79-1). Knoxville: University of Tennessee, Department of Psychology, 1979.

15

Some Latent Trait Models for Measuring Change in Qualitative Observations

GERHARD H. FISCHER

All too often, thinking and formation of concepts in the behavioral sciences have been misled by readily available statistical methods, especially of the multivariate variety. A typical example of how theorizing in psychology can be led astray by existing statistical methods is the obsolete dispute on the percentage of genetically versus environmentally determined intelligence. In this field, there has been an unwavering attempt to apply models of variance decomposition which had been developed for breeding experiments in stock farming and which are appropriate for that purpose. However, these methods are not suited for yielding any fundamental insights into the role of genetic versus environmental factors of human intelligence. Two sampling processes, which are completely beyond the control of the researcher, decide in fact the relative sizes of the estimates of the genetic and the environmental variances: First, the sample of monozygotic twins may be nonrepresentative of the genetic pool of the population. Second, the foster families must, in general, be regarded as a nonrepresentative sample of all the social environments. Hence, the estimated percentages of genetically versus environmentally determined variance are strongly biased by selection processes that, in addition, certainly are not of equal influence in data stemming from different time periods or from different countries. Thus there is no hope for stable and generalizable results.

NEW HORIZONS IN TESTING
Latent Trait Test Theory and
Computerized Adaptive Testing

ISBN 0-12-742780-5

This is not what would be expected from scientific findings; they should be independent of incidental sampling procedures and should generalize over time and over populations. Even if the variance decomposition were stable and reliable, such a result would in no way help answer the basic question: What are the effects of specified kinds of environments on cognitive development, and what can be gained by purposeful changes of children's sociocultural environments?

To answer such questions, obviously a rather different methodological approach is needed. Fischer and Formann (1981) formulated a latent trait model suitable for that purpose requiring, however, adequate test data from monozygotic twins and sufficient information about the sociocultural status of the foster families. The model assigns one effect parameter to each discrete type of social environment, thus explaining observed ability differences between twins brought up in separation in terms of environmental effects and allowing predictions about the effects of environmental changes. The results of empirical applications of this model are, in principle, sample free (i.e., they are essentially uninfluenced by nonrepresentativeness of the samples of twins and/or of environments). A further advantageous property of the model is that "intelligence" need not be represented by a unidimensional scale (such as IQ) but is considered a multidimensional construct represented by the chosen sample of items; technically, one latent ability dimension corresponds to each item.

This alternative approach to the debate on genetic versus environmental determination of intelligence is mentioned here as an illustration of the author's methodological credo: Available quantitative methods should not be hastily applied, but rather the starting point should be the conceptual structure of the given problem and then the formal models should be tailored accordingly.

This principle may be illustrated by one more example, this time taken from communications research. In 1971 an old problem in market and opinion research was posed to the author: What is the effect of an insertion in different mass media, such as television, radio, and newspapers? For the practical purpose of optimizing a campaign with a limited budget, a simple answer to this question was required (e.g., "An insertion on television is three times as effective as a comparable insertion in a local radio program"). The typically very complex results obtained by way of the traditional approaches in this area (e.g., emphasizing a multitude of moderator variables that shape the effects of a communication on the individual) were of no help for the practical decision to be made. Therefore, a relatively simple latent trait model was suggested which assigns just one effect parameter to each medium within a campaign, measuring the impact of the respective insertions in that medium on the attitudinal or behavioral traits under study (Fischer, 1972). A noteworthy property of this model is that the results are essentially independent of the sample of respondents (i.e., no need for representative sampling) and that the attitudinal or behavioral domain can be represented by a heterogeneous set of questions or items (i.e., no need for unidimensional attitude scales).

At first sight these two examples may seem quite different, but the fact is that they have many important features in common: There is some sort of "treatment" (the environmental influences on cognitive growth, or the impact of mass communications), and there are certain latent dimensions (abilities corresponding to the test items, or latent attitudes and behavior tendencies covered by the questionnaire items) that reflect the treatment effects. The interest of the researcher focuses on the latter, not on the individual differences as such. The respective results should not depend on whether the samples are representative (i.e., it should not be relevant whether the samples of twins and/or environments are representative, or whether the interviewees have been representatively sampled from the audience or not).

The demand that the results referring to the structural effect parameters should be independent of the incidental ability or attitude parameters of the persons being measured is a special case of the epistemological principle of *specific objectivity* formulated by Rasch (1967, 1968, 1972, 1977): The comparison of the parameters of interest, i.e., the structural parameters (for example, the treatment effects), should only depend on their true values but should be independent of all other irrelevant factors, such as the incidental parameters characterizing the distributional properties of the sample of observations.

Applied to the two examples presented, the request for specific objectivity of the results means that these should be independent of the samples of twins and of environments in the first case and independent of the sample of subjects in the second. These requirements are met by the models indicated, which belong to a family of *linear logistic models* closely related to the Rasch (1960, 1966) models; however, they also show some marked distinctions (e.g., the latent variables are multidimensional, and the formalism of parameter estimation and hypothesis testing is quite different).

Unfortunately, only one study dealing with effects of mass communications has employed this new approach thus far, and applications to the problem of genetic versus environmental determination of intelligence are altogether lacking. However, many problems in clinical and educational psychology are of similar structure, with educational or clinical treatments taking the place of the media or environments. A considerable number of applications in these areas have been undertaken in the past decade, and the theoretical and methodological basis of the model has been further strengthened. In view of the fact that the main purpose of such studies is to assess changes caused by treatments, the models can be conceived as methods for the measurement and assessment of change.

One important distinction between the methodology alluded to here and more conventional approaches to the measurement of change is that the data are regarded as what the observations, in fact, in most cases are: qualitative variables. By no means is there an attempt to scale or to quantify the data just to make statistical methods for quantitative data applicable. Quite the contrary, the observations are considered to be realizations of *qualitative random variables,* which are, however, governed by *quantitative latent parameters*. Change is measured in terms of these latent parameters. A beneficial consequence of this is that the well-known physicalism–subjectivism dilemma (Bereiter, 1967), which occurs in attempts to "scale" the qualitative data, is automatically avoided.

Another distinction is that the estimation of the effect parameters is sample free: As long as the model holds, the results should be the same in all subgroups of individuals, except for random errors. This implies that the model circumvents the dreaded "regression effects" (Lord, 1967), which are a troublesome concomitant phenomenon of the traditional analysis of score differences; an observed covariance between initial measures and change can be due to an interaction between initial true score and treatment effects or may just be a methodological artifact. The linear logistic model serves as the null hypothesis for testing the absence of such interactions: If the effect parameters are essentially the same in all subgroups of subjects, then the results are truly generalizable over the entire range of individual differences represented in the data. If, on the other hand, significant differences of effect parameters between certain subsamples of subjects appear, then this indicates the limits of the generalizability of the results. Hence, the model can be viewed as a *theory of generalizability of treatment effects*. This important feature will be illustrated below by means of concrete examples.

A FAMILY OF LINEAR LOGISTIC MODELS FOR QUALITATIVE DATA

There are a variety of such linear logistic latent trait models differing as to the restrictiveness of their assumptions, the kind of results deducible, and the types of data required. The most important ones are the following:

1. *The dichotomous linear logistic test model (LLTM)*. This model was originally devised for analyzing the complexity of intelligence test items in terms of the cognitive operations involved (Scheiblechner, 1972; Fischer, 1973, 1974a). A number of interesting and successful applications have been undertaken in the field of the analysis of teaching materials (Spada, 1976, 1977; Spada & Kluwe, 1981) and of test construction (Formann & Piswanger, 1979; see also Fischer & Formann, 1982), but it is also useful for measuring the effects of learning, instruction, and mental maturation (Fischer, 1974a, pp. 398–401; 1976, pp. 97–110; 1977b, pp. 203–225; 1983; see also Spada, 1973). However, it restricts the investigation to unidimensional (i.e., homogeneous) sets of items; because of this and in view of its complicated formalism, it will not be dealt with here. The interested reader is referred to the referenced literature.

2. *The dichotomous linear logistic model with relaxed assumptions (LLRA)*. Its distinctive feature, as compared with other logistic latent trait models, is the relaxation of the assumptions with regard to the dimensionality of the criterion variables (items, symptoms, or other behaviors). It has proved a very useful tool for assessing change in a variety of different situations and will be described and illustrated in this chapter.

3. *The polytomous generalization of the LLTM*. This model has been formulated by Fischer (1974a, pp. 493–500; 1977a, pp. 134–138), who has also presented the estimation equations and numerical methods for solving them. However, substantial applications are still lacking; for this reason, and because its formalism is very complicated, the details of this model will not be discussed here.

4. *The polytomous generalization of the LLRA*. This offers quite interesting possibilities for empirical analyses and hypothesis testing in those cases where the response categories cannot be reduced to dichotomies. A short outline of this generalization of the LLRA will be given below, along with some empirical applications.

THE DICHOTOMOUS LINEAR LOGISTIC MODEL WITH RELAXED ASSUMPTIONS

The Model

"Dichotomous" means that the observed criterion variables, which may be test items or clinical symptoms (or any other kind of behaviors), are binary variables. It is assumed that before and after treatment a number k of such criterion variables are observed on each subject. Then the model is defined by the equations

$$P(+|v,i,t_1) = \frac{\exp(\theta_{vi})}{1 + \exp(\theta_{vi})}, \tag{15.1}$$

$$v = 1, \ldots, n \quad \text{(subjects)},$$
$$i = 1, \ldots, k \quad \text{(criteria, items)},$$

$$P(+|v,i,t_2) = \frac{\exp(\theta_{vi} + \delta_v)}{1 + \exp(\theta_{vi} + \delta_v)}, \tag{15.2}$$

$$\delta_v = \sum_j q_{vj}\eta_j + \sum_l \sum_j q_{vj}q_{vl}\rho_{jl} + \tau. \tag{15.3}$$

Thereby $P(+|v,i,t_1)$ denotes the probability that person v gives response "+" on criterion i at time t_1 (before treatment) and $P(+|v,i,t_2)$ is the analogous probability for time t_2 (after treatment). The probability $P(+|v,i,t_1)$ depends solely on one parameter, θ_{vi}, the position of person v on the latent trait underlying criterion i. For example, let criterion i be a certain symptom of fear in clinical patients; then θ_{vi} is the latent anxiety of person v giving rise to that symptom. Thus the state of person v at time t_1 is characterized by a vectorial parameter $\theta_v = (\theta_{v1}, \ldots, \theta_{vk})$ in a k-dimensional latent space; in other words, by a set of k traits associated with the k criterion variables.

Note that the model makes no assumptions whatsoever about interdependencies or the true dimensionality of these traits; all k dimensions may be mutually independent, but they may also be statistically or functionally dependent. In particular, unidimensionality of the criteria or items is not assumed, as would be the case with the Rasch (1960, 1966) or Birnbaum (1968) models. Hence, the LLRA is maximally flexible regarding the characterization of the subjects. For example, at time t_1 person v might have a lower position on trait i than on trait j, $\theta_{vi} < \theta_{vj}$, but the reverse may hold for person w, $\theta_{wi} > \theta_{wj}$.

The characterization of the subjects at time t_2 is, in principle, analogous. It is, however, restricted by the assumption that change in each person can be described by a single scalar parameter δ_v, which according to Eq. (15.3) is a linear function of the effects η_j of the given treatments (main effects), of treatment interactions ρ_{jl}, and of a trend parameter τ, which comprises all the causes of change that are unrelated to the treatments. The constants q_{vj} are measures of the dose of treatment j as applied to person v.

One additional reservation must be made with regard to the decomposition of δ_v in Eq. (15.3): The treatment interactions ρ_{jl} are only separable from the main effects η_j if an appropriate design matrix $[(q_{vj})]$ has been chosen and if the sample of observations is large. Therefore, in all the empirical applications known so far, the interaction parameters ρ_{jl} were simply assumed to be zero [i.e., the middle term on the right-hand side of Eq. (15.3) was deleted].

The most important properties of this model, which have been studied by Fischer (1972, 1974b, 1976, 1977a, 1977b, 1983), are as follows:

1. Given appropriate data, the effect parameters η_j, τ (and the interactions, ρ_{jl}) can be estimated independently of the true values of the parameters θ_{vi}; the latter need not be known and are not estimated from the data either. This means that any proposition referring to the comparison of two treatments j and l is completely independent of the sample of subjects (i.e., specifically objective).
2. The parameter estimates have ratio scale properties, so that it is possible to arrive at statements such as "treatment j is twice as effective as treatment l."
3. It is possible to test the significance of single parameters and to test almost any conceivable meaningful composite hypothesis on the parameters by means of likelihood-ratio tests (LR tests).

The mere enumeration of the model properties does not sufficiently demonstrate the full scope of the possibilities they imply. An illustrative empirical example is therefore in order.

Sample Application

RESEARCH QUESTIONS

Rop (1977) investigated the effects of three preschool educational programs (Early Reading, Logical Thinking, and Verbal Enrichment) on the cognitive development of kindergarten children. To assess change, a battery of 64 items was given before and after the treatment period; a control group attended kindergarten but did not participate in the programs. Three primary questions were to be answered:

1. Is it possible to furnish proof that the programs accelerate cognitive development?
2. What is the generality of the effect of each of the programs (e.g., does verbal training have an effect also in the nonverbal area)?
3. What do socially and educationally disadvantaged children gain in comparison to middle-class children, that is, is early intervention a means of overcoming the deficiencies resulting from less privileged environments?

The first question, which is the easiest one, was answered in the affirmative by testing the null hypothesis that the effect parameters are zero.

The second question is far more complex: Eq. (15.3) asserts that the effect of each treatment can be measured by just one parameter δ_v per person, irrespective of the item i. Hence, if it were true that verbal enrichment had little or no effect on the nonverbal abilities (which were tested by one set of 11 nonverbal items in the test), the model could not have been true for all 64 items. Hence, the model plays the role of a H_0 against the H_1 of differential effects of the treatments in certain subgroups of items (i.e., subgroups of criterion variables). There is an analogy between this model and the well-known analysis of variance for quantitative data: An analysis of variance also begins with the global H_0 that all means are equal.

RESULTS

In Rop's study the H_0 of uniform effects of the treatments on all the ability domains represented by the items had to be rejected. As Table 15.1 shows, the 64 items had to be broken down into three subsets: (*a*) naming of objects, actions, and attributes; (*b*) verbal abilities, such as verbal fluency, enunciation, and appropriate usage of language; and (*c*) nonverbal abilities. Each of the three programs had a differential effect within each of the three domains. Hence, there were interactions between treatments and ability domains. (Note that this is another type of interaction, which has to be distinguished from the treatment interactions ρ_{jl}; the latter were not considered.)

The results in Table 15.1 show the findings of the study in a maximally generalized form: It is an essential feature of the model that it identifies *the maximal subsets of criterion variables with uniform treatment effects*. This is

TABLE 15.1
Effects of the Training Programs per Time Unit (1000 Minutes) and the Trend for Naming, Verbal Intelligence, and Nonverbal Item Groups[a]

	Item group		
Treatment	Naming	Verbal intelligence	Nonverbal intelligence
Reading	.37[b]	.15	−.04
Thinking	.51[b]	.16[b]	.25[b]
Verbal	.49[b]	.37[b]	.31[b]
Trend	.84[b]	.88[b]	.32[b]

[a] Adapted from Rop (1977).
[b] Statistically significant at $p < .01$.

a consequence of the principle of specific objectivity, namely, that the estimates of effect parameters do not depend on any irrelevant factors, such as subjects or items, as long as the model holds. In other words, only the minimum number of moderator variables that are absolutely necessary to explain the data are considered.

Rop's third question is the most intriguing one. If conventional methods of data analysis had been applied, it would have been expected that environmentally privileged children with a higher level of cognitive development, and hence with better performance at t_1, would not have increased their level of performance as much as the children with poor achievement (see the above-mentioned regression effects). The LLRA as H_0, on the contrary, asserts that the effect parameters do not depend on the person parameters θ_{vi}; in other words, if the effects of treatments were really the same for all children, the effect parameters estimated for groups of children with different ability levels should also be equal except for random error. This is again a direct consequence of the principle of specific objectivity. In Rop's study it was found, in fact, that the *treatment effects were independent of the initial level of cognitive development* and, therefore, that the preschool programs were not appropriate for bridging the gap between privileged and underprivileged children, despite their general efficacy in improving the cognitive development.

THE EPISTEMOLOGICAL BACKGROUND OF THE LINEAR LOGISTIC MODEL WITH RELAXED ASSUMPTIONS

It is obvious that the properties of the LLRA are quite advantageous and have encouraged a variety of applications. However, a better theoretical and epistemological foundation of this methodological approach seemed called for, and an answer was sought to the following question: If assessment of change is to be specifically objective, what is implied with respect to the formal structure of the model? A prerequisite for dealing with this question is that the problem of measurement of change is formalized in a sufficiently general way.

Change is detected by exposing people to a set of observational conditions such as the test items, observation of symptoms, or registration of any other kind of criterion variables. Let the behavioral disposition or state of the person at time t_1 be described by a set of k parameters $\rho_{v1,1}(\theta_{v1}), \ldots, \rho_{vk,1}(\theta_{vk})$, so that $\rho_{vi,1}$ is associated with the criterion variable i and fully describes the latent behavioral disposition of person v with respect to this criterion. In the same way, let the state of person v at time t_2 be described by the set of parameters $\rho_{v1,2}(\theta_{v1}, \delta_v), \ldots, \rho_{vk,2}(\theta_{vk}, \delta_v)$, whereby δ_v is a scalar parameter representing change. Nothing is assumed concerning the functional concatenation betewen θ_{vi} and δ_v; it is only assumed that the reaction tendency $\rho_{vi,2}$ at time t_2 is a function of the latent trait θ_{vi} at time t_1 and the change parameter δ_v. If change has occurred, the parameters $\rho_{v1,2}, \ldots, \rho_{vk,2}$ must be different from $\rho_{v1,1}, \ldots, \rho_{vk,1}$. Hence, in order to detect change, it is necessary to compare these two sets of parameters.

The formal comparison of such parameters is carried out by constructing an appropriate function of all of them, namely, a function $U(\rho_{v1,1}, \ldots, \rho_{vk,1}, \rho_{v1,2}, \ldots, \rho_{vk,2})$. From this function U, it should be possible to derive the change parameter δ_v, i.e., U should essentially be nothing but δ_v. More precisely, U should be a one-to-one function of δ_v, $U = V(\delta_v)$ and thus should be independent of the latent trait parameters θ_{vi} as well as of the chosen sample of observational situations (e.g., items, symptoms, criteria), $i = 1, \ldots, k$.

On the basis of this formalization of the measurement of change, Fischer (1977c) proved the following theorem (see also Rasch, 1972, who gives a general formalization of the concept of *specifically objective comparisons* and derives a more general theorem on latent additivity of the parameters).

Theorem. Let $\partial\rho_{vi,1}/\partial\theta_{vi} > 0$ *everywhere; let U be differentiable with respect to* $\rho_{vi,1}$ *and* $\rho_{vi,2}$*;* $\rho_{vi,1}$ *with respect to* θ_{vi}*; and* $\rho_{vi,2}$ *with respect to* θ_{vi} *and* δ_v *for* $i = 1, \ldots, k$. *Furthermore, let U be a function* $U(\rho_{v1,1}, \ldots, \rho_{vk,1}, \rho_{v1,2}, \ldots, \rho_{vk,2}) = V(\delta_v)$, *which is independent of* $\theta_{vi}, i = 1, \ldots, k$. *Then there exist monotone transformations of all the parameters, so that after transformation* $\rho_{vi,2} = \rho_{vi,1} + \delta_v$. *If, in addition, the observations are assumed to be realizations of Bernoulli variables, then, except for scale transformations, Eqs.* (15.1) *and* (15.2) *must hold.*

The meaning of the theorem is this: If there are dichotomous observations at two points of time and if it is desired to assess change in a "specifically objective" manner (i.e., if the result should not depend on the sample of person parameters θ_{vi}), then the model must be essentially of the LLRA type. Of course, some scale transformations may be applied to the parameter dimensions, entailing formal changes of the model, but any model and empirical result obtained in this manner would be equivalent to what is obtained by means of the LLRA. Hence, there would be no point in transforming the parameters and thereby departing from the specifically simply *additive structure* of the LLRA.

ESTIMATING MODEL PARAMETERS IN THE LINEAR LOGISTIC MODEL WITH RELAXED ASSUMPTIONS

Since this theorem legitimizes the LLRA as theoretically well founded, a short discourse on the technical problems of parameter estimation and hypothesis testing seems to be justified. To simplify matters, Eq. (15.3) can be rewritten

$$\delta_v = \sum_j q_{vj}\eta_j \qquad \text{with } j = 1, \ldots, p. \tag{15.4}$$

This can be done because Eq. (15.3) is linear in all the parameters; hence, parameters η_j and matrix $Q = [(q_{vj})]$ only need to be redefined appropriately, p being the number of parameters ($p = m + 1$ if there are m treatments, no interactions, and trend $\eta_{m+1} = \tau$).

Let $A_1 = [(a_{vi,1})]$ be the item-score matrix (with elements 1 if the response was "+" and 0 if "−") for time t_1 and $A_2 = [(a_{vi,2})]$ for time t_2, $v = 1, \ldots, n$ and $i = 1, \ldots, k$. Maximizing the likelihood of the data subject to the condition that the sums $a_{vi,1} + a_{vi,2} = a_{vi,*}$ are given, yields the following conditional maximum likelihood (CML) estimation equations for the parameters η_j:

$$\sum_v \sum_i q_{vj}\left\{ a_{vi,2} - \left[\sum_{t=u}^{w} \exp\left(t \sum_j q_{vj}\hat{\eta}_j\right)t\right] \Big/ \left[\sum_{t=u}^{w} \exp\left(t \sum_j q_{vj}\hat{\eta}_j\right)\right]\right\} = 0, \tag{15.5}$$

$$v = 1, \ldots, n; \qquad i = 1, \ldots, k; \qquad j = 1, \ldots, p;$$

$$u(v,i) = w(v,i) = 0 \qquad \text{if } a_{vi,1} + a_{vi,2} = 0;$$

$$u(v,i) = w(v,i) = 1 \qquad \text{if } a_{vi,1} + a_{vi,2} = 2;$$

$$u(v,i) = 0, \quad w(v,i) = 1 \qquad \text{if } a_{vi,1} + a_{vi,2} = 1.$$

A sufficient condition for this set of nonlinear equations to have a unique solution $\hat{\eta} = (\hat{\eta}_1, \ldots, \hat{\eta}_p)$ is that the matrix $Q = [(q_{vj})]$ defined above, $v = 1, \ldots, n$ and $j = 1, \ldots, p$, has full rank p and that within each group of persons with the same treatment combinations (same "doses" q_{vj} for $j = 1, \ldots, p$), on each item or symptom i, the response sequences $a_{vi,1} = 0$, $a_{vi,2} = 1$, and $a_{wi,1} = 1$, $a_{wi,2} = 0$ have been observed for at least one person v and one person w, respectively. A more complicated necessary and sufficient condition is given by Fischer (1983).

Note that since the incidental person parameters θ_{vi} have been conditioned out by means of the statistics $a_{vi,1} + a_{vi,2} = a_{vi,*}$, Eqs. (15.5) contain only the effect parameters η_j and observed quantities.

The estimation equations can easily be solved by means of the standard Newton–Raphson procedure. The second-order partial derivatives needed in this method can be found in Fischer (1974a, 1974b, 1977a). Each iteration cycle requires only very little computation, and the convergence is extremely fast; hence, the total computational expenses amount to only a fraction of those needed, for example, for estimation of item parameters in the usual latent trait models. It should be stressed that there is virtually no limitation to the number of items (symptoms) and persons; the number of treatments is restricted mainly by the practical constraints in realizing an appropriate experimental design.

The above result on the uniqueness of the solution (which, by the way, can be shown to hold also under weaker, if much more complicated, assumptions) implies an important consequence about the scale properties of the parameters: If the parameters $\hat{\eta}_j$ are unique, then by virtue of Eq. (15.3), the treatment effects are measured on a ratio scale with a unit determined by the time interval (t_1, t_2) together with the chosen unit of measurement of dosage q_{vj}.

TESTING HYPOTHESES WITH THE LINEAR LOGISTIC MODEL WITH RELAXED ASSUMPTIONS

In practical applications, estimating the effect parameters η_j is only a first step. More important is to test hypotheses on these parameters. Such tests can be carried out by means of the *likelihood ratio* (*LR*) *principle*. Its idea is to compare the (conditional) likelihood of the data under hypothesis H_1 (alternative hypothesis) with the likelihood of the data under a more restrictive hypothesis H_0 (null hypothesis). If H_0 were true, both likelihood functions would have to be more or less equal; of course, the likelihood under H_1 must always be greater than or equal to that under H_0, because H_0 is, by definition, a restriction on the parameters. If, on the other hand, H_0 were false, then the likelihood under H_0 should be considerably smaller than the one under H_1.

Let $\hat{\eta} = (\hat{\eta}_1, \ldots, \hat{\eta}_p)$ be the estimates of effect parameters under H_1 and let $L(H_1)$ be the maximized conditional likelihood of the data under H_1,

$$L(H_1) = \prod_v \prod_i L_{vi}, \tag{15.6}$$

with

$$L_{vi} = \begin{cases} \exp\left(a_{vi,2} \sum_j q_{vj}\hat{\eta}_j\right) \Big/ \left[1 + \exp\left(\sum_j q_{vj}\hat{\eta}_j\right)\right] & \\ & \text{if} \quad a_{vi,1} + a_{vi,2} = 1, \\ 1 & \text{if} \quad a_{vi,1} + a_{vi,2} = 0 \text{ or } = 2. \end{cases} \tag{15.7}$$

Furthermore, let H_0 be a null hypothesis consisting of the restrictions $\eta_j^* = \mu_j(\beta_1, \ldots, \beta_{p'})$ with $p' < p$, the matrix of partial derivatives $\partial\mu_j/\partial\beta_l$ having rank p'. Finally, let $L(H_0)$ be the likelihood of the data under H_0, whereby the maximum likelihood estimates $\hat{\eta}_j^* = \mu_j(\hat{\beta}_1, \ldots, \hat{\beta}_{p'})$ are inserted instead of $\hat{\eta}_j$ in Eqs. (15.7). Then, under H_0,

$$-2 \ln \lambda = -2 \ln[L(H_0)/L(H_1)] \tag{15.8}$$

is asymptotically χ^2 distributed with $df = p - p'$. It can easily be shown that most hypotheses relevant in practical applications can be formulated as restrictions $\eta_j^* = \mu_j(\beta_1, \ldots, \beta_{p'})$ and hence can be tested by means of this LR test. As long as the restrictions are linear contrasts, estimation equations (15.5) can be used for estimating the parameters under H_0 as well (with appropriate adjustments of the factors q_{vj}); otherwise, the estimation equations require only a minor adaptation, which need not be discussed here.

Regarding the asymptotic properties of the LR test, one qualification must be made: It is not sufficient that the total sample of persons is large (formally speaking, tends to infinity); each subgroup of persons with the same treatment combination, defined by a dose vector $(q_{v1}, \ldots, q_{vp})$, must be large. Otherwise, cases could be conceived where some treatment(s) has(have) been applied in subsamples with only very few persons, so that there would be no sufficient statistical information about these treatments.

In the following, some examples of hypotheses that are typically tested in the applications are given.

1. *The H_0 of no change* ($\eta_j = \rho_{jl} = \tau = 0$ *for* $j,l = 1, \ldots, m$). This corresponds to the H_0 of equal means in the analysis of variance. Although such a global H_0 is not likely to be true, it may be considered as a starting point for more detailed investigations.

2. *All interactions are zero* ($\rho_{jl} = 0$ *for* $j,l = 1, \ldots, m$). As mentioned before, this H_0 has never been tested in the existing applications, since it requires a large amount of data. The interaction terms have always been dropped from Eq. (15.3).

3. *Some effect parameters are zero* ($\eta_j = 0$ *for some js, and/or* $\tau = 0$). In most empirical investigations, the effect parameters were statistically significant, but in some instances one or the other treatment was ineffective. Zeman (1980), for example, found that the spelling abilities of dyslexic children were not improved significantly more in the experimental group (participants in specific training groups) than in the control group without such treatment. The observed change was therefore considered due to factors independent of the treatment (e.g., teaching in school, developmental processes). A similar case of an ineffective therapy is mentioned by Zimprich (1980), who treated children with psychosomatic diseases: Patients given only the traditional chemotherapy showed no significant improvement, whereas the group with chemotherapy and psychotherapy experienced a considerable change for the better.

A special case of insignificant effect parameters is the vanishing of the trend effect, $\tau = 0$. Such a result was found, for instance, by Heckl (1976) in the course of a study of children with speech defects. The preschool children suffering from very marked speech disorders showed no significant improvement during a 5-month period of observation, unless they underwent specific speech therapies. Since the criterion variables consisted not only of symptoms of speech defects but also comprised a large set of behavioral characteristics and intelligence test items, this finding indicates that the cognitive development of the speech handicapped, at this age, more or less comes to a standstill if no appropriate intervention takes place.

4. *Several (or all) treatments are equally effective* ($\eta_j = \eta_l$ *for some or all js and ls*). In the greater part of the applications, the differences between treatment effects were significant. There may be two reasons for this. First, very similar treatments are not usually compared, but rather substantially different ones. Second, in the LLRA the whole set of data is used to estimate only a very small number of parameters, namely, the effect parameters η_j and the trend τ. Consequently, there is much statistical information on these parameters, and small differences easily become significant. Despite that, there exist some nontrivial applications where the H_0 of equal treatment effects had to be retained. Heckl (1976), in the study already mentioned above, used three different forms of speech therapy (syntactic, semantic, and phonetic exercises). Contrary to expectation, all three proved to be equally effective. Heckl's interpretation was that the effects were brought about by the intensive devotion of the therapist to the handicapped children and by the reinforcements given to their verbal productions, relatively independent of the content of the prescribed exercises.

5. *The hypothesis of generalizability of the treatment effects over the items or symptoms.* If there are at least two subsets of items (symptoms) with different content, say, Set I and Set II, then an interesting question is whether the treatment(s) has(have) the same effect(s) on the ability or behavior domains represented by Sets I and II, respectively ($\eta_j^{(I)} = \eta_j^{(II)}$ for $j = 1, \ldots, m$). Testing such hypotheses is important because this helps to trace the limits of the generalizability of the results. One pertinent example (Rop, 1977) has already been given above; there the preschool education programs had a considerably different effect in three subsets of test items. A related study is that of Zeman (1976), who investigated the effects of early training in elementary set theory; she found a substantial transfer of the cognitive operations acquired from the material used in the learning phase to other dissimilar materials. Although the H_0 of equal training effects on all the tested ability domains had to be rejected, her results imply that the preschool education program considered was in fact a rather general vehicle for promoting cognitive development.

6. *The hypothesis of generalizability over different subsets of persons.* It is a characteristic feature of the LLRA that, because of the assignment of one latent parameter θ_{vi} to each combination of person by criterion, formally it makes no difference whether additional persons or additional criteria are added to a given set of data. Therefore differences of treatment effects between subsets of persons lend themselves to the same kind of LR test as those between subsets of criteria.

Suppose that at least two sets of persons, A and B (e.g., male versus female, older versus younger, high versus low achievers) are distinguished; then, under H_0, it is assumed that the effects $\eta_j^{(A)}$ of treatments in the first sample are equal to the parameters $\eta_j^{(B)}$ in the second sample: $\eta_j^{(A)} = \eta_j^{(B)}$ for $j = 1, \ldots, m$. An intriguing example where—contrary to expectation—this H_0 had to be retained was the study by Rop (1977) cited above, in which the children from middle-class families gained as much from the early education programs as did the underprivileged children; in this case, the treatment effects were generalizable further than anticipated.

In most other studies, however, unequal effect parameters were found in different subsamples of people; compare the following investigations: Vodopiutz (1977), who studied the effects of certain training units on complex movements in gymnastic instruction; Pendl (1976), the effects of a language laboratory on teaching a foreign language (English) in high school; Rella (1976), the results of driver improvement training in anticipating dangerous traffic situations; Platzer (1978), the effects of technical playing materials on the development of mechanical–technical understanding; Witek (1979), the effects of group-dynamic sensitivity training for business executives.

7. *The hypothesis of a linear dose-response curve.* Equation (15.4) presupposes that the total effect of a treatment is proportional to its dose. However, general experience indicates that in some cases a treatment is completely ineffective below a certain minimal dose and that satiation occurs above a certain amount of treatment. It is therefore important that the hypothesis of linearity [Eq. (15.4)] be tested against an unspecified nonlinear dose-response curve.

This can be done by means of the following parameterization: Suppose that dose is not a continuous variable, but assumes certain discrete values $u_1, \ldots, u_s$. Then it is possible to assign one effect parameter $\eta_{j1}, \ldots, \eta_{js}$ to each of these doses. To embody this set of new parameters into the model, let $b_{vj} = (b_{vj,1}, \ldots, b_{vj,s})$ be a selection vector with elements

$$b_{vj,t} = \begin{cases} 1 & \text{if subject } v \text{ has obtained dose } u_t \text{ in treatment } j \\ 0 & \text{otherwise;} \quad t = 1, \ldots, s. \end{cases} \tag{15.9}$$

The selection vector for each combination of person v by treatment j consists of zeros except for one element, which is equal to 1 and indicates the dose obtained in the respective treatment. Equation (15.4) then becomes

$$\delta_v = \sum_j \sum_t b_{vj,t}\eta_{jt}. \tag{15.10}$$

Now, consider Eq. (15.4) as H_0 and Eq. (15.10) as H_1, allowing a LR test of the linearity hypothesis. Heckl's (1976) study is one of the few where this H_0 was actually tested: A substantial difference in effect between children with fortnightly speech therapy and children with one therapeutic session per week was observed; the further benefit of two sessions per week, however, was comparatively small. Apparently, satiation occurred in the latter case.

THE POLYTOMOUS LINEAR LOGISTIC MODEL WITH RELAXED ASSUMPTIONS

Description of the Model

Although quite often the data are readily reducible to dichotomous variables, in many cases such a reduction is either not possible or makes little sense. In spite of this, designs with polytomous data have not received enough attention in the literature, owing to the lack of suitable methodology. In some of the early papers on linear logistic models for measuring change (Fischer, 1972, 1974a, 1974b, 1977b), the possibility of generalizing the LLRA to polytomous data has been recognized and the respective estimation equations have been derived. Without going into technical details, the essentials of the parameterization will be presented here.

Suppose that k polytomous variables, each of which assumes one of r qualitative or quantitative realizations, are the basis for assessment of change. A generalization of the model of Eqs. (15.1)–(15.3) is then

$$P(A_{vi}^{(h)} = 1 \mid v,i,t_1) = \exp(\theta_{vi}^{(h)}) \Big/ \left[\sum_t \exp(\theta_{vi}^{(t)}) \right], \tag{15.11}$$

$$P(A_{vi}^{(h)} = 1 \mid v,i,t_2) = \exp(\theta_{vi}^{(h)} + \delta_v^{(h)}) \Big/ \left[\sum_t \exp(\theta_{vi}^{(t)} + \delta_v^{(t)}) \right], \tag{15.12}$$

$$\delta_v^{(h)} = \sum_j q_{vj} \eta_j^{(h)} + \tau^{(h)}, \tag{15.13}$$

$$v = 1, \ldots, n; \qquad i = 1, \ldots, k;$$
$$j = 1, \ldots, m; \qquad t,h = 1, \ldots, r.$$

Thereby, $A_{vi} = (A_{vi}^{(1)}, \ldots, A_{vi}^{(r)})$ is an indicator vector-variable with realizations $a_{vi}^{(h)} = 1$ if person v's reaction to criterion i was in category h, and $a_{vi}^{(h)} = 0$ otherwise. The state of each person at t_1 is characterized by a matrix of parameters $\theta_{vi}^{(h)}$ and change is described by a vectorial parameter $\delta_v = (\delta_v^{(1)}, \ldots, \delta_v^{(r)})$; element $\delta_v^{(h)}$ measures the effect of the treatments with respect to reaction category h. Analogously, the effect of each treatment is described by a vectorial parameter $\eta_j = (\eta_j^{(1)}, \ldots, \eta_j^{(r)})$, element $\eta_j^{(h)}$ being associated with the specific effect of treatment j with respect to response category h. To be more concrete, the behavior categories of a depressive patient could be, for example, "agitated," "withdrawn," and "normal"; a certain medicative treatment might then have a very strong effect of reducing agitation and increasing withdrawal without, however, necessarily increasing the rate of normal behavior. Hence, the model is suited for describing interactions between treatments and reactions categories.

Several such qualitative categories, however, may also express different levels of an underlying latent dimension (i.e., different degrees of *one* behavioral tendency). A typical example would be the categories "very content," "rather content," "rather discontent," and "not at all content," reflecting degrees of satisfaction (e.g., with a job). The case of unidimensionality of the response categories with respect to the treatments is then formalized as follows:

$$\eta_j^{(h)} = \phi^{(h)}\eta_j \qquad (j = 1, \ldots, m; \ \ h = 1, \ldots, r). \qquad (15.14)$$

Equations (15.14) have been called the "reduction conditions"; of course, it is a purely empirical matter whether they hold or not. If they are to hold, the matrix of parameters $\eta_j^{(h)}$ must be of rank 1. The parameters $\phi^{(h)}$ are called the "category weights." [In Eqs. (15.11)–(15.14), certain normalization conditions $\theta_{vi}^{(r)} = 0$, $\eta_j^{(r)} = \tau^{(r)} = 0$ and $\phi^{(r)} = 0$ have to be fulfilled, but within the framework of this chapter a complete formal presentation of the polytomous model cannot be given.]

Parameter Estimation

As in the dichtomous LLRA, the effect parameters and the trend effects can be estimated empirically, independent of the person parameters $\theta_{vi}^{(h)}$ that characterize the state of the sample at t_1. Furthermore, hypotheses are again testable by means of LR tests. One reservation, however, must be made regarding the reduction conditions: When the parameters are estimated under the assumptions of Eqs. (15.14), the solutions of the estimation equations need not necessarily be unique.

The numerical computations for estimating parameters in the polytomous LLRA are much more complex than in its dichotomous counterpart, and some theoretical questions need additional attention (as, for example, uniqueness of the solution in case of the reduction conditions); also the amount of data required is much larger. For these reasons, only a few empirical applications have been realized so far. Nevertheless, the polytomous LLRA is a potentially powerful instrument for assessing change, as will be illustrated by the following two empirical studies.

Sample Applications

Hammer (1978) investigated the cognitive and attitudinal effects of a multimedia presentation dealing with forms of human settlement, problems of big cities, and ecology. The presentation was viewed by one sample of high school students, whereas another sample received instruction on the same topics from a teacher. The cognitive effects of both methods of instruction were measured by a questionnaire with the three response categories "correct," "partially correct," and "incorrect"; the attitudinal–emotional effects were evaluated by another questionnaire with the categories "positive," "neutral," "negative," and "don't know." As expected, the multimedia presentation proved to be generally more effective than the teacher, especially so with respect to the domain of attitudinal and emotional change; the teacher was able to impart knowledge rather than to influence attitudes or to appeal to emotions.

The second example returns to the problem of measuring effects of mass communication mentioned at the beginning: Kropiunigg (1979a, 1979b) carried out a field study on a topical problem of social and political interest in Austria connected with the reform of penal law in 1975. In Styria, one of the nine provinces of Austria, an informational campaign on this topic was carried out by the *Regionalprogramm Studio Steiermark* (radio) and by the *Kleine Zeitung Graz* (newspaper), whereby questions of probation and resocialization of convicts were covered.

Before and after the campaign, small but representative samples of the population were interviewed ($t_1 : n = 550$; $t_2 : n = 640$). The questionnaire comprised items from three attitudinal domains and one set of items for assessing familiarity with relevant facts. Since, in this survey, the people interviewed at t_1 and t_2 (in contrast to cases of the standard LLRA) were not the same, a modified version of the model for independent samples had to be used (see Fischer, 1972, 1974a, 1974b, 1977c).

This study differed from the other above-mentioned investigations in one essential respect: It was not possible to obtain generalizable propositions with respect to the effects of the media. Rather, the results supported the standard conjecture of communications theory: that effects of communications are strongly determined by a number of moderator variables (e.g., socioeconomic factors). Only the result that the radio programs were more effective than the respective articles in the daily newspaper was of some generality. Those segments of the population characterized by high contact frequency with the radio programs in question showed satiation with respect to information on the issue; a further increase in the density of the pertinent information in the newspaper, on the other hand, would still have increased the effects of the campaign. A somewhat unexpected finding was the relatively limited acceptance of the promoted ideas by women and by religious people, whereas supporters of the (ruling) socialist party showed significantly above average consent.

The principal goal of giving a simple characterization of each medium by a few effect parameters $\eta_j^{(h)}$—which in 1971 led to the development of the LLRA and other linear logistic models—was not reached in this empirical study. Perhaps the epistemological basis of these developments is not appropriate for such a complex problem of social science. But as mentioned above, the theoretical considerations and the applications in other fields indicate that it was worthwhile to venture models which attempt to derive very simplified and generalizable results from complex bodies of qualitative data.

REFERENCES

Bereiter, C. Some persisting dilemmas in the measurement of change. In C. W. Harris (Ed.), *Problems in measuring change*. Madison: The University of Wisconsin Press, 1967.

Birnbaum, A. Some latent trait models and their use in inferring an examinee's ability. In F. M. Lord & M. R. Novick, *Statistical theories of mental test scores*. Reading, Massachusetts: Addison-Wesley, 1968.

Fischer, G. H. A measurement model for the effect of mass media. *Acta Psychologica*, 1972, *36*, 207–220.

Fischer, G. H. The linear logistic model as an instrument in educational research. *Acta Psychologica*, 1973, *37*, 359–374.

Fischer, G. H. *Einführung in die Theorie psychologischer Tests*. [Introduction to the theory of psychological tests.] Bern: Huber, 1974. (a)

Fischer, G. H. Lineare logistische Modelle zur Beschreibung von Einstellungs- und Verhaltensänderungen unter dem Einfluß von Massenkommunikationen. In W. F. Kempf (Ed.), *Probabilistische Modelle in der Sozialpsychologie*. Bern: Huber, 1974. (b)

Fischer, G. H. Some probabilistic models for measuring change. In D. N. M. de Gruijter & L. J. Th. van der Kamp (Eds.), *Advances in psychological and educational measurement*. New York: Wiley, 1976.

Fischer, G. H. Some probabilistic models for the description of attitudinal and behavioral changes under the influence of mass communication. In W. F. Kempf & B. Repp (Eds.), *Mathematical models for social psychology*. Bern: Huber and New York: Wiley, 1977. (a)

Fischer, G. H. Linear logistic test models. In H. Spada & W. F. Kempf (Eds.), *Structural models of thinking and learning*. Bern: Huber, 1977. (b)

Fischer, G. H. *Some implications of specific objectivity for the measurement of change* (Research Bulletin No. 21). Vienna: University of Vienna, Institute of Psychology, 1977. (c)

Fischer, G. H. Logistic latent trait models with linear constraints. *Psychometrika*, 1983, *48*, 3–26.

Fischer, G. H., & Formann, A. K. Zur Schätzung der Erblichkeit quantitativer Merkmale. *Zeitschrift für Differentielle und Diagnostische Psychologie*, 1981, *2*, 189–197.

Fischer, G. H., & Formann, A. K. Some applications of logistic latent trait models with linear constraints on the parameters. *Applied Psychological Measurement*, 1982, *6*, 397–416.

Formann, A. K., & Piswanger, K. *Wiener Matrizentest*. Weinheim: Beltz Test, 1979.

Hammer, H. *Informationsgewinn und Motivationseffekt einer Tonbildschau und eines verbalen Lehrervortrages*. Unpublished doctoral dissertation, University of Vienna, 1978.

Heckl, U. *Therapieerfolge bei der Behandlung sprachgestörter Kinder*. Unpublished doctoral dissertation, University of Vienna, 1976.

Kropiunigg, U. *Wirkungen einer sozialpolitischen Medienkampagne*. Unpublished doctoral dissertation, University of Vienna, 1979. (a)

Kropiunigg, U. Einstellungswandel durch Massenkommunikation. *Österreichische Zeitschrift für Soziologie*, 1979, *4*, 67–71. (b)

Lord, F. M. Elementary models for measuring change. In C. W. Harris (Ed.), *Problems in measuring change*. Madison: The University of Wisconsin Press, 1967.

Pendl, P. *Effektivität des Sprachlabors an Höheren Schulen*. Unpublished doctoral dissertation, University of Vienna, 1976.

Platzer, H. *Der Einfluß technischer Spielzeuge auf das mechanisch-technische Verständnis*. Unpublished doctoral dissertation, University of Vienna, 1978.

Rasch, G. *Probabilistic models for some intelligence and attainment tests*. Copenhagen: Danmarks Paedagogiske Institut, 1960.

Rasch, G. An item analysis which takes individual differences into account. *British Journal of Mathematical and Statistical Psychology*, 1966, *19*, 49–57.

Rasch, G. An informal report on a theory of objectivity in comparisons. In L. J. Th. van der Kamp & C. A. J. Vlek (Eds.), *Psychological measurement theory*. Leyden: University of Leyden, Psychological Institute, 1967. (Proceedings of the NUFFIC international summer session in science at Het Oude Hof, The Hague, July 1966.)

Rasch, G. *A mathematical theory of objectivity and its consequences for model construction*. Paper presented at the European Meeting on Statistics, Econometrics and Management Science, Amsterdam, The Netherlands, September 1968.

Rasch, G. Objectivitet i samfundsvindenskaberne et metodeproblem. [Objectivity in the social sciences as a methodological problem.] *Nationaløkonomisk Tidsskrift*, 1972, *110*, 161–196.

Rasch, G. On specific objectivity: An attempt at formalizing the request for generality and validity of scientific statements. In M. Blegvad (Ed.), *The Danish yearbook of philosophy*. Copenhagen: Munksgaard, 1977.

Rella, E. *Trainierbarkeit des Antizipierens von Gefahrensituationen im Straßenverkehr*. Unpublished doctoral dissertation, University of Vienna, 1976.

Rop, I. The application of a linear logistic model describing the effects of pre-school curricula on cognitive growth. In H. Spada & W. F. Kempf (Eds.), *Structural models of thinking and learning*. Bern: Huber, 1977.

Scheiblechner, H. Das Lernen and Lösen komplexer Denkaufgaben. *Zeitschrift für experimentelle und angewandte Psychologie*, 1972, *19*, 476–506.

Spada, H. Die Analyse kognitiver Lerneffekte mit stichprobenunabhängigen Verfahren. In K. Frey & M. Lang (Eds.), *Kognitionspsychologie und naturwissenschaftlicher Unterricht*. Bern: Huber, 1973.

Spada, H. *Modelle des Denkens und Lernens*. Bern: Huber, 1976.

Spada, H. Logistic models of learning and thought. In H. Spada & W. F. Kempf (Eds.), *Structural models of thinking and learning*. Bern: Huber, 1977.

Spada, H., & Kluwe, H. Zwei Modelle der Denkentwicklung und ihr Bezug zur Theorie von Piaget. In R. H. Kluwe & H. Spada (Eds.), *Studien zur Denkentwicklung*. Bern: Huber, 1981.

Vodopiutz, A. *Komplexbildung im motorischen Lernen*. Unpublished doctoral dissertation, University of Vienna, 1977

Witek, J. *Die Effektivität des gruppendynamischen Sensitivity Trainings*. Unpublished doctoral dissertation, University of Vienna, 1979.

Zeman, M. *Die Wirksamkeit der mathematischen Früherziehung*. Unpublished doctoral dissertation, University of Vienna, 1976.

Zeman, M. *Kursmäßige Legasthenikertherapie an Wiens Schulen*. Vienna: Ketterl, 1980.

Zimprich, H. Behandlungskonzepte und resultate bei psychosomatischen Erkrankungen im Kindesalter. In H. Zimprich (Ed.), *Kind und Umwelt*. Vienna: Springer, 1980.

Author Index

Numbers in italics indicate the pages on which complete references can be found.

Subject Index

F

G

P

Q

R